Non-Coding RNAs

Molecular Tools for Crop Improvement

Edited by

Penna Suprasanna

Amity University, Mumbai, Maharashtra, India

Vinay Kumar

Modern College of Arts, Science and Commerce (Savitribai Phule Pune University), Pune, India

Ashish Kumar Srivastava

Bhabha Atomic Research Centre, Mumbai, India

CRC Press
Taylor & Francis Group
Boca Raton London New York

CRC Press is an imprint of the
Taylor & Francis Group, an **informa** business

Designed cover image: Shutterstock – image number – 2163064841

First edition published 2024
by CRC Press
2385 Executive Center Drive, Suite 320, Boca Raton, FL 33431

and by CRC Press
4 Park Square, Milton Park, Abingdon, Oxon, OX14 4RN

CRC Press is an imprint of Taylor & Francis Group, LLC

ISBN: 9781032425443 (hbk)
ISBN: 9781032438993 (pbk)
ISBN: 9781003369288 (ebk)

DOI: 10.1201/9781003369288

Typeset in Times
by KnowledgeWorks Global Ltd.

Contents

Foreword

Murdoch University | Food Futures Institute

Food Futures Institute
WA State Agricultural Biotechnology Centre
Centre for Crop & Food Innovation

Plant non-coding RNAs (ncRNAs) are emerging key regulators of plant growth, development, and their responses to climate change and environmental stresses. These tiny regulatory RNA molecules are taking center-stage and are being hailed as potent candidates for engineering climate-resilience in crop plants. It is, therefore, crucial to identify and functionally characterize plant ncRNAs using high-throughput, genomics, and genetic-engineering techniques. In this regard, this book, *Non-Coding RNAs: Molecular Tools for Crop Improvement,* is a complete and comprehensive coverage of highly important topics with contemporary relevance.

The chapters cover various facets of plant ncRNAs, ranging from identification and characterization, role in stress tolerance, their regulation to their exploitation for producing high-yielding plants as well as plants that can survive in harsh environments with no or very few penalties. The book provides an integrated and comprehensive blend of basic and advanced resources on the subject of plant ncRNAs for students, scholars, and plant scientists interested in, or already engaged in, research involving plant-interactions with stressful environments and dynamic climates and its post-transcriptional regulations via ncRNAs. The chapters present the current knowledge and recent advancements in plant ncRNAs, ranging from their sequencing and identification; prediction, identification, and characterization of non-coding RNAs and their targets; *in silico* tools and databases to study plant non-coding RNAs; and their application as potent targets for crop improvement.

I am delighted that the edited volume *Non-Coding RNAs: Molecular Tools for Crop Improvement* is being published. The chapters included in this book are very well-written by established and active scientists and researchers. This book describes recent advances in ncRNAs, their biology, advances in their identification and characterization, their roles in plant growth, development, and stress responses and adaptations, in addition to dedicated chapters covering how they can play pivotal roles and can be explored in crop-improvement strategies.

I congratulate the editors for unraveling this illustrious volume and hope that this will be a useful reference material for researchers, students, and agri-policymakers.

Professor Rajeev K. Varshney FRS

FOREWORD

Preface

Climate change is an inevitable phenomenon that is projected to be experienced by all the living creatures on planet earth. Climatic change will be preceded by a cascade of fluctuating events including elevated temperature, hyper-soil salinity, infrequent rainfall followed by drought, and elevated CO_2 and light, which impact food security. In such dynamic environmental fluctuations, plants are most severely affected due to their sessile nature. This necessitates the development of climate-resilient or climate-smart crops that can sustain an aberrant climate and environmental challenges with as few yield penalties as possible. Plant non-coding RNAs regulate plant responses to environmental stresses. These regulatory RNA molecules have gained increasing attention and are now considered as potent candidates for improving plant climate-resilience. Therefore, it is necessary to understand their regulatory roles, and identify and functionally characterize them using high throughput molecular, transgenic as well as genome editing techniques.

This book on plant non-coding RNAs is intended to provide an integrated and comprehensive coverage of basic and advanced topics for students, scholars, and scientists interested in, or already engaged in, research involving plant-interactions with stressful environments and their post-transcriptional regulations. The chapters are high-quality articles covering the current knowledge and recent advancements in plant non-coding RNAs, ranging from their sequencing and identification; prediction, identification, and characterization of non-coding RNAs and their targets; *in silico* tools and databases to study plant non-coding RNAs; and their utilization as candidates for engineering climate resilience in crop plants.

The book is an excellent reference source for plant scientists engaged in research related to Plant biotechnology, Crop improvement, Climate resilience, Small RNA biology, Stress tolerance and Plant genomics. The book is also suitable for both advanced undergraduate and postgraduate students specialized in biotechnology, plant sciences and agriculture. We extend our appreciation and sincere gratitude to the authors for their chapter contributions. Our wholehearted thanks are due to the entire team of Taylor & Francis for the cooperation and all the support during the publication of this book.

Penna Suprasanna
Amity University Mumbai

Vinay Kumar
Savitribai Phule Pune University

Ashish Kumar Srivastava
Bhabha Atomic Research Centre

Preface

Contributors

Uttpal Anand
Department of Life Sciences
Ben-Gurion University of the Negev
Beer-Sheva, Israel

Subham Bhakta
Bhabha Atomic Research Centre
Mumbai, India
and
Homi Bhabha National Institute
Mumbai, India

Nikhil Bhamwani
Department of Biological Sciences
Birla Institute of Technology and Science (BITS)
Pilani, India

C. Bharadwaj
ICAR-Indian Agricultural Research Institute
New Delhi, India

Ravi Kant Bhatia
Department of Bio-sciences
Himachal Pradesh University
Shimla, India

Arunima Bhattacharya
Centre National de la Recherche Scientifique
Institut National de la Santé et de la Recherche Médicale
Institut Européen de Chimie et Biologie
University of Bordeaux
Pessac, France

S. K. Chaturvedi
Rani Laxmibhai Central Agricultrual University
Jhansi, India

Sarthak Chordia
Department of Biological Sciences
Birla Institute of Technology and Science (BITS)
Pilani, India

Sandip Das
Department of Botany
University of Delhi
Delhi, India

Abhijit Dey
Department of Life Sciences
Presidency University
Kolkata, India

Satarupa Dey
Shyampur Siddheswari Mahavidyalaya
Howrah, West Bengal, India

Mimosa Ghorai
Department of Life Sciences
Presidency University
Kolkata, India

Santosh Kumar Gupta
DBT-National Institute of Plant Genome Research
New Delhi, India

Mehul Gupta
Department of Biological Sciences
Birla Institute of Technology and Science (BITS)
Pilani, India

Swati Hazra
Sharda School of Agruicultural Sciences
Sharda University
Greater Noida, India

Yuniet Hernandez
Plant RNAi Biology Group
International Centre for Genetic Engineering and Biotechnology
New Delhi, India

Niraj Kumar Jha
Department of Biotechnology
School of Engineering & Technology
Sharda University
Greater Noida, India

Shrusthti Joshi
Department of Biotechnology
P.E.S. Modern College of Arts, Science and Commerce
Savitribai Phule Pune University
Pune, India

Narendra Kadoo
CSIR-National Chemical Laboratory
Pune, India
and
Academy of Scientific and Innovative Research (AcSIR)
Ghaziabad, India

Deepak Khairnar
Department of Biotechnology,
Dr. D. Y. Patil Arts, Commerce & Science College
Savitribai Phule Pune University
Pune, India
and
National Centre for Cell Science
Savitribai Phule Pune University
Pune, India

Neeraj Kumar
ICAR-Indian Grassland and Fodder Research Institute
Jhansi, India

Nitesh Kumar
Department of Bio-sciences
Himachal Pradesh University
Shimla, India

Pradeep Kumar
North Eastern Regional Institute of Science and Technology
Nirjuli, Arunachal Pradesh, India

Sudhir Kumar
Plant RNAi Biology Group
International Centre for Genetic Engineering and Biotechnology
New Delhi, India

Vinay Kumar
Department of Biotechnology
P.E.S. Modern College of Arts, Science and Commerce
Savitribai Phule Pune University
Pune, India

Neelam Kumari
Department of Bio-sciences
Himachal Pradesh University
Shimla, India

Madhu Kamle
North Eastern Regional Institute of Science and Technology
Arunachal Pradesh, India

Milan Kumar Lal
ICAR-Central Potato Research Institute, Indian Council of Agricultural Research, Ministry of Agriculture and Farmers' Welfare,
Shimla, India

Varsha Arun Mahadik
CSIR-National Chemical Laboratory
Pune, India
and
Academy of Scientific and Innovative Research (AcSIR)
Ghaziabad, India

Sayanti Mandal
Institute of Bioinformatics Biotechnology (IBB)
Savitribai Phule Pune University
Pune, India

Sicon Mitra
Department of Biotechnology
School of Engineering & Technology
Sharda University
Greater Noida, India

Suvendu Mondal
Bhabha Atomic Research Centre
Mumbai, India
and
Homi Bhabha National Institute
Mumbai, India

Potshangbam Nongdam
Department of Biotechnology
Manipur University
Manipur, India

Devendra Kumar Pandey
Department of Biotechnology
Lovely Professional University
Punjab, India

Suraj Patil
Department of Biotechnology
P.E.S. Modern College of Arts, Science and Commerce
Savitribai Phule Pune University
Pune, India

Aryadeep Roychoudhury
School of Sciences
Indira Gandhi National Open University
New Delhi, India

Manish Roorkiwal
Khalifa Center for Genetic Engineering and Biotechnology (KCGEB), UAE University, UAE
Hyderabad, Telangana, India

Shivaraj S.M.
CSIR-National Chemical Laboratory
Pune, India

Parmeshwar K. Sahu
Department of Genetics and Plant Breeding
Indira Gandhi Krishi Vishwavidyalaya
Raipur, India

Neeti Sanan-Mishra
Plant RNAi Biology Group
International Centre for Genetic Engineering and Biotechnology
New Delhi, India

Richa Sao
Department of Genetics and Plant Breeding
Indira Gandhi Krishi Vishwavidyalaya
Raipur, India

Akanksha Sharma
Department of Bio-sciences
Himachal Pradesh University
Shimla, India

Niharika Sharma
NSW Department of Primary Industries
Orange Agricultural Institute
Orange, NSW, Australia

Priyanka Sharma
School of Bioengineering Sciences and Research
MIT-ADT University
Pune, India

Rita Sharma
Department of Biological Sciences
Birla Institute of Technology and Science (BITS)
Pilani, India

Mahipal S. Shekhawat
KM Government Institute for Postgraduate Studies and Research
Puducherry, India

Swati Singh
Department of Biotechnology
TERI School of Advanced Studies
New Delhi, India

Anjali Soni
ICAR-Indian Agricultural Research Institute
New Delhi, India

Rajeev K. Varshney
State Agricultural Biotechnology Centre
Murdoch University
Murdoch, Australia

T. Vinutha
ICAR-Indian Agricultural Research Institute
New Delhi, India

1 Small RNAs in Plants
Are These Magic Bullets for Imparting Climate Resilience in Crops?

Sandip Das[1] *and Swati Singh*[2]
[1]Department of Botany, University of Delhi, Delhi, India
[2]Department of Biotechnology, TERI School of Advanced Studies, New Delhi, India

ABBREVIATION

lncRNA: Long non-coding RNA
miRNA: microRNA
circRNA: Circular RNA

1.1 INTRODUCTION

The scourge of climate change has irreversibly altered the environment and is threatening to disrupt world food, feed and nutritional security. Shifting climate patterns are having opposite impacts on land availability for crop farming and are affecting crop adaptability to biotic and abiotic factors. Climate change is not only laying to waste vast stretches of arable land into barren fields that are economically unviable for agriculture, but also converting yet inaccessible land such as cold deserts or land covered by permafrost into farmlands. Such changing land-use patterns coupled with a rapidly increasing world population is posing a serious problem to meet the requirement for the global food demand and about the stability of the global environment (https://www.unfpa.org/data/world-population-dashboard; Gilbert 2014; Valin et al. 2014; Noya et al. 2018; IPCC 2019).

Development and deployment of climate-smart crop cultivars are thus required on an urgent basis to tackle these challenges (Wheeler and Von Braun 2013; Altieri and Nicholls 2017). Breeding programmes are designed to introgress beneficial genes or alleles responsible for yield increase, stress tolerance and adaptive advantages between closely related crop relatives and/or crop wild relatives (CWR), including primary, secondary and tertiary gene pools (Gilbert 2014; Brozynska et al. 2016; Ceoloni et al. 2017). Conventional methods of crop improvement through breeding and selection are time-consuming and labour-intensive, with the added disadvantage of linkage drag. A whole suite of forward genetic strategies (e.g., molecular breeding, marker assisted breeding, precision breeding, genomics-assisted breeding [GAB]) and selective and reverse genetic (targeted genetic engineering, gene/genome editing) strategies can be employed with a rapid development and deployment time and to produce high-yielding, climate resilient crops with minimum yield penalty (Gepts 2002; Nejat and Mantri 2017; Cobb et al. 2019; Varshney et al. 2020, 2021a, 2021b; Veillet et al. 2020; Snowdon et al. 2021).

A key requirement of targeted genetic engineering and genome/gene editing is knowledge about identity and function of molecular regulators controlling or modulating traits of interest. Recent developments in high-throughput, next generation sequencing (NGS) have improved our understanding and knowledge of gene regulatory mechanisms and epigenetics (Margulies et al. 2005; Mardis 2008). NGS-based RNA sequencing (RNA-seq) allows detection and quantification of

DOI: 10.1201/9781003369288-1

known, novel and rare protein-coding and non-coding RNA (ncRNA) transcripts (Margulies et al. 2005; Mortazavi et al. 2008; Nejat et al. 2015; Liu et al. 2017a), and helps us to understand the functional relevance of intergenic space in nuclear genome as a source of origin of ncRNA, disregarding the fact that it was traditionally referred to as junk DNA (Budak et al. 2020). In the recent past, several categories of ncRNAs have been identified with diverse functions in plants (Lelandais-Brière et al. 2010; Adenot et al. 2006; Bhogireddy et al. 2021; Nejat and Mantri 2017; Liu et al. 2019a; Yu et al. 2019; Chen et al. 2020).

In the present chapter, we discuss the impact of climate change on plant growth and yield and need for climate smart crops. We also review recent advances in knowledge of the biological functions of plant ncRNAs in biotic and abiotic stresses, and the possibility of deploying ncRNA in generation of climate-smart crops.

1.2 IMPACT OF CLIMATE CHANGE ON PLANT GROWTH AND YIELD

The frequency of extreme weather events is expected to increase globally, and the associated uncertainty has triggered anxiety as crop yield is likely to be compromised leading to an increased risk to food security, seriously affecting the universal goal of eliminating hunger by 2030 (Lesk et al. 2016; Richardson et al. 2018; Tito et al. 2018; Raza et al. 2019). With the average yield of most major crops either increasing at a slow pace, widely fluctuating or showing a downward trend worldwide, the increased frequency of drought and heavy rainfalls, temperature fluctuations, salinity and insect pest attacks is anticipated to further aggravate the problem and decrease productivity (Figure 1.1; Dhankher and Foyer 2018; https://ourworldindata.org/crop-yields; Ray et al. 2012). The outcomes of stresses on crop yield are hard to quantify, and in future, the productivity of the major crops is estimated to drop in many countries due to global warming, water shortage and other environmental impacts (Eastburn et al. 2010; Bonan and Doney 2018; Tebaldi and Lobell 2018).

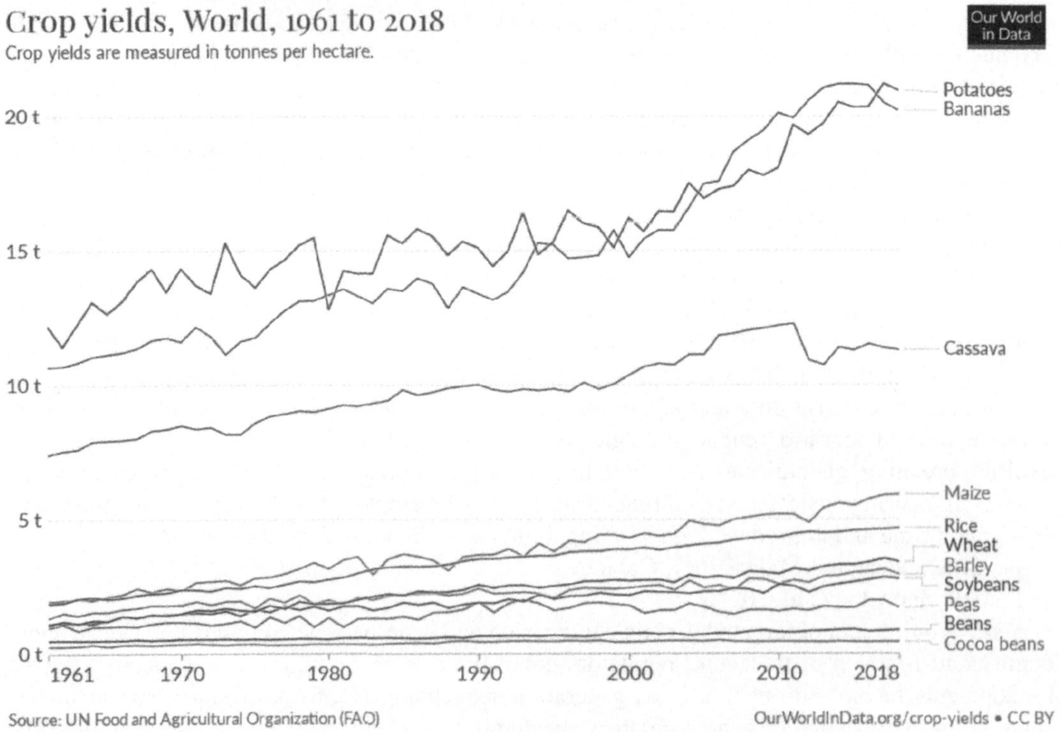

FIGURE 1.1 Trends in average yield of major crops between 1961 and 2018 based on the FAO dataset available at Our World in Data (https://ourworldindata.org/crop-yields).

Abiotic stresses such as temperature, drought and salinity play a major role in plant yield either individually or collectively. To assess their impact on plant performance, research has been widely conducted with each of these individual stresses. The effect of temperature on yield of plants has been one of the major factors responsible for reduction of crop yield. With a rise in temperature, the yield of most of the crops around the world has been evidently reduced. The effect of temperature extremes due to climate change has started becoming visible with a huge reduction in wheat production in many countries, and is likely to further reduce the crop yield by 3–10% for each 1°C rise in temperature (Asseng et al. 2015; Ray et al. 2015). Likewise, in cereals, sterility is induced in developing plants if the temperature increases to around 30°C (Saini and Aspinall 1982). Schlenker and Roberts (2009) reported that in soybean and maize, rise in temperature only up to a threshold resulted in yield increase, with further temperature increase leading to yield loss (Schlenker and Roberts 2009). A separate study reported that with every 1°C rise in temperature over and above the optimal temperature, the yield of maize decreases by 8.3% (Lobell and Field 2007; Lobell et al. 2011). In rice and sorghum, every 1°C rise over the optimal temperature causes a decrease in yield of 2.6% and 7.8%, respectively (Easterling et al. 2007).

Other than temperature, water stress such as flooding and drought conditions also jeopardizes crop yield (Dietz et al. 2021). For instance, biomass production, grain yield and flowering time were found to be severely and differentially impacted by the timing of drought (e.g., before, during and at the end of flowering) in *Pennisetum glaucum*, which is otherwise considered a hardy cereal crop (pearl millet; Winkel et al. 1997). Similarly, when subjected to drought stress during sporogenesis, wheat and rice suffered between 35% and 75% reduction in grain set (Saini and Aspinall 1981, 1982; Sheoran and Saini 1996). In rice, drought stress greatly disturbs the process of fertilization and anthesis. Water deficit can reduce the harvest index by up to 60%, thus, dramatically decreasing grain setting (Garrity and O'Toole 1994). The effect of drought has been evaluated on the yield of several crops, for example, it reduced seed yield from 2.7 t/ha to 2.3 t/ha in *Brassica napus* (Tesfamariam et al. 2010), 3.6 t/ha to 2.0 t/ha in *Glycine max* (Candogan et al. 2013), 8.2 t/ha to 5.9 t/ha in *Secale cereale* (Hubner et al. 2013), 5.9 t/ha to 4.4 t/ha in durum wheat (Johari-Pireivatlou 2010), 7.4 t/ha to 1.4 t/ha in *Zea mays* (Aydinsakir et al. 2013), weight of kinnow fruit from 148 g/fruit to 115 g/fruit (Dolker et al. 2017), fruit yield from 13.2 t/ha to 8.3 t/ha in *Vitis vinifera* (Intrigliolo et al. 2012), to name a few. Drought stress influences wheat during all developmental stages, with various reproductive stages and grain-filling stages being most critical (Pradhan et al. 2012). Wheat yield decreased from 1% to 30% during the mild drought stress at post-anthesis, while the loss increased up to 92% in the case of prolonged but mild drought stress during the flowering and grain-filling stages (Araus et al. 2002; De Oliveira et al. 2013). The yield of an important grain legume, mashbean (*Vigna mungo L.*), was reduced from 31% to 57% by drought stress during the flowering stage, and a 26% reduction was reported when drought stress was experienced during the reproductive phase (Baroowa and Gogoi 2014).

All stresses have been shown to significantly reduce plant growth and yield. For example, heat stress has been shown to cause yield loss of up to 42% in *Triticum aestivum* (Badu-Apraku et al. 1983), between 27% and 30% in *Oryza sativa* (Kumar et al. 2015), up to 90% after 14 days of heat stress in *Brassica napus* (Pokharel et al. 2021) and 73% of grain weight in *Z. mays* (Chukwudi et al. 2021).

1.3 ncRNAs AS REGULATORS

NcRNAs such as tRNA and rRNAs are essential to life forms as they perform critical housekeeping roles (Nejat and Mantri 2018). In addition to these two forms, a plethora of ncRNAs are now known to be central to gene regulation in almost all developmental and adaptive processes. Although discovered only about two decades ago, components such as "activator RNA", "nuclear confined, heterogeneous RNA of unknown function" and function of repeated DNA as source of "producer gene sequences" were proposed in 1969 (Britten and Davidson 1969). Since then, several classes of regulatory ncRNAs have been identified and reported to play important roles in a variety of biological processes (Zubko and Meyer 2007; Poething 2009; Simon and Meyers 2011; Wang and Chang 2011; Nejat and Mantri 2018; Zhang et al. 2018b Song et al. 2021).

Regulatory ncRNAs can be classified into three major subclasses on the basis of their size (Ohtani 2018; Karlik et al. 2019):

i. Long ncRNA (lncRNAs) that are >200 nt
ii. Small ncRNAs which include microRNAs (20–24 nt), small interfering RNAs (siRNAs, <200 nt), small nuclear RNA (snRNAs) and small nucleolar RNAs (snoRNAs)
iii. Circular ncRNAs (circRNAs)

1.3.1 Long Non-coding RNAs (lncRNAs)

Long non-coding RNAs (lncRNAs) are further classified based on (i) length of lncRNA transcript, (ii) association with protein-coding genes, (iii) association with DNA elements, (iv) resemblance to protein-coding RNA, (v) repeat association, (vi) association with biochemical pathway and stability of transcript, (vii) conservation of sequence and structure, (viii) expression state, (ix) cellular/sub-cellular localization and (x) function (St. Laurent et al. 2015). A simplified classification of lncRNAs is based on their source loci in the genome and includes natural antisense transcripts (NATs), long intronic ncRNAs and long intergenic ncRNA (lincRNAs) (Ron et al. 2010; Zhang and Chen 2013).

Most of the lncRNAs (such as *COLD INDUCED LONG ANTISENSE INTRAGENIC RNA* [*COOLAIR*], *COLD ASSISTED INTRONIC NON-CODING RNA* [*COLDAIR*] and *LONG-DAY-SPECIFIC MALE-FERTILITY- ASSOCIATED RNA* [*LDMAR*]) are RNA polymerase II transcribed followed by 5′-capping, splicing and 3′-polyadenylation (Swiezewski et al. 2009; Heo and Sung 2011; Ding et al. 2012a; Ariel et al. 2014; Matsui and Seki 2019). LncRNAs are also reported to be transcribed by RNA polymerase III (Pol III), RNA polymerase IV (Pol IV) and RNA polymerase V (Pol V) (Dieci et al. 2007; Wierzbicki et al. 2008, 2009; Ulitsky and Bartel 2013; Huang et al. 2021a). Over the past decade, the role of lncRNA in plants has been established in various processes such as reproduction, development, as a modulator of the splicing process and a response towards stresses; however, their specific mode of action in each of these processes still remains elusive (Heo and Sung 2011; Ding et al. 2012; Ariel et al. 2014; Bardou et al. 2014; Wang et al. 2018b; Matsui and Seki 2019; Sun et al. 2020).

1.3.2 Small Non-Coding RNA (sRNA)

The second major class of ncRNAs are small RNA (sRNAs), largely comprising microRNA (miRNAs) and small interfering RNA (siRNAs) (ca. 20–24 nt). MiRNA and siRNA differ in their sequence and structural features, biogenesis and their mode of action (Talmor-Neiman et al. 2006; Ding et al. 2012b; Borges and Martienssen 2015; Huang et al. 2016a; Chauhan et al. 2017; Brant and Budak 2018; Huang et al. 2019; Wang et al. 2019a; Yoshikawa et al. 2021). Biogenesis of miRNAs follows the general transcription principle and pathway as that of a protein-coding mRNA from a type II promoter. In a canonical pathway, RNA Pol II transcribes a primary-miRNA (pri-miRNA), which is then processed to generate precursor miRNA (pre-miRNA). In a few instances, RNA Pol III has also been implicated in miRNA biogenesis (Borchert et al. 2006; Gu et al. 2009; Diebel et al. 2010; Ha and Kim 2014; O'Brien et al. 2018). The final processing of the precursor stem-loop structure to release the mature 20–22 nt miRNA occurs either inside the nuclear membrane (plants) or in the cytosol (animals) (Axtell et al. 2011; Wang et al. 2019a). The process of miRNA biogenesis and maturation is guided by a large repertoire of enzymes, including *DICER-1*, *DICER Like* (DCLs), *Drosha*, *HASTY/Exportin-5*, *HEN1*, *HYL1* and *SERRATE*, to name a few (Borchert et al. 2006; Gu et al. 2009; Diebel et al. 2010; Ha and Kim 2014; Anand et al. 2017; O'Brien et al. 2018; Treiber et al. 2019; Wang et al. 2019a).

Biogenesis of various forms of siRNAs such as trans-acting siRNA (tasiRNA; TAS), phased-siRNA (phasiRNA) from pre-formed double-stranded RNAs transcribed by RNA-dependent RNA polymerases (RDRs/RDRPs) is carried out by transcription of the inverted repeat region by RNA

Pol II and RNA Pol IV (Huang et al. 2021a). Enzymes and proteins involved in the biogenesis and amplification of siRNAs include *DCL4, RDR6, SGS3, SDE5, AGO7, AGO1* and small RNAs such as miR173, miR390, miR161 and miR828 (Axtell et al. 2006; Allen and Howell 2010; Yoshikawa 2013; Deng et al. 2018; Carbonell 2019; Sakurai et al. 2021; Yoshikawa et al. 2021).

1.3.3 Circular RNA (circRNAs)

CircRNAs are a recently identified class of ncRNAs which are non-polyadenylated, single-stranded covalently closed loop structures of RNA without any 5′ or 3′ polarity (Chen and Yang 2015; Chu et al. 2018a, 2018b). CircRNAs originate from the back-splicing of RNA polymerase II pre-mRNA transcripts wherein a downstream splice donor is joined to an upstream splice acceptor site, thus generating a circular RNA molecule, and involves both canonical GT-AG or non-canonical GC/GG, CA/GC, GG/AG, GC/CG, CT/CC splice signals (Ye et al. 2017; Zhao et al. 2019b). CircRNAs exist in multiple isoforms, such as circular intronic RNAs (ciRNA), exon-intron-circRNA (elciRNA) or the most commonly occurring exonic circRNA (ecRNA) (Meng et al. 2017; Bose and Ain 2018). The current knowledge on the molecular mechanism underlying regulation of gene expression by circRNAs is still fragmentary; however, it has been shown that circRNAs are involved in the *cis*-regulation of transcription of parental genes by associating with Pol II (Li et al. 2015; Zhou et al. 2018). Further identification of potential functions of circRNAs in plants has been predicted using bioinformatics and these include circRNA-miRNA interaction networks in soybeans (Zhao et al. 2017), the circRNA-lincRNA-mRNA network in potatoes (Zhou et al. 2018) and the circRNA-miRNA-mRNA networks at different developmental stages in *Arabidopsis* (Chen et al. 2017; Liu et al. 2017c).

1.4 MODE OF ACTION OF ncRNAs

This section summarizes the modes and mechanisms of action of the various classes of ncRNA for the purpose of understanding how they can be deployed for trait modification for climate smart crops, and the challenges therein.

1.4.1 lncRNAs

LncRNAs have been shown to be involved in regulating transcription, RNA maturation, transport and translation and chromatin remodelling (Wapinski and Chang 2011; Böhmdorfer and Wierzbicki 2015; Wierzbicki et al. 2021). Four mechanisms have been proposed for lncRNA-mediated regulation of gene expression wherein lncRNAs act as/via (i) signal, (ii) decoy, (iii) guide and, (iv) scaffold (Wang et al. 2018b). LncRNAs participate in RNA-DNA duplex formation and mediate transcriptional regulation of gene expression in *cis* as well as in *trans*. LncRNAs interact with chromatin and can act as a "signal" to recruit chromatin remodifying complexes as has been shown for the lncRNA, AIRN, that is responsible for transcriptional regulation of the *Igf2r* locus through recruitment of histone-modifying machinery at the promoter (Latos et al. 2012). In plants, the lncRNA, *COLDAIR*, physically interacts and recruits the PRC2 complex to cause epigenetic silencing of the *FLC* locus, triggered as a vernalization response (Heo and Sung 2011).

As "decoys", lncRNAs participate in lncRNA-protein (lncRNA-RNA binding protein complexes) or lncRNA-miRNA interactions and regulate transcription of the target locus. Proteins such as chromatin-modifying proteins, transcription factors or other trans-regulatory proteins, and RNA such as miRNAs can, therefore, be sequestered by lncRNAs. For example, an lncRNA that is transcribed from a minor promoter of the dihydrofolate reductase (*DHFR*) gene acts as a "decoy", binds to the major promoter of the same gene and sequesters the transcription factor proteins necessary for the initiation of transcription (Martianov et al. 2007). An example of an lncRNA-miRNA interaction is that of the lncRNA *Induced upon Phosphate Starvation* (*IPS1*) that sequesters miR399 and regulates phosphate uptake in plants (Franco-Zorrilla et al. 2007; Lin et al. 2008; Ajmera et al. 2018).

LncRNAs can also act as "guides" wherein they physically interact with other trans-factors (proteins, siRNA) and recruit them to the target locus to regulate gene expression. For example, the lncRNA *COLDAIR* acts as a "guide" to localize the PRC2 complex to the FLC locus in *cis* and brings about change in the chromatin structure (Heo and Sung 2011). The lncRNA *HOTAIR*, on the other hand, has been proposed to act as a "guide" for the PRC2 complex, modify chromatin states in *trans*, and play an important role in tumorigenesis in cancerous cells (Gupta et al. 2010).

The fourth mode of action of lncRNAs is its as role as a "scaffold". In this model, different folds and domains of lncRNA bind simultaneously to multiple trans-factors or effector molecules and help assemble a higher order structure. The lncRNA telomerase RNA (TERC) acts as a scaffold to assemble the multiple components of the active telomerase (Grammatikakis et al. 2014). Similarly, the lncRNA nuclear paraspeckle assembly transcript 1 (*NEAT1*) is a critical component for the structure of the paraspeckle body inside the nucleus by sequestration of two proteins—PSP1 and p54 (Clemson et al. 2009). Other lncRNAs include HOTAIR, which is known to act as scaffold by interacting simultaneously with PRC2 complex (300 nt at the 5′- end), and LSD1/CoREST/REST (700 nt at 3′-end) to alter the chromatin state though histone methylation (Tsai et al. 2010).

Even though four different models have been proposed, lncRNAs are likely to employ a combination of the various modes of action and participate in RNA-DNA, RNA-RNA and RNA-protein interactions to regulate gene expression. Additionally, lncRNAs can mediate both activation and repression of transcriptional activity and gene expression. Several studies have shown that the regulation of gene expression by lncRNA is mediated via DNA methylation and acetylation, RNA-RNA interactions, nucleosome repositioning and regulation of protein-coding genes either through cis- or trans acting processes (Wilusz et al. 2009; Faust et al. 2012; Kornienko et al. 2013; Matsui and Seki 2019; Statello et al. 2021). LncRNAs are known to regulate chromatin structure and remodelling via RNA-mediated transcriptional gene silencing via RNA pol V generated lncRNA transcripts to mediate cytosine methylation and histone modification, recruitment of chromatin remodellers for nucleosome positioning, and, chromosome looping with the help of enhancer RNA (eRNAs)

1.4.2 MiRNA and siRNA

MiRNA and siRNA act by binding to a target based on sequence complementarity and triggering a post-transcriptional gene silencing (PTGS) and translational repression mechanism (Kamthan et al. 2015; Yu et al. 2017; O'Brien et al. 2018; Wang et al. 2019a). MiRNAs have been shown to be involved in cytosine methylation in a process termed "miRNA-directed DNA methylation", although only in a few cases such as that of *PHB* and *PHV* genes directed by miR165/miR166 (Bao et al. 2004; Wu et al. 2010). In a majority of the cases, miRNAs act through transcript cleavage and translational repression. The primary differentiator between the two modes of action is dependent on the extent of base-pairing between miRNA and the target mRNA, governed by a well-defined set of rules in plants. Canonically, the 5′ nucleotide of plant miRNAs is "U" and positions 2–12 of miRNA serve as seed sequence in plants, whereas positions 2–8 are deemed sufficient for target identification in animals. The seed sequence in plants that spans between positions 2–12 is most critical to target recognition, and mismatches between miRNA and the target are not tolerated. Sequence complementarity between positions 13–21 of miRNA and that of the target may have up to four discontinuous mismatches. After perfect or near-perfect target recognition, slicing and transcript cleavage occur between position 10–11 of the miRNA in plants. Imperfect pairing between the miRNA and the target, especially in the seed region around positions 9–12, is proposed to be responsible for translational repression as the mode of miRNA-mediated gene regulation (Schwab et al. 2005; Brodersen and Voinnet 2009; Ellwanger et al. 2011; Chi et al. 2012; Pandey et al. 2019).

Small interfering RNAs originate from the dsRNA template with perfect sequence complementarity and fall into 20–24 nt size classes. The 24 nt siRNAs are termed as long repeat RNA-associated sRNA or heterochromatic siRNA based on their source and association, and are involved in chromatin modification through modification of histone and DNA at the target loci (Simon and Meyer 2011; Kamthan et al. 2015; Martienssen and Moazed 2015; Xie and Yu 2015). In comparison, 20–22 nt siRNAs have been associated with gene regulation via transcript cleavage and translational repression similar to that observed for miRNAs (Kamthan et al. 2015). LncRNAs transcribed by RNA Pol V also guide siRNAs to the target loci where they are involved in the modification of the chromatin structure by recruiting multi-subunit complexes for DNA methylation via the RNA-directed DNA methylation (RdDM) pathway, also known as a plant *de-novo* RdDM process (Zheng et al. 2009; Matzke and Mosher 2014; Blevins et al. 2015). Transcriptional silencing through the RdDM pathway also is one of the other major functions performed by siRNAs in epigenetic regulation of plant growth and under environmental stress (Summanwar et al. 2020).

The two mechanisms mediated by ncRNAs and epigenetic regulation of gene expression are proposed to be a two-way mechanism by which they also regulate each other.

1.4.3 circRNA

Most of the knowledge about the mode and mechanism of action of circRNAs is derived from studies from non-plant, mammalian systems. CircRNAs have been shown to act as "sponges" for miRNA, as several of them contain miRNA binding sites and, therefore, fall in the category of miRNA sponges, target mimics or as competing endogenous RNA (ceRNA) (Franco-Zorrilla et al. 2007; Ebert and Sharp 2010; Reichel et al. 2015; Reichel and Millar 2015; Panda 2018; Bose and Ain 2018; Zhang et al. 2020b). The lack of free 5′ and 3′ ends makes circRNA resistant to cellular exoribonucleases, thus increasing their stability and half-life, and these, therefore, are more efficient as sponges as compared to linear lncRNA (Zhang et al. 2020b). CircRNAs also interact with a variety of RNA-binding proteins (RBPs), which are involved in the RBP-driven circularization process (HuR, FMRP, eIF4A3, to name a few), and may also be responsible for post-transcriptional and translational regulation; some circRNAs have also been shown to be translated and coded for proteins (Zang et al. 2020).

Deciphering the rules of complementarity and mode of action has been a critical determinant in developing tools and devising strategies to use small RNAs for trait modification (Kumar et al. 2018).

1.5 ncRNAs AND STRESSES

A large number of ncRNAs, including small ncRNA (siRNA and miRNA), lncRNAs and circRNAs, have been identified with diverse functions and appear to be critical components of a multi-layered mechanism, playing pivotal roles in regulating development and stress responses in plants. These include cell signalling, differentiation, heterosis, DNA damage repair, hormone signalling, organ development, and response to biotic and abiotic stresses across diverse plant species (Lelandais-Brière et al. 2009; Ron et al. 2010; Wahid et al. 2010; Khraiwesh et al. 2012; Sunkar et al. 2012; Mantri et al. 2013; Borges and Martienssen 2015; Wu et al. 2020).

1.5.1 ncRNAs in Abiotic Stress

1.5.1.1 lncRNAs

The role of lncRNAs in the mediation of abiotic stress tolerance has been well elucidated through different modes across many plant species. LncRNAs are known to inhibit miRNA function via target mimicry, wherein an imperfect miRNA binding site in the lncRNA mimics the miRNA binding site in the actual target. The binding of the miRNA to the mimic in the lncRNA thus prevents or minimizes interactions between miRNAs and the targets.

An example of this mode of action is of the lncRNA, *IPS1*, whose expression is induced under phosphate (Pi) deficiency. The lncRNA *IPS1* sequesters miR399, thus preventing PTGS regulation of the actual miR399 target—*PHOSPHATE2* (*PHO2*) involved in Pi homeostasis (Franco-Zorrilla et al. 2007). LncRNAs have also been shown to act as precursors of sRNA. A total of 14 lncRNAs were identified from *Populus tomentosa* under N-deficiency. Of these 14, 9 lncRNAs act as precursor for 11 known miRNAs exemplified by pto-MIR156c, e, g, h and j; pto-MIR159a, b and c; pto-MIR168b and pto-MIR169b. In addition, 5 lncRNAs serve as precursors for 14 novel miRNAs (Chen et al. 2016). The complexity of the lncRNA-mediated stress response was evident when 4 lncRNAs were found to be targeted by 29 miRNAs belonging to 5 families—*MIR156, MIR396, MIR159, MIR6462* and *MIR6457* in *P. tomentosa* (Chen et al. 2016). Similarly, two lncRNAs, *TalnRNA5* (responsive to both heat stress and powdery mildew in *T. aestivum*) and *TapmlnRNA19* (responsive to powdery mildew in *T. aestivum*), were identified as precursors for *Ta-miR2004*; and one lncRNA, *TapmlnRNA8,* responsive to powdery mildew in *T. aestivum* was identified as precursor of Ta-miR2066. Another heat-stress responsive lncRNA, *TahlnRNA27,* contains a precursor of Ta-miR2010 in wheat (Xin et al. 2011). The same study also identified 16 powdery mildew-responsive lncRNAs that act as precursors for 71 siRNAs, and 9 heat-stress-responsive lncRNAs that serve as precursors for siRNAs in wheat (Xin et al. 2011).

Plant antisense lncRNAs also mediate the stress response, wherein the antisense lncRNAs in association with sense mRNAs can form double-stranded RNA duplexes, thereby altering the gene expression pattern (Wang et al. 2014a). In poplars, five antisense lncRNAs were detected under N-deficient conditions (Chen et al. 2016); similarly, antisense lncRNAs have been observed in maize under drought stress and in wheat when infected with stripe rust fungi *Puccinia striiformis* f. sp. *Tritici* and *Arabidopsis thaliana* in response to light stimuli (Zhang et al. 2013, 2014; Wang et al. 2014c).

Other than acting through target mimicry and as antisense lncRNA, several lncRNAs have also been demonstrated to be involved in the lncRNA-mediated chromatin modification (lncR2Epi) regulatory pathway (Wang et al. 2016). An example of the lncRNA-mediated chromatin modifications (lncR2Epi) pathway is the cold-stressed *Arabidopsis,* wherein expression of two cold-induced lncRNAs, *COOLAIR* and *COLDAIR,* represses *FLOWERING LOCUS C (FLC)* expression (Michaels and Amasino 1999; Heo and Sung 2011; Csorba et al. 2014). However, both *COOLAIR* and *COLDAIR* work antagonistically in mediating methylation response to their target gene. A coordinated action of *COOLAIR,* which mediates reduction in H3K36me3 and H3K4me2 modification, and *COLDAIR,* which physically associates with polycomb repressive complex 2 (PRC2) and increases H3K27me3 modification at the *FLC* locus, is involved in flowering-time regulation (Heo and Sung 2011; Csorba et al. 2014; Marquardt et al. 2014).

1.5.1.2 miRNAs and siRNAs

It is well established that environmental stress conditions hamper plant growth, development and overall yield by altering the patterns of gene expression of not only key regulators such as small RNAs and transcription factors, but a range of other downstream targets in the gene regulatory networks. Increasing evidence confirms that miRNA levels can reprogram and control the response of a plant to their changing environmental conditions (Brant and Budak 2018; Nogoy et al. 2018).

1.5.1.2.1 Chilling/Cold Stress

Enhanced levels of miR319 through overexpression and knock-down of the targets of miR319, viz. *OsPCF5, OsPCF6, OsPCF8* and *OsPCF21* result in enhanced cold tolerance at 4°C in rice (Yang et al. 2013; Wang et al. 2014b). Other miRNAs that are differentially expressed during cold stress include miR535 (target *SPL*; *O. sativa*), miR394 (targets *LEAF CURLING RESPONSIVENESS [LCR]; A. thaliana*), miR319 (GAMyb/TC94752 as target; PCF6/TC111376; *Saccharum* sp.

Cv SP70-1143; sugarcane), miR393a (*Auxin-signalling F-box protein [AFB]; Transport Inhibitor Response protein 1 [TIR1]; Panicum virgatum L*) (Thiebaut et al. 2012; Song et al. 2016; Liu et al. 2017b; Zeng et al. 2018; Sun et al. 2020a).

1.5.1.2.2 Heat Stress

Several reports on high-temperature stress-induced small RNAs including siRNAs and miRNAs have appeared in literature (Ding et al. 2020; Das et al. 2021; Zuo et al. 2021). In a study that analysed transcriptome in various tissues during vegetative and reproductive developmental stages in *Z. mays* during heat stress, differentially expressed siRNAs and miRNAs were identified (He et al. 2019). Interestingly, heat stress caused downregulation of transposable derived 24 nt siRNA in tassels and roots. The authors identified miR159, miR166, miR396 and miR408 as the four most abundant miRNAs which are also differentially expressed under heat stress. For example, miR159 was found to be downregulated in V3-roots and R1 leaves, but upregulated in V3 stalks at different time points after administering heat stress, indicating a complex dynamic of miRNAs transcripts (He et al. 2019). In a study published recently, researchers demonstrated a phenomenon termed "thermopriming", wherein rice plants which were exposed briefly to elevated temperatures *a priori* were able to withstand subsequent heat shock. Rice plants either subjected to thermopriming or no thermopriming exhibited differential transcript abundance of osa-miR531b, osa-miR5149, osa-miR168a-5p, osa-miR1846d-5p, osa-miR5077, osa-miR156b-3p and osa-miR167e-3p along with their cognate targets and indicated that these may be critical regulatory nodes in heat-stress response (Kushawaha et al. 2021). Several other miRNAs also were found to be differentially expressed when exposed to heat stress such as osa-miR2120b-5p, osa-miR397a, osa-miR397b, osa-miR398b, osa-miR408-5p and osa-miR7693-3p that are upregulated and osamiR166k-5p, osa-miR167d-3p, osa-miR1865-3p and osa-miR5153 that are downregulated (Kushawaha et al. 2021). MiRNA expression dynamics under heat stress have been analysed in a range of plants as exemplified by reports in *Solanum lycopersicum* (miR156g, miR159a, miR167b/c/h, miR169b/c, miR170, miR171, miR172a/b, miR396; Karlova et al. 2013; Chen et al. 2015), *Populus trichocarpa* (Lu et al. 2005, 2008; Shuai et al. 2013), *P. tomentosa* (Chen et al. 2012), and *Brassica rapa* (Yu et al. 2012), to cite a limited set of examples (Table 1.1).

1.5.1.2.3 Drought Stress

The miRNA-mediated drought stress response is regulated by several miRNAs and their cognate target modules, including miR156 (*SPL/SBP*), miR159 (*MYB33/65*), miR160 (*ARF10/16/17*), miR164 (*NAC*), miR165/166 (*HD-ZIPIII*), miR167 (*ARF6/8*), miR168 (*AGO1*), miR169 (*NfγA*), miR319 (*TCP*), miR390 (*TAS3-siRNA*) and miR393 (*TIR1*), etc. (Liu et al. 2008, 2019c; Li et al. 2011; Balyan et al. 2017; Zhao et al. 2019a; Yang et al. 2019; Singroha et al. 2021). The role of miR166 was demonstrated when the expression levels of miR166 were downregulated by employing Short Tandem Target Mimic (STTM) in rice. The transgenic lines showed elevated levels of tolerance along with other natural drought tolerance phenotypes such as leaf rolling and reduced xylem diameter (Zhang et al. 2018d).

1.5.1.2.4 Salt Stress/Salinity Tolerance

A multitude of reports have implicated miRNAs in mediating salt stress responses in a variety of plants. A comparative analysis of small RNA transcriptome of radish roots revealed several novel (e.g., rsa-miR2, rsa-miR4, rsa-miR11, rsa-miR13, rsa-miR24, rsa-miR35) and known miRNAs (such as miR156, miR159, miR160, miR166, miR169, miR172, miR396, miR397, miR398 and miR399) that are differentially expressed upon being subjected to salt stress (Sun et al. 2015a). Several of these miRNAs, such as miR172c, miR397a, miR814b, rsa-miR3, rsa-miR9, rsa-miR12a, rsa-miR23 and rsa-miR48, and their targets showed temporal variation in expression kinetics, indicating their role in salt-stress response pathways (Sun et al. 2015a). In *Z. mays*, analysis of sRNA transcriptome and

TABLE 1.1
Representative Example of Non-Coding RNAs in Mediating Abiotic and Biotic Stress Responses across Various Plants

S. no.	Plant	Number/Name	Stress Response	References
		Category: lncRNAs/Abiotic Stress		
1.	*Populus tomentosa*	A total of differentially expressed 126 lncRNA; 14 of these were precursors of novel and known miRNAs.	Nitrogen deficiency stress	Chen et al. 2016
2.	*Zea mays*	664 drought stress responsive lncRNAs, of which 126 were known and 538 were novel lncRNAs; some of these are precursors of miRNA, siRNA and shRNA.	Drought stress	Zhang et al. 2014
3.	*Arabidopsis thaliana*	*HIDDEN TREASURE 1 (HID1)*	Photomorphogenesis and light sensitivity	Wang et al. 2014c
4.	*A. thaliana*	*COOLAIR* *COLDAIR*	Cold stress (Vernalization)	Csorba et al. 2014; Marquardt et al. 2014; Heo and Sung, 2011
6.	*Z. mays*	A total of 6099 lncRNAs, of which 3190 lncRNAs were differentially expressed in roots of *Z. mays* var. B73 after 0, 2, 4, 6, 8, 10, and 12 hours of water logging stress.	Waterlogging	Yu et al. 2020a
	Brassica napus	Differentially expressed lncRNA between drought tolerant (variety Q2; 5546 downregulated and 6997 upregulated) and sensitive (variety Qinyou 8; 7824 downregulated and 10,251 upregulated).	Drought stress and re-watering conditions	Tan et al. 2020
		Category: lncRNAs/biotic stress		
	Triticum aestivum	A total of 125 lncRNAs, of which some are differentially expressed upon *Blumeria graminis* infection.	*Blumeria graminis*	Xin et al. 2011
	T. aestivum	Four lncRNAs—*TalncRNA18, TalncRNA73, TalncRNA106, TalncRNA108*—that are involved in stripe rust infection.	*Puccinia striiformis* f. sp. *tritici (Pst)*	Zhang et al. 2013
	A. thaliana	A total of 159 RNAs, of which 20 lncRNA that are responsive to *Fusarium oxysporum* infection.	*Fusarium oxysporum*	Zhu et al. 2013
	Solanum lycopersicum	• The *lncRNA16397* was found to induce expression of GRX upon late blight infection. • Involvement of lncRNA33732-WRKY1 in late blight infection. • 115 upregulated and 81 downregulated out of 9011 lncRNA upon late blight infection.	*Phytopthora infestans*	Cui et al. 2017 Cui et al. 2019 Cui et al. 2020 Cao et al. 2021
	Hordeum vulgare	604 out of 12,366 lncRNAs were found to be responsive to Fusarium head blight (FHB) infection using near isogenic lines.	*Fusarium graminearum*	Huang et al. 2016b
	Oryza sativa	• 1429 lncRNA in mock inoculated resistant *O. sativa* line, 1927 lncRNA when inoculated with *M. oryzae* strain ZB13 and 1981 when inoculated with strain Zhong. • Upregulated (83) and downregulated (78) lncRNA upon blast fungus infection.	*Magnaporthe oryzae*	Jain et al. 2017 Wang et al. 2020a
	B. napus	931 differentially expressed out of 3181 lncRNAs that were identified upon *Sclerotinia* stem rot infection.	*Sclerotinia sclerotiorum*	Joshi et al. 2016

(Continued)

TABLE 1.1 (*Continued*)
Representative Example of Non-Coding RNAs in Mediating Abiotic and Biotic Stress Responses across Various Plants

S. no.	Plant	Number/Name	Stress Response	References
	O. sativa	567 *X. oryza* responsive lncRNAs; 73 lncRNAs were identified that may potentially interact with 39 transcript of JA related pathway.	*Xanthomonas oryza*	Yu et al. 2020b
	O. sativa	2518 upregulated and 2191 downregulated lncRNAs upon *D. zeae* infection.	*Dickeya zeae*	Li et al. 2018
	Solanum tuberosum	Differentially expressed lncRNA between tolerant (cv. BP1; 6139 lncRNA) and susceptible (cv Valor; 8214 lncRNA) when compared to mock inoculations.	*Pectobacterium carotovorum*	Kwenda et al. 2016a
	Actinidia deliciosa	A total of 14845 lncRNA from 12,280 loci.	*Pseudomonas syringae* pv. actinidiae	Wang et al. 2017c
	Paulownia tomentosa	• 2725 lncRNAs, of which 748 are differentially expressed. • 112 out of 3689 lncRNA that are related to phytoplasma infection.	Phytoplasma	Cao et al. 2018; Fan et al. 2018
	S. lycopersicum	A total of 1565 lncRNA identified from leaf of tomato upon infection with TYLCV.	Tomato yellow leaf curl virus (TYLCV)	Wang et al. 2015b; Zhou et al. 2019
		Category: miRNA/abiotic stress		
7.	*Helianthus annus*	miR160, miR167, miR172, miR398, miR403, miR426, miR842	Drought, high temperature, salinity and cadmium stress	Ebrahimi-Khaksefidi et al. 2015
8.	*O. sativa*	miR319	Cold stress	Yang et al. 2013; Wang et al. 2014b
9.	*O. sativa*	miR166	Drought stress	Zhang et al. 2018d; Salvador-Guirao et al. 2018b
10.	*T. aestivum*	miR156, miR160, miR164	Heat, salinity and drought stress	Pandey et al. 2014
11.	*O. sativa*	miR162a, miR166m, miR171b, miR390, miR168b, miR156l	Cadmium stress	Ding et al. 2011
12.	*O. sativa*	miR393	Salinity and alkalinity stress	Gao et al. 2011
13.	*Medicago truncatula*	miR398, miR408	Drought stress	Trinidade et al. 2010
14.	*Glycine max*	miR167d, miR169d, miR482, miR1507a, miR1508b, miR4369	Drought, salinity and alkalinity stress	Li et al. 2011
15.	*Saccharum* sp.	miRNA166III, miRNA168II, miRNA396II, miRNA398II, miRNA156V, miRNA169III miRNA159XVI	Salt stress	Carnavale Bottino et al. 2013
16.	*Triticum turgidum* ssp. *dicoccoides*	miR172b, miR1120a, miR393a	Salinity stress	Feng et al. 2017
17.	*A. thaliana*	miR396	UV-B stress	Casadevall et al. 2013

(*Continued*)

TABLE 1.1 (*Continued*)
Representative Example of Non-Coding RNAs in Mediating Abiotic and Biotic Stress Responses across Various Plants

S. no.	Plant	Number/Name	Stress Response	References
18.	*A. thaliana*	miR156	Abiotic stress: UV-B stress	Dotto et al. 2018
19.	*O. sativa*	miR169o	Nitrogen limitation	Yu et al. 2018
21.	*A. thaliana*	miR160, miR171, miR319, miR393, miR394, novel-miR66	Sulfur dioxide (SO_2) stress responsive	Li et al. 2016a
22.	*Glycine max*	miR172c	Drought and salt stress responsive	Li et al.,2015
23.	*Festuca arundinacea*	miR5813, miR9774, miR5568c, miR7758	Heat stress	Li et al. 2017a
24.	*A. thaliana*	miR156/157 and miR172	Elevated carbon dioxide (CO_2) stress responsive	May et al. 2013
25.	*T. aestivum*	miR398, miR528 and miR156	Ultraviolet radiation, continuous light and heat	Ragupathy et al. 2016
26.	*A. thaliana*	*miR398, miR390, miR319, miR159, miR156*	Ozone stress	Iyer et al. 2012
28.	*A. thaliana*	miR399	Phosphorus deficiency	Franco-Zorrilla et al. 2007
		miR408	Iron deficiency	Carrió-Seguí et al. 2019
	Brassica campestris L. ssp. *chinensis* var. *utilis*	41 conserved miRNA from 19 families	Heat stress	Ahmed et al. 2019
		Category: miRNA/Biotic Stress		
	Citrullus lanatus	*miR162, miR390, miR393, miR394, miR477, miR1222*	Cucumber green mottle mosaic virus (CGMMV) infection	Sun et al. 2017
	O. sativa	*miR156, miR159, miR164, miR166, miR169, miR171, miR396, miR827, miR1428, miR1432*	Fijivirus, rice black-streaked dwarf virus (RBSDV) infection	Sun et al. 2015b
	Curcuma longa	*miR159, miR160, miR164, miR482, miR393 and miRx2*	*Pythium aphanidermatum* infection	Chand et al. 2016
	O. sativa	*miR169*	*M. oryzae* infection	Li et al. 2017
	Cajanus cajan	*CcmiR169i, CcmiR1214, CcmiR3695, CcmiR9666b CcmiR8182*	*Fusarium udum* infection	Hussain et al. 2018
	A. thaliana	*miR158a, miR159b, miR400, miR472, miR5643a, miR5645a, miR8167a, miR869.1*, three novel *miRNAs* (*novel_85*, *novel_86* and *novel_90*)	*Bacillus velezensis FZB42* infection	Xie et al. 2017
	A. thaliana	*miR393, miR160, miR167, miR390*	*Pseudomonas* infection	Zhang et al. 2011
	B. napus	*miR156b*	*Pseudomonas brassicae* infection	Verma et al. 2014

(*Continued*)

TABLE 1.1 (*Continued*)
Representative Example of Non-Coding RNAs in Mediating Abiotic and Biotic Stress Responses across Various Plants

S. no.	Plant	Number/Name	Stress Response	References
	A. thaliana	*miR168, siRNA415*	Elicitor when infected with *P. syringae* flg22 peptide	Baldrich et al. 2014
	Ipomoea batatas	*miR408*	Response to feeding by insect pest, *Spodoptera litura*	Kuo et al. 2019
		Category: circRNA/Abiotic Stress		
29.	*A. thaliana*	A total of 1599 circRNA and 1583 heat-stress specific circRNAs.	Heat stress	Pan et al. 2017
30.	*S. lycopersicum*	A total of 163 chilling responsive out of 854 circRNAs.	Chilling injury	Zuo et al. 2016
31.	*O. sativa*	Out of a total of 12,037, 27 were found to be differentially expressed between to P-sufficiency and starvation condition.	Phosphate starvation	Ye at al., 2015
32.	*T. aestivum*	62 out of 88 circRNA were differentially expressed between stressed and well-watered control.	Dehydration stress	Wang et al. 2017
33.	*Capsicum annuum*	36 differentially expressed circRNA in fruit of bell pepper when subjected to chilling stress as compared to control.	Chilling stress	Zuo et al. 2018
34.	*Vitis vinifera*	475 differentially expressed circRNA in leaves after cold stress.	Cold stress	Gao et al. 2019
35.	*Pyrus betulifolia*	23 upregulated and 10 downregulated drought responsive out of a total of 899 circRNAs.	Drought stress	Wang et al. 2018
36.	Maize and *Arabidopsis*	2174 circRNA in maize and 1354 in Arabidopsis; Of these, 1843 in maize and 1283 are differentially expressed in maize and Arabidopsis, respectively.	Drought stress	Zhang et al. 2019
37.	*Hordeum vulgare*	62 seed transfer-cell specific circRNA originating from 48 genes.	Abiotic stress: Micro-nutrients homeostasis	Darbani et al. 2016
38.	*Citrus junos*	45 circRNA in root and 17 in leaves that are differentially expressed upon copper toxicity as compared to control.	Copper toxicity	Fu et al. 2019
39.	*Cucumis sativus*	2787 were identified; of these, 1934 were differentially expressed in roots and 44 in leaved under salt stress.	Salt stress	Zhu et al. 2019
40.	*Glycine max*	120 differentially expressed circRNA across genotypes that are tolerant and sensitive to P-levels.	Low phosphate stress	Lv et al. 2020
41.	*Brassica campestris* L. ssp. *pekinensis*	16 differentially expressed circRNA along with 151 DE lncRNA and 233 DE miRNA.	Vernalization response	Shi et al. 2021
42.	*B. campestris* L. ssp. *pekinensis*	Out of 730 circRNA, 23 and 22 were differentially expressed under calcium-deficiency when compared to optimum calcium concentration at different time points.	Calcium deficiency induced tip-burn in leaves	Wang et al. 2019c

(*Continued*)

TABLE 1.1 (*Continued*)
Representative Example of Non-Coding RNAs in Mediating Abiotic and Biotic Stress Responses across Various Plants

S. no.	Plant	Number/Name	Stress Response	References
		Category: circRNA/Biotic Stress		
43.	*S. tuberosum*	429 out of 2098 circRNAs were differentially regulated upon *Pectobacterium* infection.	*P. carotovorum* subspecies *Brasiliense* infection	Zhou et al. 2017
44.	*S. lycopersicum*	2056 lncRNAs (1767 intergenic lncRNA, 289 NAT-lncRNA) were identified; 83 lncRNA were unique to TYLCV infected samples and 32 in uninfected samples.	TYLCV infection	Wang et al. 2018
45.	*Gossypium barbadense*	686 lncRNA identified from a resistant (CSSL-1) and susceptible (CSSL-4) line. Of these, 280 were differentially expressed between the resistant and susceptible lines.	*Verticillium dahlia* infection	Xiang et al. 2018
46.	*Z. mays*	1443 and 1165 circRNA were identified from uninfected and infected samples; and 201 circRNA were shared between infected and uninfected samples. Of the 201, 155 were upregulated, and 5 downregulated in infected samples.	Maize Iranian mosaic virus (MIMV) infection	Ghorbani et al. 2018
47.	*Cucumis melo*	303 circRNA identified from PM tolerant (M1) and susceptible (B29) varieties. 17 circRNA in B29 and 23 circRNA in M1 that are differentially expressed after powdery mildew infection; 27 circRNA were predicted to be sponges for 18 different miRNAs.	*Podosphaera xanthii* race 1 (powdery mildew fungi)	Sun et al. 2021

Abbreviations: lncRNA = long non-coding RNA, miRNA = microRNA, circRNA = circular RNA.

degradome data allowed identification of known (miR164a-5p, miR164a-3p, miR167a-5p, miR167e-5p) and novel (mir_36, mir_29, mir_250, mir_316, mir_17, mir_330, mir_205, mir_189, mir_250) salt-stress responsive miRNAs from leaves and roots. Overall, a total of 1040 miRNAs were identified with 448 being common to roots and leaves (Fu et al. 2017). Such analyses in plants has led to the identification of small RNAs such as in *Gossypium hirsutum* (miRNVL5; Gao et al. 2016), *O. sativa* L. ssp. *Indica* cv. Pokkali (salt tolerant) and ssp. *Indica* cv. Badami (salt-sensitive) varieties, *Oryza coarcata*, *Suaeda maritima* (halophyte) (75 known and 200 novel; Parmar et al. 2020), *Z. mays* strains "NC286" (salt-tolerant) and "Huangzao4" (salt-sensitive) (98 miRNAs exemplified by members of zma-miR156, zma-miR159, zma-miR160, zma-miR162, zma-miR164, zma-miR166, zma-miR167, zma-miR168, zma-miR171, zma-miR319, zma-miR395, zma-miR399 families; Ding et al. 2009), to cite selected examples.

Roles of several miRNAs including miR156, miR169 and miR396 have been well elucidated in response to multiple stress conditions across different plant species. These small RNAs, therefore, act as potential candidates for further exploration of the production of crop plants that are tolerant to multiple stresses. MiR156 has been reported to be responsive to salinity, drought and heat stress in *T. aestivum* (Pandey et al. 2014), salt stress in *Saccharum* spp. (Carnavale Bottino et al. 2013), drought stress in *Glycine max* (Li et al. 2011), heavy metal stress in *O. sativa* (Ding et al. 2011; Liu et al. 2012) and UV radiation stress in *Arabidopsis* (Zhou et al. 2007). MiR396 was found to be differentially expressed when *T. aestivum* was subjected to salinity, drought, cold and heat stress (Ding et al. 2011; Liu et al. 2012; Pandey et al. 2014; Akdogan et al. 2016) and to heavy metal stress in *O. sativa* (Barrera-Figueroa et al. 2012); miR169 also exhibits varied expression under salt (Carnavale Bottino et al. 2013) and drought (Gentile et al. 2015) stress in *Saccharum* spp., heat stress in *T. aestivum* (Pandey et al. 2014) and UV radiation in *A. thaliana* (Zhou et al. 2007) and *Populus tremula* (Jia et al. 2009). Several other miRNAs have been characterized as responsive to abiotic stresses including drought (Zhang et al. 2018d), salinity (Feng et al. 2017), exposure to UV radiation (Casadevall et al. 2013; Dotto et al. 2018), cold (Megha et al. 2018), nutrient deficiency (Yu et al. 2018), sulfur dioxide stress (Li et al. 2016a), heat (Li et al. 2017a), metal stress (Gupta et al. 2014), elevated CO_2 levels (May et al. 2013), continuous light (Matsunaga et al. 2015; Ragupathy et al. 2016) and oxidative stress (Iyer et al. 2012) from different plant species.

1.5.1.3 CircRNAs

Abiotic stresses, such as nutrient depletion, high light, heat, chilling and frost, drought and salt have been shown to influence steady-state transcription levels of circRNAs. In spite of several reports, mechanistic aspects of gene regulation mediated via circRNAs and their significance in abiotic stress response and amelioration still need to be elucidated. In one of the initial reports, computational analysis and mapping of publicly available RNA sequence datasets of *O. sativa* root and *A. thaliana* leaves and the generation of transcriptome data from *O. sativa* root subjected to phosphate sufficiency and deficiency allowed the identification of 12,037 circRNAs from *O. sativa* and 6012 from *A. thaliana* (Ye et al. 2017). A comparative analysis showed variable origins of circRNAs in the two species. For instance, ecRNAs comprised 50.5% (6074 out of 12,037) in *O. sativa* and 85% (5152 out of 6012) in *A. thaliana*; similarly, circRNAs derived from two or more genes made up 34.4% in *O. sativa* and 13.4% in *A. thaliana*, thus suggesting complex and variable dynamics of circRNAs. Among all the circRNAs identified in the two species, reciprocal BLAST identified more than 700 orthologous genes. Across nearly 300 orthologs, the circRNA was derived from a similar position (Ye et al. 2017).

In other published reports, 36 differentially expressed circRNAs (28 upregulated and 8 downregulated) were identified in bell pepper in response to chilling stress (Zuo et al. 2018); 163 chilling-responsive circRNAs in tomato fruit, of which 102 circRNAs acted as miRNA sponges; 1759 differentially expressed circRNAs from tomato leaves, of which 383 acted as sponges for 266 miRNAs (Zuo et al. 2016; Yang et al. 2020); 467 circRNAs, of which 156 were common to drought, heat and drought + heat stresses. Among the circRNAs, 21 were unique to drought, 17 to heat and 36 to drought + heat stresses in tomato leaves (Zhou et al. 2020). Likewise, 475 differentially expressed circRNAs were identified in grape leaves subjected to cold stress, and overexpression of one circRNA, *circATS1*, conferred cold tolerance in *A. thaliana* (Gao et al. 2019). The grape Vv-circATS1, derived from glycerol-3-P acyltransferase, was found to improve cold tolerance in *Arabidopsis* by regulating the expression of stimulus-responsive genes, such as *CSD2*, *PRXCA*, *PME41*, *LOX3* and *WRKY48* (Gao et al. 2019).

Under dehydration-stressed conditions, differentially expressed circRNAs have been detected in wheat (Wang et al. 2017a), pear (Wang et al. 2018a), maize and *Arabidopsis* (Zhang et al. 2019a). Similar differentially expressed circRNAs were identified in barley (Darbani et al. 2016), *Brassica rapa* (Wang et al. 2019c), and soybean (Lv et al. 2020) under nutrient stress; *Citrus junos* (Fu et al. 2019) under copper toxicity and *Cucumis sativus* under salt stress (Zhu et al. 2019). In a significant finding, 1599 novel and 1583 heat-stress-induced circRNAs were identified in *A. thaliana* and heat stress was found to influence biogenesis of circRNAs, leading to altered circRNA length, number of circularized exons and alternative circularization events, revealing the complexities of circRNA-mediated gene regulation in plant responses to abiotic stresses (Pan et al. 2018).

1.5.2 ncRNAs in Biotic Stress

1.5.2.1 lncRNAs

LncRNAs that may play a role in mediating biotic stress have been observed in a range of species. In one of the earliest studies, 71 putative lncRNAs that may mediate response to *Blumeria graminis* f. sp. *tritici* (*Bgt*), an obligate biotrophic fungus responsible for causing powdery mildew disease in wheat, were documented using microarrays and SBS (sequencing-by-synthesis) technologies (Xin et al. 2011). These were in addition to 77 lncRNAs that were heat-stress responsive and 23 lncRNAs that were common to heat-stress and powdery mildew disease caused by *Blumeria graminis* f. sp. *tritici* (Xin et al. 2011). Four lncRNAs (*TalncRNA18, TalncRNA73, TalncRNA106* and *TalncRNA108*) were differentially expressed in response to stripe rust caused by *Puccinia striiformis* f. sp. *tritici (Pst)* in wheat (Zhang et al. 2013). Twenty *Fusarium oxysporum*-specific lncTARs (transcriptionally active regions) were observed in *Arabidopsis* out of 159 novel lncTARs (Zhu et al. 2014). In tomatoes, a total of 688 differentially expressed lncRNAs were identified between *Solanum pimpinellifolium* L3708, known to be resistant to *Phytopthora infestans,* and *S. lycopersicum* Zaofen No. 2, known as *P. infestans* susceptible line. Overexpression of lncRNA16397 resulted in induction of expression of *S. lycopersicum glutaredoxin* (*SlGRX*) genes, and co-expression of lncRNA and cDNA of *SlGRX* led to a reduction in reactive oxygen species (ROS) accumulation, leading to reduced injury to the cell membrane, eventually providing an increased resistance against *P. infestans* (Cui et al. 2017). Subsequently, a number of lncRNA including 115 upregulated and 81 downregulated lncRNAs were identified and the lncRNA-miRNA regulatory network was analysed during tomato-*P. infestans* interaction (Cui et al. 2020). The lncRNA33772-respiratory burst oxidase module was also shown to be involved in tomato-*P. infestans* interaction via *WRKY1* (Cui et al. 2019). In *Hordeum vulgare* (barley), a total of 12,366 lncRNAs were identified, of which 604 lncRNAs were found to be responsive to *Fusarium* head blight (FHB) infection caused by *Fusarium graminearum* (Huang et al. 2016b). In two separate studies, lncRNAs that are differentially expressed upon *Magnaporthe oryzae* infection in rice have been identified. In one study, *in-silico* analysis of the gene ontology omnibus (GEO) dataset (GSE62911) allowed identification of 1429 lncRNAs in mock, and in two rice lines (1927 lncRNAs in ZB13 and 1981 lncRNAs in the Zhong dataset) resistant to *M. oryzae* (Jain et al. 2017). In another study, RNA-seq revealed several upregulated (83) and downregulated (78) lncRNAs in rice after *M. oryzae* infection, and based on the analysis of the data, the authors propose that lncRNAs may be involved in Jasmonic Acid (JA)-mediated defence against *M. oryzae* in rice (Wang et al. 2020a). In a separate study, a total of 3181 lncRNAs were identified from *Brassica napus,* of which 931 lncRNAs were differentially expressed in response to infection by *Sclerotinia sclerotiorum* that caused stem rot disease (Joshi et al. 2016). Similarly, lncRNAs have been identified in response to bacterial phytopathogens *Xanthomonas oryzae* infection (567 lncRNAs; Yu et al. 2020) and *Dickeya zeae* causing foot-rot disease (4709 lncRNAs; Li et al. 2018) identified in rice. A correlation could be established between 73 out of 567 lncRNAs and 39 genes of the JA pathway, and *ALEX1,* an lncRNA, was found to be involved in resistance against *X. oryzae* (Yu et al. 2020). Out of the 4709 lncRNAs in response to *D. zeae* infection in rice, 2518 lncRNAs were upregulated and 2191 were downregulated, and several lncRNAs were predicted to be sponges for Osa-miR156 and Osa-miR396 (Li et al. 2018). Similarly, differentially expressed lncRNAs have been identified from species such as in potato against the cause of soft-rot, *Pectobacterium carotovorum* (Kwenda et al. 2016a, 2016b), kiwifruit plants infected with *Pseudomonas syringae* pv. *actinidiae* (Wang et al. 2017c) and against *Wickerhamomyces anomalus,* which is a yeast antagonist (Zhao et al. 2021), in *Juglans regia* (walnut) against *Colletotrichum gloeosporioides* (Feng et al. 2021) and in *Brassica rapa* against *Hyaloperonospora brassicae,* which causes downy-mildew (Zhang et al. 2021). LncRNAs also have been postulated to play an important role in the manifestation and response to viral infections (Nejat and Mantri 2018). Virus-Induced Gene Silencing (VIGS) has been established in tomatoes with a role of ~1560 lncRNAs associated with *Tomato Yellow Leaf Curl Virus* (TYLCV) (Wang and Luan 2015; Zhou et al. 2019).

1.5.2.2 miRNAs and siRNAs

Other than lncRNAs, small RNAs such as siRNA and miRNA have been identified as key regulators that mediate responses against various biotic stresses such as viral, fungal and bacterial phytopathogens (Wang et al. 2019d). In *Citrullus lanatus,* several miRNAs (miR162, miR390, miR393, miR394, miR477, miR1222) have been identified in response to cucumber green mottle mosaic virus (Liang et al. 2019). A comparative analysis of miRNA profiles from susceptible (ROC22) and tolerant (YA05-179) varieties of sugarcane allowed identification of a large number of already known and novel miRNAs that are differentially expressed in response to smut disease caused by *Sporisorium scitamineum* (Su et al. 2017). These miRNAs include 264 known and 137 novel in RCK (ROC22-control), 263 known and 140 novel in RT (ROC22-treatment with *S. scitamineum*), 260 known and 111 novel in YACK (YA05-179 control) and 262 known and 119 novel miRNAs in YAT (YA05-179- treatment with *S. scitamineum*). The authors shortlist five miRNAs (miR5671, miR5054, miR5783, miR5221, miR6478) and their targets as candidates for mediating pathogen response (Su et al. 2017).

In response to black-streaked dwarf virus (RBSDV) infection, multiple miRNAs—14 in leaves, and 16 in roots—that are differentially expressed and 15 novel miRNAs have been identified in rice (Sun et al. 2015b). Degradome analysis allowed identification of novel targets for miR159, miR164, miR167, miR169, miR172, miR396, miR528, miR529, miR530, miR812, miR814, miR818, miR1430, miR1432 and several others, indicating the yet undiscovered complexities of miRNA target-mediated adaptive and developmental responses (Sun et al. 2015b). Similarly, 28 conserved (members of miR166, miR167, miR169, miR399 families) and 14 novel (miRp1, miRp2, miRp3, miRp4, miRp5, miRp6, miRp7, miRp8, miRp9, miRp10, miRp11, miRp12, miRp13, miRp14) miRNAs were differentially expressed in response to RBSDV in *Z. mays,* which causes maize rough dwarf disease (MRDD) in maize (Zhou et al. 2016).

Small RNAs, such as miRNAs, that play vital roles during plant-pathogen interactions have been identified in several crop plants in response to attack by pathogens. These miRNA or small RNAs may act through TGS and PTGS mechanisms, thereby mediating defence responses and contributing to plant immunity. Several miRNAs such as miR472, miR482, miR1510, miR2109, miR160, miR393, miR390, miR400, miR398 and a number of other small RNA are proposed to be involved in Effector Triggered Immunity (ETI) and PAMP-Triggered Immunity (PTI) responses in rice (Baldrich and San Segundo 2016). MiRNAs have been identified in response to *Pythium aphanidermatum* infection in *Curcuma longa* (28 miRNAs; Chand et al. 2016), *M. oryzae* infection in rice (Li et al. 2017b; Zhang et al. 2018a, 2019c) and *Fusarium udum* infection in *Cajanus cajan* (five novel miRNAs: miR169i-3p, miR1214, miR3695, miR9666b-3p and miR8182; Hussain et al. 2018). In another study, elicitation of transcriptional regulation of miR168 was observed in *A. thaliana* when exposed to *Fusarium oxysporum*. In addition, the study also demonstrated the role of heterochromatin-derived siRNA415 in RDdM and miR823 in response to *F. oxysporum* infection (Baldrich et al. 2014). In a candidate gene-centred approach, functional analysis of the polycistronic miRNA cluster, miR166h-miR166k, showed that miR166k-5p is involved in increasing immunity against *M. oryzae* (rice blast) and *Fusarium fujikuroi* (bakane) through transcriptional regulation of *ethylene-insensitive (EIN-2)* in rice (Salvador-Guirao et al. 2018b).

Recent studies have identified miRNAs responsive to *Bacillus velezensis* in *A. thaliana* (Xie et al. 2017) and *Z. mays* (Xie et al. 2019) and *Bacillus amyloliquefaciens* in *A. thaliana* (Xie et al. 2018). In a study, differential expression of members of miRNA gene families was observed when *A. thaliana* was infected with virulent and non-virulent strains of *P. syringae* pv. tomato (*Pst*), *Pst DC3000 hrcC* (15 miRNA families), *Pst DC3000 EV* (27 miRNA families) and *Pst DC3000 avrRpt2* (20 miRNA families) when small RNA profiling was done 6 hours and 14 hours post-infection

(Zhang et al. 2011a). In a separate investigation, the role of AGO2/miR393* in mediating innate immunity via *P. syringae* pv. tomato (*Pst*) was established (Zhang et al. 2011b). A comparative analysis of the roots of *B. napus* infected with *Plasmodiophora brassicae* (causal organism of club-root disease) at 0 days post-infection (dpi), 10 dpi and 20 dpi identified a total of 10 differentially expressed miRNA at 10 dpi (most upregulated) and 34 miRNAs (21 upregulated, 13 downregulated) at 20 dpi (Verma et al. 2014). Another comparative analysis of root transcriptome of near-isogenic lines (NILs) of *B. napus* that are resistant (409R) and susceptible (409S) to *P. brassicae* led to identification of a total of 120 miRNAs, of which 48 were conserved and 72 were novel in nature. Out of these 120, 18 miRNAs (9 each upregulated and downregulated) were differentially expressed in roots of 409R before and after *P. brassicae* infection, whereas such a pattern was observed for only one miRNA (novel_1, upregulated) in 409S roots (Li et al. 2021).

1.5.2.3 CircRNAs

Recent studies have also identified circRNAs in mediating biotic stress responses in plants. A total of 2098 circRNAs, of which 429 circRNAs are differentially expressed between resistant (cultivar BP1) and susceptible (cultivar Valor), of potato in response to *P. carotovorum* subsp. *brasiliense (Pcb)* infection have been identified (Zhou et al. 2018). Analysis revealed that 931 out of these 2098 circRNAs were intergenic in origin, and cultivar-specific circRNAs (761 from Valor, and 694 from BP1) could be identified (Zhou et al. 2018).

CircRNAs also act as target mimics to sequester miRNAs, similar to lncRNAs. TYLCV is known to infect a number of crops such as tomatoes, beans, peppers and cucurbits and causes significant yield loss (Kil et al. 2016). In a recent study, 184 circRNAs and 2056 lncRNAs were identified from JS-CT-9210, a susceptible strain of tomato, before and after being infected with TYLCV. Thirty-two of these circRNAs were identified from untreated, and 83 circRNAs were identified from the TYLCV-treated JS-CT-9210 strain. The removal of circRNA slcirc108 and its parent gene Solyc07g043420.2.1 resulted in increased TYLCV infection, and partial resistance was achieved after reinjection of the parent gene by agro-infiltration, indicating that circRNAs can modulate expression levels of their parent genes (Wang et al. 2018b). Similarly, comparative analysis of a *Gossypium barbadense* chromosome segment substitution line in *G. hirsutum* background, CSSL-1 resistant to Verticillium wilt with that of susceptible line CSSL-4, after *Verticillium dahlia* infection allowed identification of 686 novel circRNAs in cotton; 280 of these were differentially expressed between the two lines as un-inoculated or inoculated state (Xiang et al. 2018).

RNA-seq of the tissue-culture-derived leaf, root and stem of *Actinidia chinensis* (kiwifruit) cultivar Hongyang, tissue-culture-derived leaf of cultivar Jinyan, *A. eriantha* and *A. argute* when infected with *P. syringae* pv. *actinidiae* (Psa) (causal organism of Bacterial Canker) allowed identification of a total of 3582 circRNAs from all the libraries that were analysed. One thousand two hundred thirty circRNAs were identified from the tissue dataset of the *A. chinensis* cultivar Hongyang, and 2914 circRNAs from the taxon dataset (*A. chinensis* cultivars Hongyang, Jinyan, *A. eriantha,* and *A. argute*) infected with *P. syringae* pv. *Actinidiae* were identified and analysed at two time points at 2 days post-inoculation and 14 dpi (Wang et al. 2017b).

1.6 ncRNAs AS POTENTIAL TARGETS FOR PRODUCTION OF CLIMATE-SMART CROPS

The role of ncRNAs as master riboswitches for regulating gene expression in response to environmental stress conditions makes them prime candidates and potential targets for trait modification to engineer climate resilience in crops (Agrawal et al. 2015; Rajwanshi et al. 2014; Zhang 2015; Chaudhary et al. 2021; Huang et al. 2021b). This strategy, however, has multiple challenges, including pleiotropy and unintended non-specific effects. Analysis of the published literature reveals that

(i) several miRNAs are involved in a single stress; (ii) a single miRNA is involved in multiple stresses; and (iii) several miRNAs across different species tend to have conserved mature 21 nt sequence, whereas a significant proportion of miRNAs are species- or lineage-specific and, thus, are of limited distribution across taxonomic boundaries. Discrete genomic location or *MIRNA* genes as a source of precursor miRNAs permits application of reverse genetics strategies for manipulation of miRNA loci, such as generation of overexpression, knock-out, knock-down and gene-edited lines (Teotia and Tang 2017; Zhou et al. 2017; Zhang et al. 2018c; Lian et al. 2021). Additionally, transcriptional regulation of *MIRNA* genes mediated via RNA polymerase II confers spatio-temporal regulation of the expression pattern. Each of these facts are to be considered while devising trait manipulation using miRNAs. For instance, multi-stress-responsive miRNAs hold advantages and are emerging targets for engineering plants, as such plants may withstand multiple stresses that may manifest simultaneously or in succession. Identification of novel and known miRNAs using NGS and bioinformatics platforms has also aided in characterization of miRNAs that might act as potential targets for engineering stress tolerance in transgenic plants. Over the past decade, stress tolerance of a variety of plant species has been improved through overexpression or downregulation of miRNA and numerous reports have appeared. The following section provides a summary of some of these reports.

Overexpression of the mir408 precursor from *A. thaliana* in *Cicer arietinum* resulted in increased expression of miR408 with a corresponding decrease of the target gene *Plantacyanin*. When the miR408 overexpression lines were subjected to drought stress, an increase in expression of the transcription factors (TF) *DREB1/2A, Rd17, Rd29a* and *Rd29b* was observed; phenotype analysis of transgenic plants and comparison with plants transformed with empty vector control revealed a significantly higher adaptive response of miR408 overexpression transgenic lines to drought stress (Hajyzadeh et al. 2015). Overexpression of miR408 in *A. thaliana* background on one hand led to enhanced tolerance to multiple stresses including salinity (100 mM NaCl), cold (-0.5°C for 12 hours) and oxidative (grown on 20 μM methyl viologen) stress, and on the other hand reduced tolerance to drought (administered by regulating schedule of plant watering) and osmotic (20% PEG) stress (Ma et al. 2015). Overexpression of miR408 also led to an overall increase in photosynthetic rate, plant biomass as measured through leaf area, petiole and silique length, size of flower and seed yield, whereas overexpression of miR408-target, i.e. *Plantacyanin* and *Laccase 13,* had an opposite effect (Song et al. 2018). This increase in biomass and plant size was attributed to an increase in cell size rather than cell number (Song et al. 2018). The role of miR408 in iron and copper homeostasis was also evaluated through analysis of knock-out and overexpression lines, and indicated the multiple regulatory roles of miR408 in iron homeostasis in *A. thaliana* (Carrió-Seguí et al. 2019). Apart from multiple abiotic stresses, the expression state of miR408 was also shown to be altered upon wounding stress in *Ipomoea batatas* cv. Tainung 57, and transgenic plants overexpressing miR408 exhibited reduced resistance to feeding by the insect *Spodoptera litura* (sweetpotato; Kuo et al. 2019).

Overexpression of miR393a from rice (*Osa-miR393a*) in *Agrostis stolonifera* (creeping bent grass) also increased tolerance to multiple stresses. The transgenic plants of creeping bent grass have a lower density of stomata and increased cuticle density, which correlated to improved tolerance against drought (administered by limiting water supply), enhanced expression of small heat-shock proteins genes that provide tolerance against heat (40°C/35°C) and increased uptake of potassium, leading to salinity tolerance (250 mM NaCl; Zhao et al. 2019a). Compared to overexpression, knock-down of miR393 in poplar variety 84K developed using *Populus alba* × *P. glandulosa* led to the promotion of plant growth as measured by plant height, stem thickness, number of internodes, stem anatomy and lignin content (Chu et al. 2021).

The complexity of miRNA-mediated regulation of the stress response was again demonstrated when miR399f was constitutively overexpressed in *A. thaliana*. These transgenic plants are endowed with enhanced tolerance against salinity (150 mM NaCl) as well as abscisic acid (ABA; 1 μM) treatment, but have increased sensitivity to drought stress (watering withheld for 12 days) (Baek et al. 2016).

MiR399 is also known to regulate phosphate uptake and homeostasis, as well as anther development and male fertility in mini-citrus *Fortunella hindsii* (Wang et al. 2020b).

MIR172 is a multi-gene miRNA family with each member playing a distinct role, and several members playing overlapping and redundant roles in the development and adaptation in plants (Lian et al. 2021; Ó'Maoiléidigh et al. 2021). Overexpression of one of the members, miR172c from *Glycine max* (*gma-miR172c*) in *A. thaliana*, resulted in increased tolerance to multiple stresses such as salt stress (250 mM NaCl) and water deficit stress (administered by 10% w/v PEG6000) as measured by higher germination rate, cotyledon greening and reduced leaf water loss, reflecting better tolerance capabilities; overexpression of miR172c was also correlated to enhanced sensitivity to ABA (100 μM), which, in turn, induced early flowering (Li et al. 2016b). Along similar lines, Pan et al. (2016) demonstrated that salt and drought, but not cold stress, induced the expression of miR172a in *Glycine max*. Overexpression of miR172a via *Agrobacterium rhizogenes* was used to generate overexpressing hairy root (OHR), and improved salt tolerance (200 mM NaCl) was observed in *G. max pre-miR172a*-OHR plants. Overexpression of miR172a was accompanied by downregulation of *Salt Suppressed AP2 domain-containing (SSACs)* target genes for miR172 in the transgenic plants (Pan et al. 2016).

An miRNA, miR319d, identified from *Solanum habrochaites*, a wild species of cultivated tomato, when overexpressed increased tolerance to two opposite spectra of temperature stress, viz. chilling and heat stress in the cultivated species *S. lycopersicum* (Shi et al. 2019). Transgenic plants overexpressing *Osa-miR319b* in *O. sativa* demonstrated 80% to 90% recovery compared to the 30% of wild-type plants when subjected to cold stress (4°C for 7 days). The expression of the predicted targets of *Osa-miR319b,* namely, *OsPCF6* and *OsTCP21,* was found to be induced upon cold stress, and cold-stress-mediated induction of these genes was found to be reduced in transgenic plants of *Osa-miR319b*. In contrast to these, the expression levels of DREB1A/B/C, DREB2A and TPP1/2 was found to increase in transgenic plants (Wang et al. 2014b). MiRNA319a from *O. sativa*, Osa-miR319a, when overexpressed in *Agrostis stolonifera* (creeping bentgrass), was shown to confer increased tolerance to salt and drought stress (Zhou and Luo 2014).

In another study, miR396b from the trifoliate orange *Poncirus trifoliata* was overexpressed in *Citrus limon* for improved cold tolerance (Zhang et al. 2016). The transformed lines showed higher levels of polyamines, reduced accumulation of ROS along with reduced transcripts of 1-aminocyclopropane-1-carboxylic acid oxidase (ACO) and ethylene content, which can be hypothesized as a cause for recovering growth in cold stress (Zhang et al. 2016).

The role of lncRNA in mediating multiple stresses can be illustrated by a case study which analysed a knock-out mutant and overexpression of *DROUGHT-INDUCED* lncRNA (DRIR). The T-DNA knock-out-*drir*^{D} was found to be a dominant mutant with higher expression of *DRIR* than the wild-type *A. thaliana*. Knock-out and overexpression analysis of this lncRNA DRIR in *A. thaliana* revealed increased tolerance towards drought and salt stress. Phenotypic analysis revealed greater sensitivity towards ABA, exhibited as differences in stomatal closure and seedling growth by both the mutant and overexpression transgenic seedlings when compared with untransformed wild-type *A. thaliana*. Genome-wide transcriptome analysis demonstrated that the expression of genes involved in ABA-regulated stress tolerance, *P5CS1*, *RD29A*, *RD29B*, *AtrbohB*, *FUT4*, *NIP1*, *TIP4*, *ANNAT7*, *NAC3* and *WRKY8,* was significantly increased in the mutant and overexpression transgenic lines conferring salt and dehydration tolerance (Qin et al. 2017).

Another lncRNAs (lncRNA973) associated with salt stress in cotton was identified through RNA-seq. Overexpression of lncRNA973 from cotton into *A. thaliana* resulted in an increased seed germination rate, fresh weight and root length of the transgenic plants in response to salt stress when compared to wild-type. Similarly, knock-down of lncRNA973 resulted in wilted plants with drooped and yellowing leaves under salt stress conditions, indicating diminished tolerance to salt stress (Zhang et al. 2019b).

1.7 ncRNA-BASED TECHNOLOGIES AND THEIR DEPLOYMENT FOR TRAIT MANIPULATION

NcRNA-based trait modification is based on PTGS via RNA interference (RNAi). Over the past few years, several small RNA-based gene modification tools based on the principles of target recognition and mechanism of action have been developed, including:

a. RNA interference
b. Artificial miRNA
c. Target mimicry

1.7.1 RNAi

The most common strategy for trait modification and imparting climate resilience has been to overexpress candidate miRNAs or small RNAs (besides other target genes) to generate transgenic plants. Overexpression vectors for PTGS are broadly classified as a co-suppression vectors, antisense vectors, hairpin constructs and intron-spliced hairpin constructs based on the principle to transcribe dsRNA or hairpin-structured RNA to induce RNAi (Wesley et al. 2001; Figure 1.2). These constructs rely on generating sense and antisense transcripts against genes or regions of interest. The dsRNA/inverted repeat or hairpin RNA that is transcribed is further processed to generate 21–22 nt dsRNA by the enzymes that are involved in biogenesis of small RNA, which then leads to RNAi mediated via PTGS of the gene of interest (Figure 1.2).

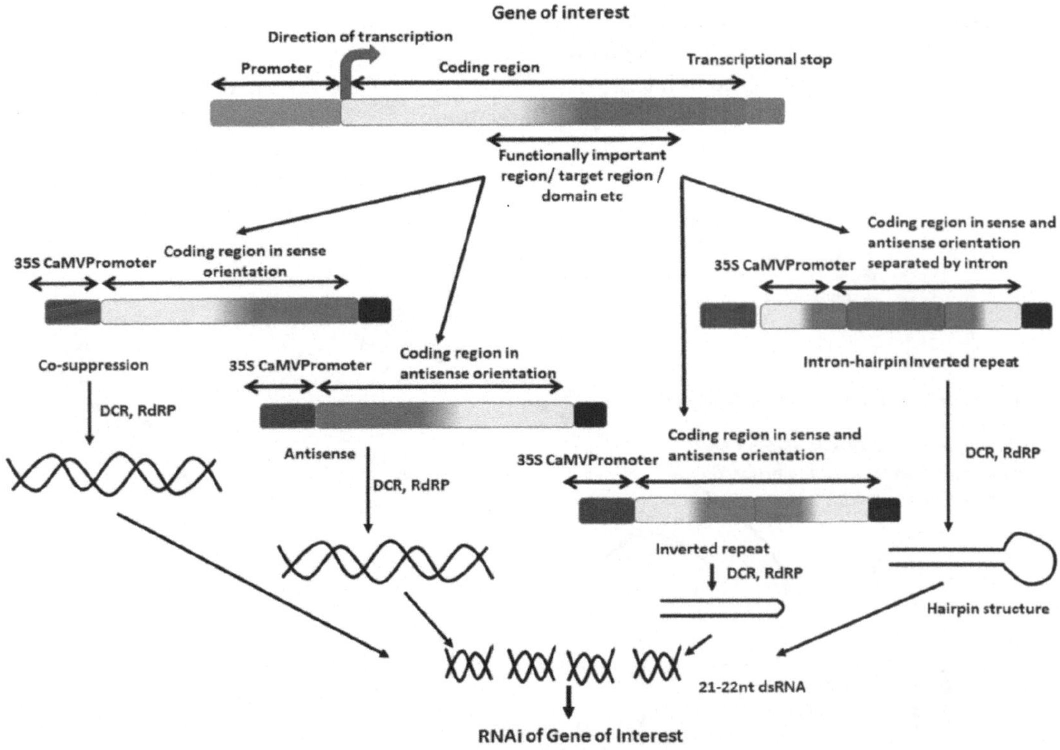

FIGURE 1.2 Diagrammatic representation of the basis of co-suppression, antisense, inverted repeat and intron-hairpin-inverted repeat constructs for RNAi-mediated gene silencing.

1.7.2 Artificial miRNA (amiRNA)

Small RNA-based gene silencing, one of the most commonly employed approaches in trait improvement of crop plants, suffers from the unintended side effects of recognition and cleavage of non-target mRNA. To overcome this, artificial miRNA (amiRNA) technology was developed initially in mammalian cells, and subsequently expanded to plants (Zeng et al. 2002; Parizotto et al. 2004; Schwab et al. 2006; Zhang et al. 2018c). The amiRNA strategy relies on analysing the target gene sequence to identify a 21 nt mature amiRNA:miRNA* duplex (Schwab et al. 2006; Singh and Das 2021). The amiRNA:amiRNA* duplex is then used to replace miRNA:miRNA* in the natural or native miRNA precursor backbone with the help of overlapping PCR or site-directed mutagenesis (Schwab et al. 2006). The backbone of several miRNA genes such as MIR319 and MIR390 in *A. thaliana* (Schwab et al. 2006; Carbonell et al. 2014), MIR528 in *O. sativa* (Warthmann et al. 2008), MIR164 in *T. aestivum* (Gasparis et al. 2017) and MIR156d in *Malus domestica* (Charrier et al. 2019) has been used for constructing amiRNA-mediated gene silencing vectors in plants. These constructs, when introduced into plants, produce a single dsRNA duplex of amiRNA:amiRNA* from each precursor transcript, which is instrumental in bringing about PTGS of the target locus without affecting the non-target transcripts (Schwab et al. 2006; Niu et al. 2006; Wagaba et al. 2016; Faisal et al. 2017; Singh and Das 2021; Figure 1.3).

1.7.3 Target Mimicry

Apart from the commonly employed strategy of overexpression of miRNAs/ncRNAs and the silencing of target genes via amiRNAs, strategies have also been developed to generate loss-of-function mutants of small RNAs. One such ncRNA-based technology that has a potential for trait manipulation in plants is through target mimicry. As discussed previously, lncRNAs modulate gene expression via acting as a (i) signal, (ii) decoy, (iii) guide, and (iv) scaffold by participating in lncRNA-DNA, lncRNA-RNA and lncRNA-protein interactions (Wang et al. 2018b). Of the four modes of action, lncRNAs have been shown to act as "decoys", "target mimics" or "sponges" and sequester small

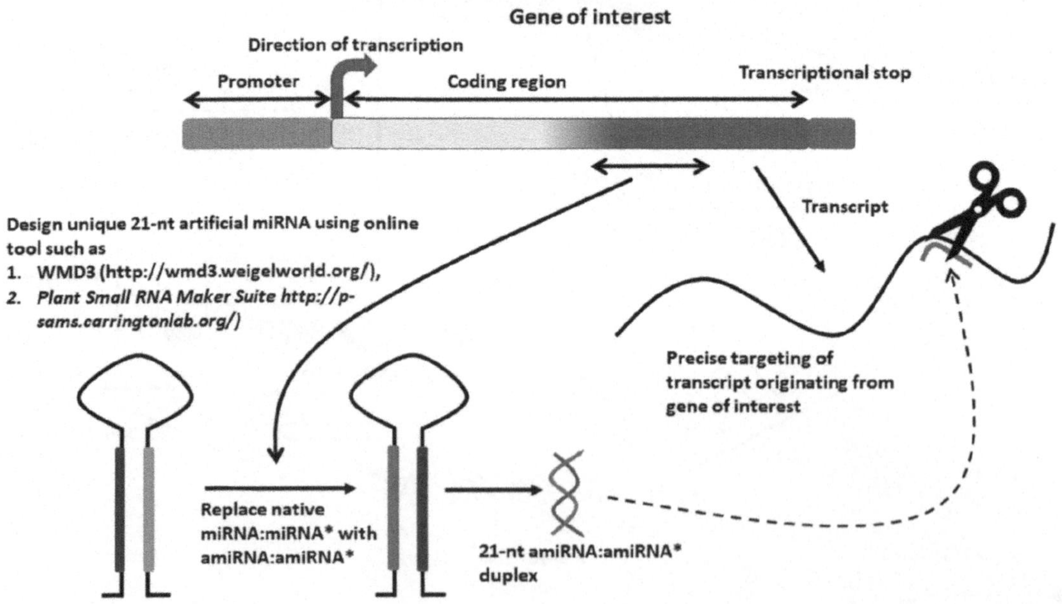

FIGURE 1.3 Diagrammatic representation of the basis of design of the artificial miRNA construct and PTGS-mediated gene silencing.

ncRNAs such as miRNAs and siRNAs (Franco-Zorrilla et al. 2007). LncRNAs, therefore, are also termed as endogenous target mimics (eTMs; Wu et al. 2013; Wang et al. 2015b; Karakulah et al. 2016; Deng et al. 2020). Similarly, circRNAs are also known to contain binding sites for miRNAs and act as ceRNA as sponges or target mimics (Franco-Zorrilla et al. 2007; Ebert and Sharp 2010; Reichel et al. 2015; Panda 2018; Bose and Ain 2018; Zhang et al. 2020b). Based on this principle of action of the lncRNAs and circRNAs, strategies such as artificial target mimics (aTM) and STTM, miRNA sponges have been developed (Franco-Zorrilla et al. 2007; Tang et al. 2012; Tang and Tang 2013; Reichel et al. 2015; Peng et al. 2018). All these strategies rely on sequestering miRNA/ncRNA by overexpressing a competitive transcript with an miRNA binding site (Figure 1.4). The transcript could be a modified gene, lncRNA, synthetic sequence or circRNA, each with one (or multiple as in STTM) miRNA binding site(s) that incorporates mismatches at the 10/11 position to prevent target cleavage. The lncRNA *IPS1* was found to harbour a mutated binding site for miR399 that is induced upon phosphate starvation. The mutation at the miRNA cleavage site prevents PTGS of *IPS1* while sequestering miR399 and leading to upregulation of the miR399 target transcript *PHO2* (Franco-Zorrilla et al. 2007). The authors coined the term "target mimicry" to explain the novel mode of regulation of miRNA activity and demonstrated the universality and utility of target mimicry for modulating miRNA expression by incorporating mutated miRNA binding sites for miR156b (MIM156b) and miR319b (MIM319b) (Franco-Zorrilla et al. 2007). Target mimicry has been used to generate loss-of-function mutants of members of miRNA families, miR156 (MIN156), miR157 (MIM157), miR159 (MIM159), miR164 (MIM164), miR167 (MIM167), miR172 (MIM172), miR319 (MIM319) and miR393 (MIM393) that show altered development, to cite a few examples (Todesco et al. 2010). "miR773 (MIM773) of eTMs (Wu et al. 2013)" with "target mimicry against miR773 (MIM773) was found to have an enhanced response to the necrotrophic fungal pathogen *Plectosphaerella cucumerina* and two hemibiotrophs, *Fusarium oxysporum* and *Colletotrichum higginsianum* (Salvador-Guirao et al. 2018a); similarly, target mimicry against miR396 (MIM396)

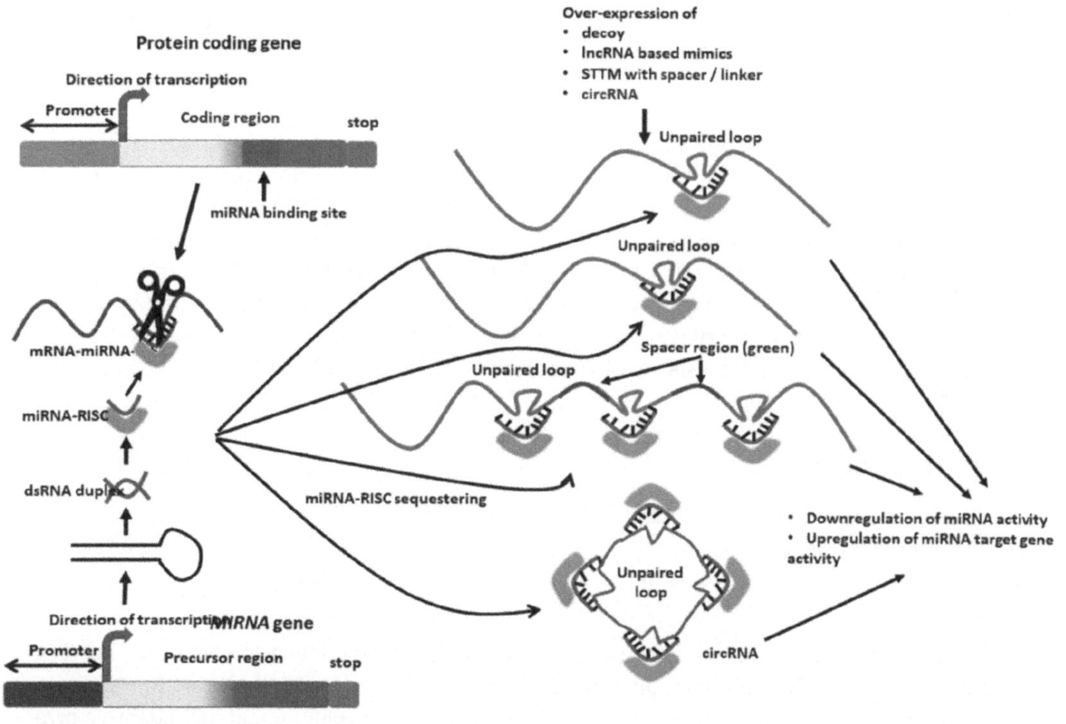

FIGURE 1.4 Diagrammatic representation of the design of decoy, mimics, STTM and miRNA sponges based on expression of modified lncRNA, synthetic RNA with spacer/linker and modified circRNA.

showed an increased tolerance of MIM396 plants to *Plectosphaerella cucumerina*, *Botrytis cinerea* (both necrotrophic fungi) and *Fusarium oxysporum* f. sp. *conglutinans* and *Colletotrichum higginsianum* (both hemibiotrophic fungi) (Soto-Suárez et al. 2017). Finally computational tools has been used to identify endogenous target mimics (eTMs) that originate from intergenic space or non-coding regions and targeted against multiple miRNAs in *A. thaliana* and *O. sativa* (Wu et al. 2013).

STTM were subsequently developed by designing a "shorter non-coding RNAs (sncRNA)" of 108 nt RNA with two-copies of miR165/miR166 binding sites with mismatches, separated by a short linker sequence (Yan et al. 2012). STTM technology has now been used to modify several developmental traits and analyse response to abiotic and biotic stresses in several plants, including crops. STTM-mediated downregulation of members of the miR482 and miR2118 families was shown to enhance resistance against strains of *P. infestans* and *P. syringae* via negative regulation of nucleotide-binding domain leucine-rich repeat (*NLR*) genes in *S. lycopersicum* (Canto-Pastor et al. 2019). STTM has also been used to study the: involvement of miR171 and miR390 in hypoxia in roots in *A. thaliana* and wild tomato—*Solanum habrochaites* (Hou et al. 2019); role of miR9678 in production and regulation of a lncRNA-*WSGAR* (*Wheat Seed Germination Associated RNA*) and biogenesis of phasiRNA from the lncRNA and impact on seed germination in *T. aestivum* (Guo et al. 2018); role of miR1514a in regulation of two NAC transcription factors, Phvul.010g121000 and Phvul.010g120700, and biogenesis of phasiRNA under drought stress in *Phaseolus vulgaris* (Sosa-Valencia et al. 2017); molecular regulation of leaf development and drought tolerance mediated via miR160, miR165/166 in *A. thaliana* (Yang et al. 2019); effect of downregulation of miR172, miR165/miR166 via Tobacco rattle virus (TRV)-based STTM to develop virus-based miRNA silencing (VbMS) in *S. lycopersicum* and *Nicotiana benthamiana* (Sha et al. 2014); miR393 in biomass production in poplars (Chu et al. 2021); interplay of miR160 and auxin signalling in root architecture in *Solanum tuberosum* (Yang et al. 2021); development of a new STTM vector system based on a maize-infecting strain of cucumber mosaic virus strain ZMBJ (ZMBJ-CMV) to downregulate miR159, miR165/miR166 to study development in *Z. mays* (Liu et al. 2019b); TRV-STTM-mediated silencing of miR164d, miR396b, Novel-miR1 and Novel-miR7 to study resistance against *Corynespora cassiicola* in *Cucumis sativus* L. (Wang et al. 2019b); and modulation of miR167b-3p based on potato virus X (PVX)-STTM to study infection and tolerance against tobacco curly shoot virus in *Nicotiana benthamiana* (Wu et al. 2021), to cite a few recent examples.

Another form of artificial target mimicry is the MiRNA sponge, which is produced from transgenes containing multiple miRNA binding sites located in tandem repeats. These sponges have been used to target and inhibit entire families of miRNAs (Reichel et al. 2015; Thomson and Dinger 2016).

The deployment of circRNA as miRNA sponges, although, has been informatically predicted and demonstrated in several published work as in *Pyrus betulifolia* (Wang et al. 2018), *Brassica rapa* L. ssp. *pekinensis* (Wang et al. 2019c) and *O. sativa* (Huang et al. 2021); their deployment in trait manipulation has not yet been extensively reported except in a few cases where CRISPR-Cas9-mediated editing of the circRNA locus Os06circ02797 showed the circRNA to be a sponge for miR408 (Zhou et al. 2021).

1.8 CONCLUSION

The deluge of sequence information and functional evidence into the role of ncRNA in mediating biotic and abiotic stress tolerance has increased manifold in recent years. Based on availability, as well as a lack of literature, the following concluding remarks can be made:

a. As has been widely reported for protein-coding genes, regulatory networks of ncRNA including small RNA, lncRNA and circRNA are lacking, including their role in transcriptional gene silencing. Given the enormous complexity of developmental and adaptive processes, it is extremely challenging and difficult to pinpoint the exact role of any molecular regulator, including ncRNAs.

b. CRISPR-Cas9-based gene editing has been employed to introduce indels at miRNA processing sites in *MIR* genes, thus delaying miRNA biogenesis (Rojo et al. 2018; Tiwari et al. 2021), to impede miRNA-mRNA target pairing, to complete deletion and knock-in of *MIR* genes and/or promoters (Zhou et al. 2017) and to impart abiotic stress tolerance in *O. sativa* (Yue et al. 2020). The application of CRISPR-Cas9, one of the most efficient genome-editing tools for editing either an entire miRNA gene family or targeting specific miRNA within the gene family, has thus opened up newer avenues for modulating expression of ncRNAs.
c. Recent advances in gene-delivery methods, such as delivery of small dsRNA inside a plant cell with the help of an organic carrier molecules (ribonucleoprotein cross linkers) or nanoparticles (biopolymer of chitosan, silicon, carbon and clay nanosheets) (Joga et al. 2016; Cunningham et al. 2018; McLoughlin et al. 2018; Hendrix et al. 2021), are promising to widen the scope of application of small RNAs beyond the plant species that are otherwise amenable to *Agrobacterium*-mediated genetic transformation.
d. Development of software and tools, and the cataloguing and availability of several ncRNA-related databases such as miRFAN (Liu et al. 2012), PncStress (database of ncRNA involved in stress response; Wu et al. 2020), PendoTMBase (Plant endoTarget mimicry; Deng et al. 2020), PeTMbase (database of endogenous target mimic in plants; Karakulah et al. 2016), GreenCircRNA (Zhang et al. 2020a), PCirc (Yin et al. 2021), PlantcircBase (http://ibi.zju.edu.cn/plantcircbase/; Chu et al. 2017), miRSponge (Wang et al. 2015a) and CropCircDB (Wang et al. 2019b) are likely to be exploited by researchers to study plant development and responses to biotic and abiotic stresses.
e. Several confounding reports of small RNAs under various stresses have appeared including: (i) one small RNA (miRNA) in multiple stresses, (ii) multiple small RNA (miRNA) in a single stress, (iii) conserved small RNAs and species-specific small RNA (miRNA), (iv) lack of information on diversity and conservation of lncRNA and circRNA across species and (v) absence of information on correlation between genome complexity, ploidy and diversity of small RNA sequence and structure. It is unclear whether the information generated in a model species can be extrapolated to crop plants, or plants with complex genome features.

Small RNAs, owing to their central role as master regulators, have great, but complicated, potential. They hold enormous promise for trait modification, including imparting climate resilience; however, this hinges on identifying small RNAs that are unique or specific to one or few stress responses as well as have negligible roles in organ or plant development.

Plants are likely to encounter multiple stresses simultaneously, and, therefore, identification of candidate small RNAs that are involved in mediating such multiple stresses is prudent. Additionally, small RNAs with additive effects are potential candidates for imparting resilience against multiple stresses (Taylor et al. 2014; Shriram et al. 2016; Xu et al. 2019). The choice of small RNA to impart stress resilience must avoid or minimize introducing developmental defects, such as a pleiotropic effect, because of the role of small RNA in both development and adaptive processes. Precise spatio-temporal expression and inducible expression of candidate small RNAs, or that of decoys/sponges coupled with gene-editing, may hold promise to address the challenges in future research programmes.

ACKNOWLEDGEMENTS

The research in the laboratory of SD is funded by the Department of Biotechnology, Govt. of India; University Grants Commission, Govt. of India through an extra-mural research grant; and Delhi University through research and development support, Institute of Eminence (IoE) grants, and Infrastructure grants. SS is grateful to Council of Scientific & Industrial Research (CSIR), Govt. of India, for the award of the RA Fellowship.

We would also like to apologize to those authors of excellent research and review articles that could not be cited due to space constraints.

REFERENCES

Adenot X, Elmayan T, Lauressergues D, Boutet S, Bouche N, Gasciolli V, Vaucheret H (2006) DRB4-dependent TAS3 trans-acting siRNAs control leaf morphology through AGO7. Curr Biol 16:927–932

Agrawal A, Rajamani V, Reddy VS, et al. (2015) Transgenic plants over-expressing insect-specific microRNA acquire insecticidal activity against *Helicoverpa armigera*: An alternative to Bt-toxin technology. Transgenic Res 24:791–801

Ahmed W, Xia Y, Zhang H, Li R, Bai G, Siddique KHM, Guo P (2019) Identification of conserved and novel miRNAs responsive to heat stress in flowering Chinese cabbage using high-throughput sequencing. Sci Rep 9(1):14922

Ajmera I, Shi J, Giri J, Wu P, Stekel DJ, Lu C, Hodgman TC (2018) Regulatory feedback response mechanisms to phosphate starvation in rice. Syst Biol Appl 4:4

Akdogan G, Tufekci ED, Uranbey S, Unver T (2016) miRNA-based drought regulation in wheat. Funct Integr Genomics 16:221–233

Allen E, Howell MD (2010) miRNAs in the biogenesis of trans-acting siRNAs in higher plants. Semin Cell Dev Biol 21:798–804

Altieri MA, Nicholls CI (2017) The adaptation and mitigation potential of traditional agriculture in a changing climate. Clim Change 140:33–45

Anand S, Singh NK, Das S (2017) Small RNAs: I: Role as developmental and adaptive regulators in plants. In: Sajad Majeed Zargar, Vandna Rai (eds.) Plant OMICS and Crop Breeding. Apple Academic Press/CRC Press, New Jersey, USA. pp. 115–161

Araus J, Slafer G, Reynolds M, Royo C (2002) Plant breeding and drought in C3 cereals: What should we breed for? Ann Bot 89:925–940

Ariel F, Jegu T, Latrasse D, Romero-Barrios N, Christ A, Benhamed M, Crespi M (2014) Noncoding transcription by alternative RNA polymerases dynamically regulates an auxin-driven chromatin loop. Mol Cell 55(3):383–396

Asseng S, Ewert F, Martre P, Rötter RP, Lobell D, Cammarano D, Kimball B, Ottman M, Wall G, White JW (2015) Rising temperatures reduce global wheat production. Nat Clim Change 5:143

Axtell MJ, Jan C, Rajagopalan R, Bartel DP (2006) A two-hit trigger for siRNA biogenesis in plants. Cell 127(3):565–577

Axtell MJ, Westholm JO, Lai EC (2011) Vive la différence: Biogenesis and evolution of microRNAs in plants and animals. Genome Biol 12:1–13

Aydinsakir K, Erdal S, Buyuktas D, Bastug R, Toker R (2013) The influence of regular deficit irrigation applications on water use, yield, and quality components of two corn (*Zea mays* L) genotypes Agricultural Water Management 128:65–71

Badu-Apraku B, Hunter RB, Tollenaar M (1983). Effect of temperature during grain filling on whole plant and grain yield in maize (*Zea mays* L.). Can J Plant Sci 63:357–363

Baek D, Chun HJ, Kang S, Shin G, Park SJ, Hong H, Kim C, Kim DH, Lee SY, Kim MC, Yun DJ (2016) A role for *Arabidopsis* miR399f in salt, drought, and ABA signaling. Mol Cells 39(2):111–118

Baldrich P, Kakar K, Siré C, Moreno AB, Berger A, García-Chapa M, López-Moya JJ, Riechmann JL, San Segundo B (2014) Small RNA profiling reveals regulation of *Arabidopsis* miR168 and heterochromatic siRNA415 in response to fungal elicitors. BMC Genomics 15:1083

Baldrich P, San Segundo B (2016) MicroRNAs in rice innate immunity. Rice 9:6

Balyan S, Kumar M, Mutum RD, Raghuvanshi U, Agarwal P, Mathur S, Raghuvanshi S (2017) Identification of miRNA-mediated drought responsive multi-tiered regulatory network in drought tolerant rice, Nagina 22. Sci Rep 27:15446

Bao N, Lye KW, Barton MK (2004). MicroRNA binding sites in *Arabidopsis* class III HD-ZIP mRNAs are required for methylation of the template chromosome. Dev Cell 7(5):653–662

Bardou F, Ariel F, Simpson CG, Romero-Barrios N, Laporte P, Balzergue S, Brown JWS, Crespi M (2014) Long noncoding RNA modulates alternative splicing regulators in *Arabidopsis*. Dev Cell 30:166–176

Baroowa B, Gogoi N (2014) Biochemical changes in black gram and green gram genotypes after imposition of drought stress. J Food Legum 27:350–353

Barrera-Figueroa BE, Gao L, Wu Z, et al. (2012) High throughput sequencing reveals novel and abiotic stress-regulated microRNAs in the inflorescences of rice. BMC Plant Biol 12:132

Bhogireddy S, Mangrauthia SK, Kumar R, Pandey AK, Singh S, Jain A, Budak H, Varshney RK, Kudapa H (2021) Regulatory non-coding RNAs: A new frontier in regulation of plant biology. Funct Integr Genomics 21(3–4):313–330

Blevins T, Podicheti R, Mishra V, Marasco M, Wang J, Rusch D, Tang H, Pikaard CS (2015) Identification of pol IV and RDR2-dependent precursors of 24 nt siRNAs guiding de novo DNA methylation in *Arabidopsis*. Elife 4:e09591

Böhmdorfer G, Wierzbicki AT (2015) Control of chromatin structure by long noncoding RNA. Trends Cell Biol 25(10):623–632

Bonan GB, Doney SC (2018) Climate, ecosystems, and planetary futures: The challenge to predict life in earth system models. Science 359(80):359(6375):eaam8328

Borchert GM, Lanier W, Davidson BL (2006) RNA polymerase III transcribes human microRNAs. Nat Struct Mol Biol 13(12):1097–101

Borges F, Martienssen RA (2015) The expanding world of small RNAs in plants. Nat Rev Mol Cell Biol 16:727–741

Bose R, Ain R (2018) Regulation of transcription by circular RNAs. In: J. Xiao (ed.) Circular RNAs, Advances in Experimental Medicine and Biology, vol. 1087 Springer Nature Singapore Pte Ltd. pp. 81–94

Brant EJ, Budak H (2018) Plant small non-coding RNAs and their roles in biotic stresses, Front Plant Sci 9:1038–1038.

Britten RJ, Davidson EH (1969) Gene regulation for higher cells: A theory. Science 165:349–357

Brodersen P, Voinnet O (2009) Revisiting the principles of microRNA target recognition and mode of action. Nat Rev Mol Cell Biol 10:141–148

Brozynska M, Furtado A, Henry RJ (2016) Genomics of crop wild relatives: Expanding the gene pool for crop improvement. Plant Biotechnol J 14:1070–1085

Budak H, Kaya SB, Cagirici HB (2020) Long non-coding RNA in plants in the era of reference sequences. Front Plant Sci 11:276

Candogan BN, Sincik M, Buyukcangaz H, Demirtas C, Goksoy AT, Yazgan S. (2013) Yield, quality and crop water stress index relationships for deficit irrigated soybean [*Glycine max* (L.) Merr.] In subhumid climatic conditions. Agric Water Manage 118:113–121

Canto-Pastor A, Santos BAMC, Valli AA, Summers W, Schornack S, Baulcombe DC (2019) Enhanced resistance to bacterial and oomycete pathogens by short tandem target mimic RNAs in tomato. Proc Natl Acad Sci USA 116(7):2755–2760

Cao Y, Fan G, Zhai X, Dong Y (2018) Genome-wide analysis of lncRNAs in *Paulownia tomentosa* infected with phytoplasmas, Acta Physiol Plant 40:49–49

Cao W, Gan L, Wang C, Zhao X, Zhang M, Du J, Zhou S, Zhu C (2021) Genome-wide identification and characterization of potato long non-coding RNAs associated with *Phytophthora infestans* resistance. Front Plant Sci 12:619062

Carbonell A (2019) Secondary small interfering RNA-based silencing tools in plants: An update. Front Plant Sci 10:687

Carbonell A, Takeda A, Fahlgren N, Johnson SC, Cuperus JT, Carrington JC (2014) New generation of artificial MicroRNA and synthetic trans-acting small interfering RNA vectors for efficient gene silencing in *Arabidopsis*. Plant Physiol 165:15–29

Carnavale Bottino M, Rosario S, Grativol C, et al. (2013) High-throughput sequencing of small RNA transcriptome reveals salt stress regulated MicroRNAs in sugarcane. PLoS ONE 8:e59423

Carrió-Seguí À, Ruiz-Rivero O, Villamayor-Belinchón L, Puig S, Perea-García A, Peñarrubia L (2019) The altered expression of microRNA408 influences the *Arabidopsis* response to iron deficiency. Front Plant Sci 10:324

Casadevall R, Rodriguez RE, Debernardi JM, et al. (2013) Repression of growth regulating factors by the MicroRNA396 inhibits cell proliferation by UV-B radiation in *Arabidopsis* leaves. Plant Cell 25:3570–3583

Ceoloni C, Kuzmanović L, Ruggeri R, et al. (2017) Harnessing genetic diversity of wild gene pools to enhance wheat crop production and sustainability: Challenges and opportunities. Diversity 9: https://doi.org/10.3390/d9040055

Chand SK, Nanda S, Rout E, Mohanty J, Mishra R, Joshi RK (2016) Identification and characterization of microRNAs in turmeric (*curcuma Longa* L.) Responsive to infection with the pathogenic fungus *Pythium aphanidermatum*. Physiol Mol Plant Pathol 93:119–128

Charrier A, Vergne E, Joffrion C, Richer A, Dousset N, Chevreau E (2019) An artificial miRNA as a new tool to silence and explore gene functions in apple. Transgenic Res 28:611–626

Chaudhary S, Grover A, Sharma PC (2021) MicroRNAs: Potential targets for developing stress- tolerant crops. Life 11:289

Chauhan S, Yogindran S, Rajam MV (2017) Role of miRNAs in biotic stress reactions in plants. Indian J Plant Physiol 22:514–529

Chen H, Chen X, Chen D, Li J, Zhang Y, Wang A (2015) A comparison of the low temperature transcriptomes of two tomato genotypes that differ in freezing tolerance: *Solanum lycopersicum* and *Solanum habrochaites*. BMC Plant Biol 15:1–16

Chen G, Cui J, Wang L, et al. (2017) Genome-wide identification of circular RNAs in *Arabidopsis thaliana.* Front Plant Sci 8:1678–1678

Chen L, Ren Y, Zhang Y, Xu J, Sun F, Zhang Z, Wang Y (2012) Genome-wide identification and expression analysis of heat-responsive and novel microRNAs in *Populus tomentosa*. Gene 504:160–165

Chen M, Wang C, Bao H, Chen H, Wang Y (2016) Genome-wide identification and characterization of novel lncRNAs in populus under nitrogen deficiency. Mol Genet Genomics 291:1663–1680

Chen LL, Yang L (2015) Regulation of circRNA biogenesis, RNA Biol 12:381–388

Chen L, Zhu QH, Kaufmann K (2020) Long non-coding RNAs in plants: Emerging modulators of gene activity in development and stress responses. Planta 252:1–14

Chi SW, Hannon GJ, Darnell RB (2012) An alternative mode of microRNA target recognition. Nat Struct Mol Biol 19:321–327

Chu Q, Bai P, Zhu X, Zhang X, Mao L, Zhu QH, Fan L, Ye CY (2018a) Characteristics of plant circular RNAs. Brief Bioinform 15:135–143

Chu L, He X, Shu W, Wang L, Tang F (2021) Knockdown of miR393 promotes the growth and biomass production in poplar. Front Plant Sci 12:714907

Chu Q, Shen E, Fan L (2018b) Emerging roles of plant circular RNAs. J Plant Cell Dev 1:1–14

Chu Q, Zhang X, Zhu X, Liu C, Mao L, Ye C, Zhu Q, Fan L (2017) PlantcircBase: A database for plant circular RNAs. Mol Plant 10(8):1126–1128

Chukwudi UP, Kutu FR, Mavengahama S (2021) Heat stress effect on the grain yield of three drought-tolerant maize varieties under varying growth conditions. Plants 10:1532

Clemson CM, Hutchinson JN, Sara SA, Ensminger AW, Fox AH, Chess A, Lawrence JB (2009) An architectural role for a nuclear noncoding RNA: NEAT1 RNA is essential for the structure of paraspeckles. Mol Cell 33:717–726

Cobb JN, Biswas PS, Platten JD (2019) Back to the future: Revisiting MAS as a tool for modern plant breeding. Theor Appl Genetics 132:647–667

Csorba T, Questa JI, Sun Q, Dean C (2014) Antisense COOLAIR mediates the coordinated switching of chromatin states at FLC during vernalization. Proc Natl Acad Sci USA 111:16160–16165

Cui J, Jiang N, Hou X, Wu S, Zhang Q, Meng J, Luan Y (2020) Genome-wide identification of lncRNAs and analysis of ceRNA networks during tomato resistance to *Phytophthora infestans*. Phytopathology 110:456–464

Cui J, Jiang N, Meng J, et al. (2019) LncRNA33732-respiratory burst oxidase module associated with WRKY1 in tomato- *Phytophthora infestans* interactions. Plant J 97:933–946

Cui J, Luan Y, Jiang N, Bao H, Meng J (2017) Comparative transcriptome analysis between resistant and susceptible tomato allows the identification of lncRNA16397 conferring resistance to *Phytophthora infestans* by co-expressing glutaredoxin. Plant J 89:577–589

Cunningham FJ, Goh NS, Demirer GS, et al. (2018) Nanoparticle-mediated delivery towards advancing plant genetic engineering. Trends Biotechnol 36:882–897

Darbani B, Noeparvar S, Borg S (2016) Identification of circular RNAs from the parental genes involved in multiple aspects of cellular metabolism in barley. Front Plant Sci 7:776

Das R, Mukherjee A, Basak S, Kundu P (2021) Plant miRNA responses under temperature stress. Plant Gene 28:100317

De Oliveira ED, Bramley H, Siddique KH, Henty S, Berger J, Palta JA (2013) Can elevated CO_2 combined with high temperature ameliorate the effect of terminal drought in wheat? Funct Plant Biol 40:160–171

Deng J, Li Q, Huang L, Tang W, Zheng K, Yan J, Wu W (2020) PendoTMBase: A database for plant endogenous target mimics. Interdiscip Sci 12(4):526–529

Deng P, Muhammad S, Cao M, Wu L (2018) Biogenesis and regulatory hierarchy of phased small interfering RNAs in plants. Plant Biotechnol J 16:965–975

Dhankher OP, Foyer CH (2018) Climate resilient crops for improving global food security and safety. Plant Cell Environ 41:877–884

Diebel KW, Smith AL, van Dyk LF (2010) Mature and functional viral miRNAs transcribed from novel RNA polymerase III promoters. RNA 16(1):170–185

Dieci G, Fiorino G, Castelnuovo M, Teichmann M, Pagano A (2007) The expanding RNA polymerase III transcriptome. Trends Genet 23(12):614–622

Dietz KJ, Zörb C, Geilfus CM (2021) Drought and crop yield. Plant Biol (Stuttg) 23(6):881–893

Ding Y, Chen Z, Zhu C (2011) Microarray-based analysis of cadmium-responsive microRNAs in rice (*Oryza sativa*). J Exp Bot 62:3563–3573

Ding Y, Huang L, Jiang Q, Zhu C (2020) MicroRNAs as important regulators of heat stress responses in plants. J Agric Food Chem 68(41):11320–11326

Ding J, Lu Q, Ouyang Y, Mao H, Zhang P, Yao J, Xu C, Li X, Xiao J, Zhang Q (2012a) A long noncoding RNA regulates photoperiod-sensitive male sterility, an essential component of hybrid rice. Proc Natl Acad Sci USA 109(7):2654–2659

Ding D, Zhang L, Wang H, Liu Z, Zhang Z, Zheng Y (2009) Differential expression of miRNAs in response to salt stress in maize roots. Ann Bot 103(1):29–38

Ding J, Zhou S, Guan J (2012b) Finding MicroRNA targets in plants: Current status and perspectives. Genomics, Proteomics Bioinf 10:264–275

Dolker D, Bakshi P, Dorjey S, Choudhary P, Kour K, Singh M (2017) Effect of deficit irrigation scheduling on yield and quality of kinnow mandarin fruits. Int J Curr Microbiol Appl Sci 6:261–269

Dotto M, Gómez MS, Soto MS, Casati P (2018) UV-B radiation delays flowering time through changes in the PRC2 complex activity and miR156 levels in *Arabidopsis thaliana*. Plant Cell Environ 41:1394–1406

Eastburn DM, Degennaro MM, Delucia EH, Dermody O, McElrone AJ (2010) Elevated atmospheric carbon dioxide and ozone alter soybean diseases at SoyFACE. Globle Change Biol 16:320–330.

Easterling WE, Aggrawal PK, Batima P, Brander KM, Erda L, Howden SM, Kirilenko A, Morton J, Soussana J-F, Schmidhuber J, Tubiello FN (2007) Food, fibre and forest products. In: In: Parry ML, Canziani OF, Palutikof JP, Van Der Linden PJ, Hanson CE (eds) Climate Change 2007: Impacts, Adaptation and Vulnerability. Contribution of Working Group II to the Fourth Assessment Report of the Intergovernmental Panel on Climate Change. Cambridge University Press, Cambridge, UK. Pp 273–313

Ebert MS, Sharp PA (2010) MicroRNA sponges: Progress and possibilities. RNA 16(11):2043–2050

Ellwanger DC, Buttner FA, Mewes HW, Stumpflen V (2011). The sufficient minimal set of miRNA seed types. Bioinformatics 27:1346–1350

Faisal A, Biswas S, Zerin T, Rahman T, Seraj Z (2017) Downregulation of the DST transcription factor using artificial microRNA to increase yield, salt and drought tolerance in rice. Am J Plant Sci 8:2219–2237

Fan G, Cao Y, Wang Z (2018) Regulation of long noncoding RNAs responsive to phytoplasma infection in *Paulownia tomentosa*. Int J Genomics 2018: Article ID 3174352

Faust T, Frankel AD, D'Orso I (2012) Transcription control by long non-coding RNAs. Transcription 3(2):78–86

Feng S, Fang H, Liu X, Dong Y, Wang Q, Yang KQ (2021) Genome-wide identification and characterization of long non-coding RNAs conferring resistance to *Colletotrichum gloeosporioides* in walnut (*Juglans regia*). BMC Genomics 22:1–17

Feng K, Nie X, Cui L, Deng P, Wang M, Song W (2017) Genome-wide identification and characterization of salinity stress-responsive miRNAs in wild emmer wheat (*Triticum turgidum* ssp. *dicoccoides*). Genes (Basel) 8(6):156

Franco-Zorrilla JM, Valli A, Todesco M, Mateos I, Puga MI, Rubio-Somoza I, Leyva A, Weigel D, García JA, Paz-Ares J (2007) Target mimicry provides a new mechanism for regulation of microRNA activity. Nat Genet 39:1033–1037

Fu XZ, Zhang XY, Qiu JY, et al. (2019) Whole-transcriptome RNA sequencing reveals the global molecular responses and ceRNA regulatory network of mRNAs, lncRNAs, miRNAs and circRNAs in response to copper toxicity in Ziyang Xiangcheng (*Citrus junos* sieb. Ex tanaka). BMC Plant Biol 19:509

Fu R, Zhang M, Zhao Y, He X, Ding C, Wang S, Feng Y, Song X, Li P, Wang B (2017) Identification of salt tolerance-related microRNAs and their targets in maize (*Zea mays* L.) Using high-throughput sequencing and degradome analysis. Front Plant Sci 8:864

Gao P, Bai X, Yang L, Lv D, Pan X, Li Y, Cai H, Ji W, Chen Q, Zhu Y. osa-MIR393: a salinity- and alkaline stress-related microRNA gene. Mol Biol Rep. 2011 Jan; 38(1):237–42

Gao Z, Li J, Luo M, Li H, Chen Q, Wang L, Song S, Zhao L, Xu W, Zhang C, Wang S, Ma C (2019) Characterization and cloning of grape circular RNAs identified the cold resistance-related vv-*circATS1*. Plant Physiol 180(2):966–985

Gao S, Yang L, Zeng HQ, Zhou ZS, Yang ZM, Li H, Sun D, Xie F, Zhang B (2016) A cotton miRNA is involved in regulation of plant response to salt stress. Sci Rep 6:19736

Garrity DP, O'Toole JC (1994) Screening rice for drought resistance at the reproductive phase. Field Crops Res 39:99–110

Gasparis S, Kała M, Przyborowski M, Orczyk W, Nadolska-Orczyk A (2017) Artificial microRNA based specific gene silencing of grain hardness genes in polyploid cereals appeared to be not stable over transgenic plant generations. Front Plant Sci 7:2017

Gentile A, Dias LI, Mattos RS, Ferreira TH, Menossi M (2015) MicroRNAs and drought responses in sugarcane. Front Plant Sci 6:58

Gepts P (2002) A comparison between crop domestication, classical plant breeding, and genetic engineering. Crop Sci 42:1780–1790

Ghorbani A, Izadpanah K, Peters JR, Dietzgen RG, Mitter N (2018) Detection and profiling of circular RNAs in uninfected and maize Iranian mosaic virus-infected maize. Plant Sci 274:402–409

Gilbert N (2014) Cross-bred crops get fit faster. Nature 513:292

Grammatikakis I, Panda AC, Abdelmohsen K, Gorospe M (2014) Long noncoding RNAs(lncRNAs) and the molecular hallmarks of aging. Aging (Albany NY) 6(12):992–1009

Gu TJ, Yi X, Zhao XW, Zhao Y, Yin JQ (2009) Alu-directed transcriptional regulation of some novel miRNAs. BMC Genomics 10:563

Guo G, Liu X, Sun F, Cao J, Huo N, Wuda B, Xin M, Hu Z, Du J, Xia R, Rossi V, Peng H, Ni Z, Sun Q, Yao Y (2018) Wheat miR9678 affects seed germination by generating phased siRNAs and modulating abscisic acid/gibberellins signaling. Plant Cell 30:796–814

Gupta RA, Shah N, Wang KC, et al. (2010) Long non-coding RNA *HOTAIR* reprograms chromatin state to promote cancer metastasis. Nature 464:1071–1076

Gupta OP, Sharma P, Gupta RK, Sharma I (2014) MicroRNA mediated regulation of metal toxicity in plants: Present status and future perspectives. Plant Mol Biol 84:1–18

Hajyzadeh M, Turktas M, Khawar KM, Unver T (2015) miR408 overexpression causes increased drought tolerance in chickpea. Gene 555(2):186–193

Ha M, Kim VN (2014) Regulation of microRNA biogenesis. Nat Rev Mol Cell Biol 15:509–524

He J, Jiang Z, Gao L, You C, Ma X, Wang X, Xu X, Mo B, Chen X, Liu L (2019) Genome-wide transcript and small RNA profiling reveals transcriptomic responses to heat stress. Plant Physiol 181(2):609–629

Hendrix B, Zheng W, Bauer MJ, Havecker ER, Mai JT, Hoffer PH, Sanders RA, Eads BD, Caruano-Yzermans A, Taylor DN, Hresko C, Oakes J, Iandolino AB, Bennett MJ, Deikman J (2021) Topically delivered 22 nt siRNAs enhance RNAi silencing of endogenous genes in two species. Planta 254(3):60

Heo JB, Sung S (2011) Vernalization-mediated epigenetic silencing by a long intronic noncoding RNA. Science 331:76–79

Hou Y, Jiang F, Zheng X, Wu Z (2019) Identification and analysis of oxygen responsive microRNAs in the root of wild tomato (*Solanum habrochaites*). BMC Plant Biol 19(1):100.

http://www.fao.org/faostat/en/#data

https://ourworldindata.org/crop-yields

Huang Y, Li L, Smith KP, Muehlbauer GJ (2016b) Differential transcriptomic responses to *Fusarium graminearum* infection in two barley quantitative trait loci associated with *fusarium* head blight resistance. BMC Genomics 17:387

Huang CY, Wang H, Hu P, Hamby R, Jin H (2019) Small RNAs - big players in plant-microbe interactions. Cell Host Microbe 26(2):173–182

Huang K, Wu XX, Fang CL, Xu ZG, Zhang HW, Gao J, Zhou CM, You LL, Gu ZX, Mu WH, Feng Y, Wang JW, Zhang Y (2021a) Pol IV and RDR2: A two-RNA-polymerase machine that produces double-stranded RNA. Science 374(6575):1579–1586

Huang J, Yang M, Lu L, Zhang X (2016a) Diverse functions of small RNAs in different plant-pathogen communications. Front Microbiol 7:1–17

Huang X, Zhang H, Guo R, Wang Q, Liu X, Kuang W, Song H, Liao J, Huang Y, Wang Z (2021b) Systematic identification and characterization of circular RNAs involved in flag leaf senescence of rice. Planta 253(2):26

Hubner M, Wilde P, Schmiedchen B, Dopierala P, Gowda M, Reif JC, Miedaner T (2013) Hybrid rye performance under natural drought stress in Europe. Theor Appl Genet 126:475–482

Hussain K, Mungikar K, Kulkarni A, Kamble A (2018) Identification, characterization and expression analysis of pigeonpea miRNAs in response to *Fusarium* wilt. Gene 653(5):57–64

Intrigliolo DS, P´erez D, Risco D, Yeves A, Castel JR. (2012) Yield components and grape composition responses to seasonal water deficits in tempranillo grapevines. Irrig Sci 30:339–349

IPCC (2019) Summary for Policymakers. In: Climate Change and Land: an IPCC special report on climate change, desertification, land degradation, sustainable land management, food security, and greenhouse gas fluxes in terrestrial ecosystems [P.R. Shukla, J. Skea, E. Calvo Buendia, V. Masson-Delmotte, H.-O. Pörtner, D. C. Roberts, P. Zhai, R. Slade, S. Connors, R. van Diemen, M. Ferrat, E. Haughey, S. Luz, S. Neogi, M. Pathak, J. Petzold, J. Portugal Pereira, P. Vyas, E. Huntley, K. Kissick, M. Belkacemi, J. Malley, (eds.)]. In press

Iyer NJ, Jia X, Sunkar R, Tang G, Mahalingam R (2012) microRNAs responsive to ozone-induced oxidative stress in *Arabidopsis thaliana*. Plant Signal Behav 7(4):484–491

Jain P, Sharma V, Dubey H, Singh PK, Kapoor R, Kumari M, Singh J, Pawar DV, Bisht D, Solanke AU, Mondal TK, Sharma TR (2017) Identification of long non-coding RNA in rice lines resistant to rice blast pathogen *Maganaporthe oryzae*. Bioinformation 13(8):249–255

Jia X, Ren L, Chen QJ, et al. (2009) UV-B-responsive microRNAs in *Populus tremula*. J Plant Physiol 166:2046–2057

Joga MR, Zotti MJ, Smagghe G, Christiaens O (2016) RNAi efficiency, systemic properties, and novel delivery methods for pest insect control: What we know so far. Front Physiol 7:1–14

Johari-Pireivatlou M. (2010) Effect of soil water stress on yield and proline content of four wheat lines. Afr J Biotechnol 9:36–40

Joshi RK, Megha S, Basu U, Rahman MH, Kav NNV (2016) Genome wide identification and functional prediction of long non-coding RNAs responsive to *Sclerotinia sclerotiorum* infection in *Brassica napus*. PLoS ONE 11: e0158784

Kamthan A, Chaudhuri A, Kamthan M, Datta A (2015) Small RNAs in plants: Recent development and application for crop improvement. Front PlantSci 6:208

Karakülah G, Yücebilgili Kurtoğlu K, Unver T (2016) PeTMbase: A database of plant endogenous target mimics (eTMs). PLoS ONE 11(12): e0167698

Karlik E, Ari S, Gozukirmizi N (2019) LncRNAs: Genetic and epigenetic effects in plants. Biotechnol Biotechnol Equip 33:429–439

Karlova R, van Haarst JC, Maliepaard C, van de Geest H, Bovy AG, Lammers M, Angenent GC, de Maagd RA (2013) Identification of microRNA targets in tomato fruit development using high-throughput sequencing and degradome analysis. J Exp Bot 64:1863–1878

Khaksefidi E, Mirlohi R, Khalaji SF, et al. (2015) Differential expression of seven conserved microRNAs in response to abiotic stress and their regulatory network in *Helianthus annuus*. Front Plant Sci 6:741

Khraiwesh B, Zhu JK, Zhu J (2012) Role of miRNAs and siRNAs in biotic and abiotic stress responses of plants. Biochim Biophys Acta 1819:137–148

Kil E-J, Kim S, Lee Y-J, Byun H-S, Park J, Seo H, Kim C-S, Shim J-K, Lee J-H, Kim J-K, Lee K-Y, Choi HS, Lee s (2016) Tomato yellow leaf curl virus (TYLCV-IL): A seed-transmissible geminivirus in tomatoes. Sci Rep 6:19013

Kornienko AE, Guenzl PM, Barlow DP, Pauler FM (2013) Gene regulation by the act of long non-coding. BMC Biol 11:1–14

Kumar V, Khare T, Shriram V, Wani SH (2018) Plant small RNAs: The essential epigenetic regulators of gene expression for salt-stress responses and tolerance, Plant Cell Rep 37:61–75

Kumar N, Kumar N, Shukla A, Shankhdhar SC, Shankhdhar D (2015) Impact of terminal heat stress on pollen viability and yield attributes of rice (*Oryza sativa* L.). Cereal Res Commun 43:616–626

Kuo YW, Lin JS, Li YC, Jhu MY, King YC, Jeng ST (2019) MicroR408 regulates defense response upon wounding in sweet potato. J Exp Bot 70(2):469–483

Kushawaha AK, Khan A, Sopory SK, Sanan-Mishra N (2021) Priming by high temperature stress induces MicroRNA regulated heat shock modules indicating their involvement in thermopriming response in rice. Life (Basel) 11(4):291

Kwenda S, Birch PR, Moleleki LN (2016a) Genome-wide identification of potato long intergenic noncoding RNAs responsive to *Pectobacterium carotovorum* subspecies *brasiliense* infection. BMC Genomics 17:614–614

Kwenda S, Motlolometsi TV, Birch PR, Moleleki LN (2016b) RNA-seq profiling reveals defense responses in a tolerant potato cultivar to stem infection by *Pectobacterium carotovorum* ssp. brasiliense. Front Plant Sci 7:1905

Latos PA, Pauler FM, Koerner MV, Şenergin HB, Hudson QJ, Stocsits RR, Allhoff W, Stricker SH, Klement RM, Warczok KE, Aumayr K, Pasierbek P, Barlow DP (2012) Airn transcriptional overlap, but not its lncRNA products, induces imprinted Igf2r silencing. Science 338(6113):1469–1472

Lelandais-Brière C, Naya L, Sallet E, Calenge F, Frugier F, Hartmann C, Gouzy J, Crespi M (2009) Genome-wide Medicago truncatula small RNA analysis revealed novel microRNAs and isoforms differentially regulated in roots and nodules. Plant Cell 21(9):2780–2796

Lelandais-Brière C, Sorin C, Declerck M, Benslimane A, Crespi M, Hartmann C (2010) Small RNA diversity in plants and its impact in development. Curr Genomics 11:14–23

Lesk C, Rowhani P, Ramankutty N (2016) Influence of extreme weather disasters on global crop production. Nature 529:84–87

Liang C, Liu H, Hao J, Li J, Luo L (2019) Expression profiling and regulatory network of cucumber microRNAs and their putative target genes in response to cucumber green mottle mosaic virus infection. Arch Virol 164:1121–1134

Lian H, Wang L, Ma N, Zhou C-M, Han L, Zhang T-Q, Wang J-W (2021) Redundant and specific roles of individual MIR172 genes in plant development. PLoS Biol 19(2): e3001044

Li H, Dong Y, Yin H, Wang N, Yang J, Liu X, Wang Y, Wu J, Li X (2011) Characterization of the stress associated microRNAs in *Glycine max* by deep sequencing. BMC Plant Biol 11:170

Li H, Hu T, Amombo E, Fu J (2017a) Genome-wide identification of heat stress-responsive small RNAs in tall fescue (*Festuca arundinacea*) by high-throughput sequencing. J Plant Physiol 213:157–165

Li Z, Huang C, Bao C, Chen L, Lin M, Wang X, Zhong G, Yu B, Hu W, Dai L, Zhu P, Chang Z, Wu Q, Zhao Y, Jia Y, Xu P, Liu H, Shan G (2015) Exon-intron circular RNAs regulate transcription in the nucleus. Nat Struct Mol Biol 22(3):256–264

Li Y, Jeyakumar JMJ, Feng Q, Zhao Z-X, Fan J, Khaskheli MI, Wang WM (2019) The roles of rice microRNAs in rice-*Magnaporthe oryzae* interaction. Phytopathol Res 1:33

Li WQ, Jia YL, Liu FQ, Wang FQ, Fan FJ, Wang J, Zhu JY, Xu Y, Zhong YG, Yang J (2018) Genome-wide identification and characterization of long non-coding RNAs responsive to *Dickeya zeae* in rice. RSC Adv 8:34408–34417

Lin SI, Chiang SF, Lin WY, Chen JW, Tseng CY, Wu PC, Chiou TJ (2008) Regulatory network of microRNA399 and PHO2 by systemic signaling. Plant Physiol 147(2):732–746

Li Q, Shah N, Zhou X, Wang H, Yu W, Luo J, Liu Y, Li G, Liu C, Zhang C, Chen P (2021) Identification of MicroRibonucleic acids and their targets in response to *Plasmodiophora brassicae* infection in *Brassica napus*. Front Plant Sci 12:734419

Liu H, Jin T, Liao R, Wan L, Xu B, Zhou S, Guan J (2012) miRFANs: An integrated database for *Arabidopsis thaliana* microRNA function annotations. BMC Plant Biol 12:68

Liu G, Liu J, Pei W, Li X, Wang N, Ma J, Zang X, Zhang J, Yu S, Wu M, Yu J (2019a) Analysis of the MIR160 gene family and the role of MIR160a_A05 in regulating fiber length in cotton. Planta 250(6):2147–2158

Liu X, Liu S, Wang R, Chen X, Fan Z, Wu B, Zhou T (2019b) Analyses of MiRNA functions in maize using a newly developed ZMBJ-CMV-2bN81-STTM vector. Front. Plant Sci 10:1277

Liu D, Mewalal R, Hu R, Tuskan GA, Yang X (2017a) New technologies accelerate the exploration of non-coding RNAs in horticultural plants. Hortic Res 4:17031

Liu HH, Tian X, Li YJ, Wu CA, Zheng CC (2008) Microarray-based analysis of stress-regulated microRNAs in *Arabidopsis thaliana*. RNA 14:836–843

Liu Q, Yan S, Yang T, Zhang S, Chen YQ, Liu B (2017b) Small RNAs in regulating temperature stress response in plants. J Integr Plant Biol 59(11):774–791

Liu T, Zhang L, Chen G, Shi T (2017c) Identifying and characterizing the circular RNAs during the lifespan of *Arabidopsis* leaves, Front Plant Sci 8:1278–1278

Liu X, Zhang X, Sun B, et al. (2019c) Genomewide identification and comparative analysis of drought-related micro-RNAs in two maize inbred lines with contrasting drought tolerance by deep sequencing. PLoS ONE 14:e0219176

Li W, Wang T, Zhang Y, Li Y (2016b) Overexpression of soybean miR172c confers tolerance to water deficit and salt stress, but increases ABA sensitivity in transgenic *Arabidopsis thaliana*. J Exp Bot 67:175–194

Li L, Xue M, Yi H (2016a) Uncovering microRNA-mediated response to SO_2 stress in *Arabidopsis thaliana* by deep sequencing. J Hazard Mater 316:178–185

Li Y, Zhao SL, Li JL, Hu XH, Wang H, Cao XL, Xu YJ, Zhao ZX, Xiao ZY, Yang N, Fan J, Huang F, Wang WM (2017b) Osa-miR169 negatively regulates rice immunity against the blast fungus *Magnaporthe oryzae*. Front Plant Sci 8:2

Lobell DB, Bänziger M, Magorokosho C, Vivek B (2011) Nonlinear heat effects on African maize as evidenced by historical yield trials. Nat Clim Change 1:42–45

Lobell DB, Field CB (2007) Global scale climate-crop yield relationships and the impacts of recent warming. Environ Res Lett 2:014002

Lu S, Sun YH, Chiang VL (2008) Stress-responsive microRNAs in Populus. Plant J 55:131–151

Lu S, Sun YH, Shi R, Clark C, Li L, Chiang VL (2005) Novel and mechanical stress-responsive MicroRNAs in *Populus trichocarpa* that are absent from Arabidopsis. Plant Cell 17(8):2186–203

Lucero L, Ferrero L, Fonouni-Farde C, Ariel F (2021) Functional classification of plant long noncoding RNAs: A transcript is known by the company it keeps. New Phytol 229:1251–1260

Lv L, Yu K, Lü H, Zhang X, Liu X, Sun C, Xu H, Zhang J, He X, Zhang D (2020) Transcriptome-wide identification of novel circular RNAs in soybean in response to low-phosphorus stress. PLoS ONE 15(1):e0227243

Ma C, Burd S, Lers A (2015) miR408 is involved in abiotic stress responses in *Arabidopsis*. Plant J 84(1):169–187

Mantri N, Baskar N, Ford R, Pang ECK, Pardeshi V (2013) The role of micro-ribonucleic acids in legumes with a focus on abiotic stress response. Plant Genome 6:3

Mardis ER (2008) Next-generation DNA sequencing methods. Annu Rev Genomics Hum Genet 9:387–402

Margulies M, Egholm M, Altman WE, et al. (2005) Genome sequencing in microfabricated high-density picolitre reactors. Nature 437:376–380

Marquardt S, Raitskin O, Wu Z, Liu F, Sun Q, Dean C (2014) Functional consequences of splicing of the antisense transcript COOLAIR on FLC transcription. Mol Cell 54:156–165

Martianov I, Ramadass A, Serra Barros A, Chow N, Akoulitchev A (2007). Repression of the human dihydrofolate reductase gene by a noncoding interfering transcript. Nature.445:666–670

Martienssen R, Moazed D (2015) RNAi and heterochromatin assembly. Cold Spring Harb Perspect Biol 7:a019323

Matsui A, Seki M (2019) The involvement of long noncoding RNAs in response to plant stress. Methods Mol Biol 1933:151–171

Matsunaga W, Ohama N, Tanabe N, et al. (2015) A small RNA mediated regulation of a stress-activated retrotransposon and the tissue specific transposition during the reproductive period in Arabidopsis. Front Plant Sci 6:1–12

Matzke MA, Mosher RA (2014) RNA-directed DNA methylation: An epigenetic pathway of increasing complexity, Nat Rev Genet 15:394–408

May P, Liao W, Wu Y, Shuai B, McCombie WR, Zhang MQ, Liu QA (2013) The effects of carbon dioxide and temperature on microRNA expression in *Arabidopsis* development. Nat Commun 4:2145

McLoughlin AG, Wytinck N, Walker PL, Girard IJ, Rashid KY, de Kievit T, Fernando WGD, Whyard S, Belmonte MF (2018) Identification and application of exogenous dsRNA confers plant protection against *Sclerotinia sclerotiorum* and *Botrytis cinerea*. Sci Rep 8(1):7320

Megha S, Basu U, Kav NNV (2018) Regulation of low temperature stress in plants by microRNAs. Plant Cell Environ 41:1–15

Meng X, Li X, Zhang P, Wang J, Zhou Y, Chen M (2017) Circular RNA: An emerging key player in RNA world. Brief Bioinform 18(4):547–557

Michaels SD, Amasino RM (1999) FLOWERING LOCUS C encodes a novel MADS domain protein that acts as a repressor of flowering. Plant Cell 11:949–956

Mortazavi A, William BA, McCue K, Schaeffer L, Wold B (2008) Mapping and quantifying mammalian transcriptomes by RNA-seq. Nat Methods 5:621–628

Nejat N, Cahill DM, Vadamalai G, Ziemann M, Rookes J, Naderali N (2015) Transcriptomics-based analysis using RNA-seq of the coconut (*Cocos nucifera*) leaf in response to yellow decline phytoplasma infection. Mol Genet Genomics 290(5):1899–1910

Nejat N, Mantri N (2017) Plant immune system: Crosstalk between responses to biotic and abiotic stresses the missing link in understanding plant defence. Curr Issues Mol Biol 23:1

Nejat N, Mantri N (2018) Emerging roles of long non-coding RNAs in plant response to biotic and abiotic stresses. Crit Rev Biotechnol 38:93–105

Niu QW, Lin SS, Reyes JL, Chen KC, Wu HW, Yeh SD, Chua NH (2006) Expression of artificial microRNAs in transgenic *Arabidopsis thaliana* confers virus resistance. Nat Biotechnol 24(11):1420–1428

Nogoy FM, Niño MC, Song JY, Jung JY, Kang KK, Nou I, Cho Y-G (2018) Plant microRNAs in molecular breeding. Plant Biotechnol Rep 12:15–25

Noya I, González-García S, Bacenetti J, Fiala M, Moreira MT (2018) Environmental impacts of the cultivation-phase associated with agricultural crops for feed production. J Clean Prod 172:3721–3733

O'Brien J, Hayder H, Zayed Y, Peng C (2018) Overview of MicroRNA biogenesis, mechanisms of actions, and circulation. Front Endocrinol 9:402

Ó'Maoiléidigh DS, van Driel AD, Singh A, et al. (2021) Systematic analyses of the MIR172 family members of *Arabidopsis* define their distinct roles in regulation of APETALA2 during floral transition. PLoS Biol 19(2): e3001043

Ohtani M (2018) Plant snRNP biogenesis: A perspective from the nucleolus and cajal bodies. Front Plant Sci 8:2184

Panda AC (2018) Circular RNAs act as miRNA sponges. Adv Exp Med Biol 1087:67–79

Pandey R, Joshi G, Bhardwaj AR, Agarwal M, Katiyar-Agarwal S (2014) A comprehensive genome-wide study on tissue-specific and abiotic stress-specific miRNAs in *Triticum aestivum*. PLoS ONE 9(4):e95800

Pandey P, Srivastava PK, Pandey SP (2019) Prediction of plant miRNA targets. In: Stefan de Folter (ed.), Plant MicroRNAs: Methods and Protocols, Methods in Molecular Biology, vol. 1932. Humana Press, New York, NY, pp. 99–107. https://doi.org/10.1007/978-1-4939-9042-9_7

Pan T, Sun XQ, Liu YX, et al. (2018) Heat stress alters genome-wide profiles of circular RNAs in *Arabidopsis*. Plant Mol Biol 96:217–229

Pan WJ, Tao JJ, Cheng T, Bian XH, Wei W, Zhang WK, Ma B, Chen SY, Zhang JS (2016) Soybean miR172a improves salt tolerance and can function as a long-distance signal. Mol Plant 9(9):1337–1340

Parizotto EA, Dunoyer P, Rahm N, Himber C, Voinnet O (2004) In vivo investigation of the transcription, processing, endonucleolytic activity, and functional relevance of the spatial distribution of a plant miRNA. Genes Dev 18:2237–2242

Parmar S, Gharat SA, Tagirasa R, Chandra T, Behera L, Dash SK, Shaw BP (2020) Identification and expression analysis of miRNAs and elucidation of their role in salt tolerance in rice varieties susceptible and tolerant to salinity. PLoS ONE 15(4): e0230958

Peng T, Qiao M, Liu H, Teotia S, Zhang Z, Zhao Y, Wang B, Zhao D, Shi L, Zhang C, Le B, Rogers K, Gunasekara C, Duan H, Gu Y, Tian L, Nie J, Qi J, Meng F, Huang L, Chen Q, Wang Z, Tang J, Tang X, Lan T, Chen X, Wei H, Zhao Q, Tang G (2018) A Resource for inactivation of MicroRNAs using short tandem target mimic technology in model and crop plants. Mol Plant 11(11):1400–1417

Poething RS (2009) Small RNAs and developmental timing in plants. Curr Opin Genet Dev 19:374–378

Pokharel M, Stamm M, Hein NT, Jagadish KSV (2021) Heat stress affects floral morphology, silique set and seed quality in chamber and field grown winter canola. J Agro Crop Sci 00:1–16

Pradhan GP, Prasad PVV, Fritz AK, Kirkham MB, Gill BS (2012) Effects of drought and high temperature stress on synthetic hexaploid wheat. Funct Plant Biol 39(3):190–198

Qin T, Zhao H, Cui P, Albesher N, Xiong L (2017) A nucleus-localized long non-coding RNA enhances drought and salt stress tolerance. Plant Physiol 175(3):1321–1336

Ragupathy R, Ravichandran S, Mahdi MSR, et al. (2016) Deep sequencing of wheat sRNA transcriptome reveals distinct temporal expression pattern of miRNAs in response to heat, light and UV. Sci Rep 6:1–15

Rajwanshi R, Chakraborty S, Jayanandi K, Deb B, Lightfoot DA (2014). Orthologous plant microRNAs: Microregulators with great potential for improving stress tolerance in plants. Theor Appl Genet 127(12):2525–43

Ray DK, Gerber JS, Macdonald GK, West PC (2015) Climate variation explains a third of global crop yield variability. Nat Commun 6:5989

Ray D, Ramankutty N, Mueller N, West PC, Foley JA (2012) Recent patterns of crop yield growth and stagnation. Nat Commun 3:1293

Raza A, Razzaq A, Mehmood SS, Zou X, Zhang X, Lv Y, Xu J (2019) Impact of climate change on crops adaptation and strategies to tackle its outcome: A review. Plants (Basel) 8(2):34

Reichel M, Li Y, Li J, Millar AA (2015) Inhibiting plant microRNA activity: Molecular SPONGEs, target MIMICs and STTMs all display variable efficacies against target microRNAs. Plant Biotechnol J 13(7): 915–926

Reichel M, Millar AA (2015) Specificity of plant microRNA target MIMICs: Cross-targeting of miR159 and miR319. J Plant Physiol 180:45–48

Richardson KJ, Lewis KH, Krishnamurthy PK, Kent C, Wiltshire AJ, Hanlon HM (2018) Food security outcomes under a changing climate: Impacts of mitigation and adaptation on vulnerability to food insecurity. Clim Change 147:327–341

Rojo FP, Nyman RKM, Johnson AAT, Navarro MP, Ryan MH, Erskine W, Kaur P (2018) CRISPR-cas systems: Ushering in the new genome editing era. Bioengineered 9(1):214–221

Ron M, Saez MA, Williams LE, Fletcher JC, McCormick S (2010) Proper regulation of a sperm-specific cis-nat-siRNA is essential for double fertilization in *Arabidopsis*. Genes Dev 24:1010–1021

Saini HS, Aspinall D (1981) Effect of water deficit on sporogenesis in wheat (*Triticum aestivum* L.). Ann Bot 48:623–633

Saini HS, Aspinall D (1982) Abnormal sporogenesis in wheat (*Triticum aestivum* L.) Induced by short periods of high temperature. Ann Bot 49:835–846

Sakurai Y, Baeg K, Lam AYW, Shoji K, Tomari Y, Iwakawa HO (2021) Cell-free reconstitution reveals the molecular mechanisms for the initiation of secondary siRNA biogenesis in plants. Proc Natl Acad Sci USA 118(31):e2102889118

Salvador-Guirao R, Baldrich P, Weigel D, Rubio-Somoza I, San Segundo B (2018a) The MicroRNA miR773 is involved in the *Arabidopsis* immune response to fungal pathogens. Mol Plant Microbe Interact 31(2):249–259

Salvador-Guirao R, Hsing Y-I, Segundo S B (2018b) The polycistronic miR166k-166h positively regulates rice immunity via post-transcriptional control of EIN2. Front Plant Sci 9:337

Schlenker W, Roberts MJ (2009) Nonlinear temperature effects indicate severe damages to u.S. Crop yields under climate change. Proc Natl Acad Sci USA 106:15594–15598

Schwab R, Ossowski S, Riester M, Warthmann N, Weigel D (2006) Highly specific gene silencing by artificial microRNAs in Arabidopsis. Plant Cell 18(5):1121–1133.

Schwab R, Palatnik JF, Riester M, Schommer C, Schmid M, Weigel D (2005) Specific effects of microRNAs on the plant transcriptome. Dev Cell 8:517–527

Sha A, Zhao J, Yin K, Tang Y, Wang Y, Wei X, Hong Y, Liu Y (2014) Virus-based microRNA silencing in plants. Plant Physiol 164(1):36–47

Sheoran IS, Saini HS (1996) Drought-induced male sterility in rice: Changes in carbohydrate levels and enzyme activities associated with the inhibition of starch accumulation in pollen. Sex Plant Reprod 9:161–169

Shi X, Jiang F, Wen J, Wu Z (2019) Overexpression of *Solanum habrochaites* microRNA319d (sha-miR319d) confers chilling and heat stress tolerance in tomato (*S. lycopersicum*). BMC Plant Biol 19:214

Shi F, Xu H, Liu C, Tan C, Ren J, Ye X, Feng H, Liu Z (2021) Whole-transcriptome sequencing reveals a vernalization-related ceRNA regulatory network in chinese cabbage (*Brassica campestris* L. ssp. *pekinensis*). BMC Genomics 22(1):819

Shriram V, Kumar V, Devarumath RM, Khare TS, Wani SH (2016) MicroRNAs as potential targets for abiotic stress tolerance in plants. Front Plant Sci 7:817

Shuai P, Liang D, Zhang Z, Yin W, Xia X (2013) Identification of drought-responsive and novel *Populus trichocarpa* microRNAs by high-throughput sequencing and their targets using degradome analysis. BMC Genomics 14:233

Simon SA, Meyers BC (2011) Small RNA-mediated epigenetic modifications in plants. Curr Opin Plant Biol 14:148–155

Singh A, Das S (2021) Slicing Messengers by Artificial Designs: Artificial MicroRNA Induced Gene Silencing in Polyploid Plants for Functional Genomics and Trait Modification. In: G. Tang et al. (eds.), RNA-Based Technologies for Functional Genomics in Plants, Concepts and Strategies in Plant Sciences, https://doi.org/10.1007/978-3-030-64994-4_6

Singroha G, Sharma P, Sunkur R (2021) Current status of microRNA-mediated regulation of drought stress responses in cereals. Physiol Plant 172:1808–1821

Snowdon RJ, Wittkop B, Chen TW, Stahl A (2021) Crop adaptation to climate change as a consequence of long-term breeding. Theor Appl Genet 134:1613–1623

Song L, Fang Y, Chen L, Wang J, Chen X (2021) Role of non-coding RNAs in plant immunity. Plant Commun 2(3): 100180

Song JB, Gao S, Wang Y, Li BW, Zhang YL, Yang ZM (2016) miR394 and its target gene LCR are involved in cold stress response in *Arabidopsis*. Plant Gene 5:56–64

Song Z, Zhang L, Wang Y, Li H, Li S, Zhao H, Zhang H (2018) Constitutive expression of miR408 improves biomass and seed yield in *Arabidopsis*. Front. Plant Sci 8:2114

Sosa-Valencia G, Palomar M, Covarrubias AA, Reyes JL (2017) The legume miR1514a modulates a NAC transcription factor transcript to trigger phasiRNA formation in response to drought. J Exp Bot 68:2013–2026

Soto-Suárez M, Baldrich P, Weigel D, Rubio-Somoza I, San Segundo B (2017) The *Arabidopsis* miR396 mediates pathogen-associated molecular pattern-triggered immune responses against fungal pathogens. Sci Rep 7:44898

St. Laurent G, Wahlestedt C, Kapranov P (2015) The landscape of long noncoding RNA classification. Trends Genet 31:239–251

Statello L, Guo CJ, Chen LL, Huarte M (2021) Gene Regulation by long non-coding RNAs and its biological functions. Nat Rev Mol Cell Biol 22(2):96–118

Su Y, Zhang Y, Huang N, Liu F, Su W, Xu L, Ahmad W, Wu Q, Guo J, Que Y (2017) Small RNA sequencing reveals a role for sugarcane miRNAs and their targets in response to *Sporisorium scitamineum* infection. BMC Genomics 18(1):325

Summanwar A, Basu U, Rahman H, Kav NNV (2020) Non-coding RNAs as emerging targets for crop improvement. Plant Sci 297:110521

Sun J, Dong Y, Wang C, Xiao S, Jiao Z, Gao C (2021) Identification and characterization of melon circular RNAs involved in powdery mildew responses through comparative transcriptome analysis. PeerJ 9:e11216

Sun Z, He Y, Li J, Wang X, Chen J (2015b) Genome-wide characterization of rice black streaked dwarf virus-responsive MicroRNAs in rice leaves and roots by small RNA and degradome sequencing, Plant Cell Physiol 56(4): 688–699

Sun Z, Huang K, Han Z, et al. (2020b) Genome-wide identification of *Arabidopsis* long noncoding RNAs in response to the blue light. Sci Rep 10:1–10

Sun M, Shen Y, Yang J, Cai X, Li H, Zhu Y, Jia B, Sun X (2020a) miR535 negatively regulates cold tolerance in rice. Mol Breeding 40:14

Sun X, Xu L, Wang Y, Yu R, Zhu X, Luo X, Gong Y, Wang R, Limera C, Zhang K, Liu L (2015) Identification of novel and salt-responsive miRNAs to explore miRNA-mediated regulatory network of salt stress response in radish (*Raphanus sativus* L.). BMC Genomics 16(1):197

Sunkar R, Li YF, Jagadeeswaran G (2012) Functions of microRNAs in plant stress responses. Trends Plant Sci 17:196–203

Swiezewski S, Liu F, Magusin A, Dean C (2009) Cold-induced silencing by long antisense transcripts of an *Arabidopsis* polycomb target. Nature 462:799–802

Talmor-Neiman M, Stav R, Klipcan L, Buxdorf K, Baulcombe DC, Arazi T (2006) Identification of trans-acting siRNAs in moss and an RNA-dependent RNA polymerase required for their biogenesis. Plant J 48:511–521

Tan X, Li S, Hu L, Zhang C (2020) Genome-wide analysis of long non-coding RNAs (lncRNAs) in two contrasting rapeseed (*Brassica napus* l.) Genotypes subjected to drought stress and re-watering. BMC Plant Biol 20:81

Tang G, Tang X (2013) Short tandem target mimic: A long journey to the engineered molecular landmine for selective destruction/blockage of microRNAs in plants and animals. J Genet Genomics 40(6):291–296

Tang G, Yan J, Gu Y, Qiao M, Fan R, Mao Y, Tang X (2012) Construction of short tandem target mimic (STTM) to block the functions of plant and animal microRNAs. Methods 58(2):118–125

Taylor RS, Tarver JE, Hiscock SJ, Donoghue PCJ (2014) Evolutionary history of plant microRNAs. Trends Plant Sci 19:175–182

Tebaldi C, Lobell D (2018) Estimated impacts of emission reductions on wheat and maize crops. Clim Change 146:533–545

Teotia S, Tang G (2017) Silencing of stress-regulated miRNAs in plants by short tandem target mimic (STTM) approach. In: R. Sunkar (eds.) Plant Stress Tolerance. Methods in Molecular Biology, vol. 1631. Humana Press, New York, NY

Tesfamariam EH, Annandale JG, Steyn JM (2010) Water stress effects on winter canola growth and yield. Agron J 102:658–666

Thiebaut F, Rojas CA, Almeida KL, Grativol C, Domiciano GC, Lamb CR, Jde E, Hemerly A, Ferreira AS PC (2012) Regulation of miR319 during cold stress in sugarcane. Plant Cell Environ 35(3):502–512

Thomson DW, Dinger ME (2016) Endogenous microRNA sponges: Evidence and controversy. Nat Rev Genet 17:272–283

Tito R, Vasconcelos HL, Feeley KJ (2018) Global climate change increases risk of crop yield losses and food insecurity in the tropical Andes. Glob Chang Biol 24(2):e592–e602

Tiwari P, Khare T, Shriram V, Bae H, Kumar V (2021) Plant synthetic biology for producing potent phyto-antimicrobials to combat antimicrobial resistance. Biotechnol Adv 48:107729

Todesco M, Rubio-Somoza I, Paz-Ares J, Weigel D (2010) A collection of target mimics for comprehensive analysis of microRNA function in *Arabidopsis thaliana*. PLoS Genet 6(7):e1001031

Treiber T, Treiber N, Meister G (2019) Regulation of microRNA biogenesis and its crosstalk with other cellular pathways. Nat Rev Mol Cell Biol 20:5–20

Trindade I, Capitão C, Dalmay T, Fevereiro MP, Santos DM. miR398 and miR408 are up-regulated in response to water deficit in Medicago truncatula. Planta. 2010 Feb;231(3):705–16

Tsai MC, Manor O, Wan Y, Mosammaparast N, Wang JK, Lan F, Shi Y, Segal E, Chang HY (2010) Long noncoding RNA as modular scaffold of histone modification complexes. Science 329:689–693

Ulitsky I, Bartel DP (2013) lincRNAs: Genomics, evolution, and mechanisms. Cell 154(1):26–46

Valin H, Sands RD, van der Mensbrugghe D, et al. (2014) The future of food demand: Understanding differences in global economic models. Agric Econ 45:51–67

Varshney RK, Bohra A, Roorkiwal M, Barmukh R, Cowling WA, Chitikineni A, Lam HM, Hickey LT, Croser JS, Bayer PE, Edwards D, Crossa J, Weckwerth W, Millar H, Kumar A, Bevan MW, Siddique KHM (2021a) Fast-forward breeding for a food-secure world. Trends Genet 37(12):1124–1136

Varshney RK, Bohra A, Yu J, Graner A, Zhang Q, Sorrells ME (2021b) Designing future crops: Genomics-assisted breeding comes of age. Trends Plant Sci 26(6):631–649

Varshney RK, Sinha P, Singh VK, et al. (2020) 5Gs for crop genetic improvement. Curr Opin Plant Biol 56:190–196

Veillet F, Durand M, Kroj T, Cesari S, Gallois JL (2020) Precision breeding made real with CRISPR: Illustration through genetic resistance to pathogens. Plant Commun 1(5):100102

Verma SS, Rahman MH, Deyholos MK, Basu U, Kav NNV (2014) Differential expression of miRNAs in *Brassica napus* root following infection with *Plasmodiophora brassicae*. PLoS ONE 9:e86648

Wagaba H, Patil BL, Mukasa S, Alicai T, Fauquet CM, Taylor NJ (2016) Artificial microRNA-derived resistance to cassava brown streak disease. J Virol Methods 231:38–43

Wahid F, Shehzad A, Khan T, Kim YY (2010) MicroRNAs: Synthesis, mechanism, function, and recent clinical trials. Biochim Biophys Acta - Mol Cell Res 1803:1231–1243

Wang KC, Chang HY (2011) Molecular mechanisms of long noncoding RNAs. Mol Cell 43(6):904–914

Wang H, Chung PJ, Liu J, et al. (2014a) Genome-wide identification of long noncoding natural antisense transcripts transcripts and their responses to light in *Arabidopsis*. Genome Res 24:444–453

Wang R, Fang YN, Wu XM, Qing M, Li CC, Xie KD, Deng XX, Guo WW (2020b) The miR399-CsUBC24 module regulates reproductive development and male fertility in Citrus. Plant Physiol 183(4):1681–1695

Wang Y, Fan X, Lin F, He G, Terzaghi W, Zhu D, Deng XW (2014c) *Arabidopsis* noncoding RNA mediates control of photomorphogenesis by red light. Proc Natl Acad Sci USA 111(28):10359–10364

Wang LL, Jin JJ, Li LH, Qu SH (2020a) Long non-coding RNAs responsive to blast fungus infection in rice. Rice 13(1):77

Wang J, Lin J, Wang H, Li X, Yang Q, Li H, Chang Y (2018a) Identification and characterization of circRNAs in *Pyrus betulifolia* bunge under drought stress. PLoS ONE 13(7):e0200692

Wang Z, Liu Y, Li D, Li L, Zhang Q, Wang S, Huang H (2017b) Identification of circular RNAs in kiwifruit and their species-specific response to bacterial canker pathogen invasion. Front Plant Sci 8:413

Wang Z, Liu Y, Li L, Li D, Zhang Q, Guo Y, Wang S, Zhong C, Huang H (2017c) Whole transcriptome sequencing of *Pseudomonas syringae* pv. *actinidiae*-infected kiwifruit plants reveals species-specific interaction between long non-coding RNA and coding genes. Sci Rep 7(1):4910

Wang W, Luan Y (2015) The advance of tomato disease-related microRNAs. Plant Cell Rep 34:1089–1097

Wang J, Mei J, Ren G (2019a) Plant microRNAs: Biogenesis, homeostasis, and degradation. Front Plant Sci 10:1–12

Wang J, Meng X, Yuan C, Harrison AP, Chen M (2016) The roles of cross-talk epigenetic patterns in *Arabidopsis thaliana*. Brief Funct Genomics 15(4):278–287

Wang S-t, Sun X-I, Hoshino Y, Yu Y, Jia B, Sun Z, Sun M, Duan X, Zhu Y (2014b) MicroRNA319 positively regulates cold tolerance by targeting OsPCF6 and OsTCP21 in rice (*Oryza sativa* L.). PLoS ONE 9(3): e91357

Wang K, Wang C, Guo B, Song K, Shi C, Jiang X, Wang K, Tan Y, Wang L, Wang L, Li J, Li Y, Cai Y, Zhao H, Sun X (2019b) CropCircDB: A comprehensive circular RNA resource for crops in response to abiotic stress, Database, 2019:baz053. https://doi.org/10.1093/database/baz053

Wang W, Wang J, Wei Q, Li B, Zhong X, Hu T, Hu H, Bao C (2019c) Transcriptome-wide identification and characterization of circular RNAs in leaves of Chinese cabbage (*Brassica rapa* L. Ssp. *pekinensis*) in response to calcium deficiency-induced tip-burn. Sci Rep 9(1):14544

Wang J, Yang Y, Jin L, Ling X, Liu T, Chen T, Ji Y, Yu W, Zhang B (2018b) Re-analysis of long non-coding RNAs and prediction of circRNAs reveal their novel roles in susceptible tomato following TYLCV infection. BMC Plant Biol 18(1):104

Wang Y, Yang M, Wei S, et al. (2017a) Identification of circular RNAs and their targets in leaves of *Triticum aestivum* L. Under dehydration stress. Front Plant Sci 7:1–10

Wang J, Yu W, Yang Y, Li X, Chen T, Liu T, Ma N, Yang X, Liu R, Zhang B (2015b) Genome-wide analysis of tomato long non-coding RNAs and identification as endogenous target mimic for microRNA in response to TYLCV infection. Sci Rep 5:16946

Wang X, Yu G, Zhao J, Cui N, Yu Y, Fan H (2019d) Functional identification of *corynespora cassiicola*-responsive miRNAs and their targets in cucumber. Front Plant Sci 10:668

Wang P, Zhi H, Zhang Y, Liu Y, Zhang J, Gao Y, Guo M, Ning S, Li X (2015a) miRSponge: A manually curated database for experimentally supported miRNA sponges and ceRNAs. Database (Oxford):bav098

Wapinski O, Chang HY (2011). Long noncoding RNAs and human disease. Trends Cell Biol 21(6):354–361

Warthmann N, Chen H, Ossowski S, Weigel D, Hervé P (2008) Highly specific Gene silencing by artificial miRNAs in rice. PLoS ONE 3(3):e1829

Wesley SV, Helliwell CA, Smith NA, Wang MB, Rouse DT, Liu Q, Gooding PS, Singh SP, Abbott D, Stoutjesdijk PA, Robinson SP, Gleave AP, Green AG, Waterhouse PM (2001) Construct design for efficient, effective and high-throughput gene silencing in plants. Plant J 27(6):581–590

Wheeler T, Von Braun J (2013) Climate change impacts on global food security. Science 341:508–513

Wierzbicki AT (2012) The role of long non-coding RNA in transcriptional gene silencing. Curr Opin Plant Biol 15(5):517–522

Wierzbicki AR, Blevins T, Swiezewski S (2021) Long noncoding RNAs in plants. Annu Rev Plant Biol 72:245–271

Wierzbicki AT, Haag JR, Pikaard CS (2008) Noncoding transcription by RNA polymerase pol IVb/pol v mediates transcriptional silencing of overlapping and adjacent genes. Cell 135:635–648

Wierzbicki AT, Ream TS, Haag JR, Pikaard CS (2009) RNA polymerase v transcription guides ARGONAUTE4 to chromatin. Nat. Genet 41:630–634

Wilusz JE, Sunwoo H, Spector DL (2009) Long noncoding RNAs: Functional surprises from the RNA world. Genes Dev 23:1494–1504

Winkel T, Renno J-F, Payne W (1997) Effect of the timing of water deficit on growth, phenology and yield of pearl millet (*Pennisetum glaucum* (L.) R. Br.) Grown in sahelian conditions. J Exp Bot48:1001–1009

Wu HJ, Wang ZM, Wang M, Wang XJ (2013) Widespread long noncoding RNAs as endogenous target mimics for microRNAs in plants. Plant Physiol 161(4):1875–1884

Wu W, Wu Y, Hu D, Zhou Y, Hu Y, Chen Y, Chen M (2020) PncStress: A manually curated database of experimentally validated stress-responsive non-coding RNAs in plants. Database (Oxford) 2020:baaa001

Wu R, Wu G, Wang L, Wang X, Liu Z, Li M, Tan W, Qing L (2021) Tobacco curly shoot virus down-regulated the expression of nbe-miR167b-3p to facilitate its infection in *Nicotiana benthamiana*. Front Microbiol 12:791561

Wu L, Zhou H, Zhang Q, Zhang J, Ni F, Liu C, Qi Y (2010). DNA methylation mediated by a microRNA pathway. Mol Cell 38(3): 465–475

Xiang L, Cai C, Cheng J, Wang L, Wu C, Shi Y, Luo J, He L, Deng Y, Zhang X, Yuan Y, Cai Y (2018) Identification of circular RNAs and their targets in *Gossypium* under *Verticillium* wilt stress based on RNA-seq. PeerJ 6:e4500

Xie S, Jiang H, Ding T, Xu Q, Chai W, Cheng B (2018) *Bacillus amyloliquefaciens* FZB42 represses plant miR846 to induce systemic resistance via a jasmonic acid-dependent signalling pathway. Mol Plant Pathol 19(7): 1612–1623

Xie S, Jiang H, Xu Z, Xu Q, Cheng B (2017) Small RNA profiling reveals important roles for miRNAs in *Arabidopsis* response to *bacillus velezensis* FZB42. Gene 629(20): 9–15

Xie M, Yu B (2015) siRNA-directed DNA methylation in plants. Curr Genomics 16(1):23–31

Xie S, Yu H, Li E, Wang Y, Liu J, Jiang H (2019) Identification of miRNAs involved in *Bacillus velezensis* FZB42-activated induced systemic resistance in maize. Int J Mol Sci 20(20):5057

Xin M, Wang Y, Yao Y, et al. (2011) Identification and characterization of wheat long non-protein coding RNAs responsive to powdery mildew infection and heat stress by using microarray analysis and SBS sequencing. BMC Plant Biol 11:61

Xu J, Hou QM, Khare T, et al. (2019) Exploring miRNAs for developing climate-resilient crops: A perspective review. Sci Total Environ 653:91–104

Yan J, Gu Y, Jia X, Kang W, Pan S, Tang X, Chen X, Tang G (2012) Effective small RNA destruction by the expression of a short tandem target mimic in *Arabidopsis*. Plant Cell 24(2):415–427

Yang C, Li D, Mao D, Liu X, Ji C, Li X, Zhao X, Cheng Z, Chen C, Zhu L (2013) Overexpression of microRNA319 impacts leaf morphogenesis and leads to enhanced cold tolerance in rice (*Oryza sativa* L.). Plant Cell Environ 36(12):2207–2218

Yang X, Liu Y, Zhang H, Wang J, Zinta G, Xie S, Zhu W, Nie W-F (2020) Genome-wide identification of circular RNAs in response to low-temperature stress in tomato leaves. Front Genet 11:591806

Yang T, Wang Y, Teotia S, Wang Z, Shi C, Sun H, Gu Y, Zhang Z, Tang G (2019) The interaction between miR160 and miR165/166 in the control of leaf development and drought tolerance in *Arabidopsis*. Sci Rep 9(1):2832

Yang J, Zhang N, Zhang J, Jin X, Zhu X, Ma R, Li S, Lui S, Yue Y, Si H (2021) Knockdown of MicroRNA160a/b by STTM leads to root architecture changes via auxin signaling in *Solanum tuberosum*. Plant Physiol Biochem 166:939–949

Ye CY, Zhang X, Chu Q, Liu C, Yu Y, Jiang W, Zhu QH, Fan L, Guo L (2017) Full-length sequence assembly reveals circular RNAs with diverse non-GT/AG splicing signals in rice. RNA Biol 14(8):1055–1063

Yin S, Tian X, Zhang J, Sun P, Li G (2021) PCirc: Random forest-based plant circRNA identification software. BMC Bioinf 22:10

Yoshikawa M (2013) Biogenesis of trans-acting siRNAs, endogenous secondary siRNAs in plants. Genes Genet Syst 88(2):77–84

Yoshikawa M, Han YW, Fujii H, Aizawa S, Nishino T, Ishikawa M (2021) Cooperative recruitment of RDR6 by SGS3 and SDE5 during small interfering RNA amplification in *Arabidopsis*. Proc Natl Acad Sci USA 118(34): e2102885118

Yu C, Chen Y, Cao Y, Chen H, Wang J, Bi YM, Tian F, Yang F, Rothstein SJ, Zhou X, He C (2018) Overexpression of miR169o, an overlapping MicroRNA in response to both nitrogen limitation and bacterial infection, promotes nitrogen use efficiency and susceptibility to bacterial blight in rice. Plant Cell Physiol 59(6):1234–1247

Yu Y, Jia T, Chen X (2017) The 'how' and 'where' of plant microRNAs. New Phytol 216(4):1002–1017

Yu F, Tan Z, Fang T, Tang K, Liang K, Qiu F (2020a) A comprehensive transcriptomics analysis reveals long non-coding RNA to be involved in the key metabolic pathway in response to waterlogging stress in maize. Genes (Basel) 11(3):267

Yu X, Wang H, Lu Y, de Ruiter M, Cariaso M, Prins M, van Tunen A, He Y (2012) Identification of conserved and novel microRNAs that are responsive to heat stress in *Brassica rapa*. J Exp Bot 63(2):1025–1038

Yu Y, Zhang Y, Chen X, Chen Y (2019) Plant noncoding RNAs: Hidden players in development and stress responses. Annu Rev Cell Dev Biol 35:407–431

Yu Y, Zhou YF, Feng YZ, He H, Lian JP, Yang YW, Lei MQ, Zhang YC, Chen YQ (2020b) Transcriptional landscape of pathogen-responsive lncRNAs in rice unveils the role of ALEX1 in jasmonate pathway and disease resistance. Plant Biotechnol J 18(3):679–690

Yue E, Cao H, Liu B (2020) Osmir535, a potential genetic editing target for drought and salinity stress tolerance in *Oryza sativa*. Plants 9:1–10

Zang J, Lu D, Xu A (2020) The interaction of circRNAs and RNA binding proteins: An important part of circRNA maintenance and function. J Neurosci Res 98(1):87–97

Zeng Y, Wagner EJ, Cullen BR (2002) Both natural and designed micro RNAs can inhibit the expression of cognate mRNAs when expressed in human cells. Mol Cell 9:1327–1335

Zeng X, Xu Y, Jiang J, Zhang F, Ma L, Wu D, Wang Y, Sun W (2018) Identification of cold stress responsive microRNAs in two winter turnip rape (*Brassica rapa* L.) By high throughput sequencing. BMC Plant Biol 18(1): 52

Zhang B (2015) MicroRNA: A new target for improving plant tolerance to abiotic stress. J Exp Bot 66(7):1749–1761

Zhang X, Bao Y, Shan D, Wang Z, Song X, Wang Z, Wang J, He L, Wu L, Zhang Z, Niu D, Jin H, Zhao H (2018a) *Magnaporthe oryzae* induces the expression of a MicroRNA to suppress the immune response in rice. Plant Physiol 177(1):352–368

Zhang Y-C, Chen Y-Q (2013) Long noncoding RNAs: New regulators in plant development. Biochem Biophys Res Commun 436 (2013) 111–114

Zhang H, Chen X, Wang C, Xu Z, Wang Y, Liu X, Kang Z, Ji W (2013) Long non-coding genes implicated in response to stripe rust pathogen stress in wheat (*Triticum aestivum* L.). Mol Biol Rep 40(11):6245–6253

Zhang X, Dong J, Deng F, Wang W, Cheng Y, Song L, Hu M, Shen J, Xu Q, Shen F (2019b) The long non-coding RNA lncRNA973 is involved in cotton response to salt stress. BMC Plant Biol 19(1):459

Zhang P, Fan Y, Sun X, Chen L, Terzaghi W, Bucher E, Li L, Dai M (2019a) A large-scale circular RNA profiling reveals universal molecular mechanisms responsive to drought stress in maize and *Arabidopsis*. Plant J 98(4):697–713

Zhang W, Gao S, Zhou X, Chellappan P, Chen Z, Zhou X, Zhang X, Fromuth N, Coutino G, Coffey M, Jin H (2011a) Bacteria-responsive microRNAs regulate plant innate immunity by modulating plant hormone networks. Plant Mol Biol 75(1–2):93–105

Zhang W, Han Z, Guo Q, et al. (2014) Identification of maize long non-coding RNAs responsive to drought stress. PLoS ONE 9:e98958

Zhang J, Hao Z, Yin S, Li G (2020a) GreenCircRNA: A database for plant circRNAs that act as miRNA decoys, Database 2020:baaa039

Zhang P, Li S, Chen M (2020b) Characterization and function of circular RNAs in plants. Front Mol Biosci 7:91

Zhang B, Su T, Li P, Xin X, Cao Y, Wang W, Zhao X, Zhang D, Yu Y, Li D, Yu S, Zhang F (2021) Identification of long noncoding RNAs involved in resistance to downy mildew in Chinese cabbage. Hortic Res 8:44

Zhang X, Wang W, Wang M, Zhang HY, Liu JH (2016) The miR396b of *Poncirus trifoliata* functions in cold tolerance by regulating ACC oxidase Gene expression and modulating ethylene-polyamine homeostasis. Plant Cell Physiol 57(9):1865–1878

Zhang Y, Yun Z, Gong L, Qu H, Duan X, Jiang Y, Zhu H (2018b) Comparison of miRNA evolution and function in plants and animals. MicroRNA 7(1):4–10

Zhang N, Zhang D, Chen SL, et al. (2018c) Engineering artificial microRNAs for multiplex gene silencing and simplified transgenic screen. Plant Physiol 178:989–1001

Zhang J, Zhang H, Srivastava AK, Pan Y, Bai J, Fang J, Shi H, Zhu JK (2018d) Knockdown of rice MicroRNA166 confers drought resistance by causing leaf rolling and altering stem xylem development. Plant Physiol 176(3):2082–2094

Zhang X, Zhao H, Gao S, Wang WC, Katiyar-Agarwal S, Huang HD, Raikhel N, Jin H (2011b) *Arabidopsis* argonaute 2 regulates innate immunity via miRNA393(*)-mediated silencing of a golgi-localized SNARE Gene, MEMB12. Mol Cell 42(3):356–366

Zhao W, Cheng Y, Zhang C, et al. (2017) Genome-wide identification and characterization of circular RNAs by high throughput sequencing in soybean. Sci Rep 7:1–11

Zhao W, Chu S, Jiao Y (2019b) Present scenario of circular RNAs (circRNAs) in plants. Front Plant Sci 10:379

Zhao Q, Yang Q, Wang Z, Sui Y, Wang Q, Liu J, Zhang H (2021) Analysis of long non-coding RNAs and mRNAs in harvested kiwifruit in response to the yeast antagonist, *Wickerhamomyces anomalus*. Comput Struct Biotechnol J 19:5589–5599

Zhao J, Yuan S, Zhou M, Yuan N, Li Z, Hu Q, Bethea FG Jr, Liu H, Li S, Luo H (2019) Transgenic creeping bentgrass overexpressing osa-miR393a exhibits altered plant development and improved multiple stress tolerance. Plant Biotechnol J 17(1): 233–251

Zheng B, Wang Z, Li S, et al. (2009) Intergenic transcription by RNA polymerase II coordinates pol IV and pol v in siRNA-directed transcriptional Gene silencing in *Arabidopsis*. Genes Dev 23:2850–2860

Zhou Y, Cho WK, Byun HS, et al. (2019) Genome-wide identification of long non-coding RNAs in tomato plants irradiated by neutrons followed by infection with tomato yellow leaf curl virus. PeerJ 2019:e6286

Zhou J, Deng K, Cheng Y, Zhong Z, Tian L, Tang X, Tang A, Zheng X, Zhang T, Qi Y, Zhang Y (2017) CRISPR-Cas9 based genome editing reveals new insights into MicroRNA function and regulation in rice. Front Plant Sci 8:1598

Zhou M, Luo H (2014) Role of microRNA319 in creeping bentgrass salinity and drought stress response, Plant Signaling Behavior 9(4):e28700

Zhou X, Wang G, Zhang W (2007) UV-B responsive microRNA genes in *Arabidopsis thaliana*. Mol Syst Biol 3:21

Zhou Y, Xu Z, Duan C, Chen Y, Meng Q, Wu J, Hao Z, Wang Z, Li M, Yong H, Zhang D, Zhang S, Weng J, Li X (2016) Dual transcriptome analysis reveals insights into the response to rice black-streaked dwarf virus in maize, J Exp Botany 67 (15):4593–4609

Zhou J, Yuan M, Zhao Y, Quan Q, Yu D, Yang H, Tang X, Xin X, Cai G, Qian Q, Qi Y, Zhang Y (2021) Efficient deletion of multiple circle RNA loci by CRISPR-Cas9 reveals Os06circ02797 as a putative sponge for OsMIR408 in rice. Plant Biotechnol J 19(6):1240–1252

Zhou R, Yu X, Ottosen CO, Zhang T, Wu Z, Zhao T (2020) Unique miRNAs and their targets in tomato leaf responding to combined drought and heat stress. BMC Plant Biol 20(1):107

Zhou R, Zhu Y, Zhao J, Fang Z, Wang S, Yin J, Chu Z, Ma D (2018) Transcriptome-wide identification and characterization of potato circular RNAs in response to *pectobacterium carotovorum* subspecies brasiliense infection, Int J Mol Sci 19:71–71

Zhu YX, Jia JH, Yang L, Xia YC, Zhang HL, Jia JB, Zhou R, Nie PY, Yin JL, Ma DF, Liu LC (2019) Identification of cucumber circular RNAs responsive to salt stress. BMC Plant Biol 19:164

Zhu QH, Stephen S, Taylor J, Helliwell CA, Wang MB (2014) Long noncoding RNAs responsive to *Fusarium oxysporum* infection in *Arabidopsis thaliana*. New Phytol 201 574–584

Zubko E, Meyer P (2007) A natural antisense transcript of the *Petunia hybrida* sho Gene suggests a role for an antisense mechanism in cytokinin regulation. Plant J 52:1131–1139

Zuo ZF, He W, Li J, Mo B, Liu L (2021) Small RNAs: The essential regulators in plant thermotolerance. Front Plant Sci 12:726762

Zuo J, Wang Q, Zhu B, Luo Y, Gao L (2016) Deciphering the roles of circRNAs on chilling injury in tomato. Biochem Biophys Res Commun 479(2):132–138

Zuo J, Wang Y, Zhu B, Luo Y, Wang Q, Gao L (2018) Analysis of the coding and non-coding RNA transcriptomes in response to bell pepper chilling. Int J Mol Sci 19(7):2001

2 Regulatory Role(s) of Plant Small Non-Coding RNAs in Relation to Trait Improvement in Crops

Sicon Mitra[1], Niraj Kumar Jha[1], Devendra Kumar Pandey[3], Milan Kumar Lal[4], Sayanti Mandal[5], Mahipal S. Shekhawat[6], Mimosa Ghorai[2], Potshangbam Nongdam[7], Satarupa Dey[8], Pradeep Kumar[9], Madhu Kamle[10], and Santosh Kumar Gupta[11]

[1]Department of Biotechnology, School of Engineering & Technology (SET), Sharda University, Greater Noida, India

[2]Department of Life Sciences, Presidency University, Kolkata, India

[3]Department of Biotechnology, Lovely Professional University, Phagwara, Punjab, India

[4]ICAR-Central Potato Research Institute, Indian Council of Agricultural Research, Ministry of Agriculture and Farmers' Welfare, Shimla, India

[5]Department of Chemistry and Biochemistry, School of Basic Sciences & Research, Sharda University, Greater Noida, Uttar Pradesh, India

[6]Plant Biotechnology Unit, KM Government Institute for Postgraduate Studies and Research, Puducherry, India

[7]Department of Biotechnology, Manipur University, Canchipur, Manipur, India

[8]Department of Botany, Shyampur Siddheswari Mahavidyalaya, Howrah, West Bengal, India

[9]North Eastern Regional Institute of Science and Technology (NERIST), Nirjuli, Andhra Pradesh, India

[10]Applied Microbiology Laboratory, Department of Forestry, North Eastern Regional Institute of Science and Technology, Nirjuli, Andhra Pradesh, India

[11]Plant Biotechnology & Radioisotope Facility, DBT-National Institute of Plant Genome Research, Aruna Asaf Ali Marg, New Delhi, India

2.1 INTRODUCTION

Small RNAs (sRNAs) assist in regulating other genes, whereas non-coding RNAs do not code for any particular protein. These can be termed as microRNAs (miRNAs), owing to being identified as small double-stranded RNAs (dsRNAs). *Caenorhabditis elegans*, for instance, is an ideal model

DOI: 10.1201/9781003369288-2

organism that showed RNA interference, in addition the silencing role of these sRNAs [Lee et al., 1993]. MicroRNA (mRNA) is degraded in the presence of regulatory proteins that further carry out gene silencing. In some cases, small interfering RNAs (siRNAs) denote RNA silencing, which then provides defense against viral pathogens. These classes of sRNAs also enhance the expression of desired products, thereby assisting in crop improvement. In some plants, proteins such as RNA-dependent RNA polymerase (RDR), Dicer-like (DCL), and Argonautes (AGO) carry out RNA silencing. The first type of protein, which is present in fungi, worms, and amphioxus, synthesizes siRNA in the presence of the DCL protein via transcription of single-stranded RNA (ssRNAs) into dsRNAs [Wassenegger and Krczal, 2006].

Contrarily, the remaining two types of protein are present in every organism that carries out RNA silencing. DCL proteins are RNase (type III) that process hairpin RNA or dsRNA into miRNA or siRNA, respectively. AGO proteins are endonucleases, which are responsible for forming the RNA-induced silencing complex (RISC) with miRNAs or siRNAs. This chapter emphasizes different types of sRNAs, biogenesis, and roles in crop improvement. Specifically, protection against viral pathogens, restoration of male fertility, tolerance against drought, cold temperature, and salinity are elaborated.

2.2 CLASSIFICATIONS OF SMALL NON-CODING RNAs (sncRNAs)

Classification of sncRNAs includes miRNAs, siRNAs, small nuclear RNAs (snRNAs), small nucleolar RNAs (snoRNAs), heterochromatic siRNAs (hetsiRNAs), Hairpin-derived siRNAs (hp-siRNAs), and natural antisense siRNAs (natsiRNAs). Sub-classification of siRNAs further consists of endogenous trans-acting siRNA (tasiRNA), small scan RNA (scnRNA), and repeat-associated siRNA (rasiRNA).

2.3 BIOGENESIS OF TYPES OF sRNAs

In the process of biogenesis of different classes of sncRNAs, the following steps are followed. In coding mRNAs, dsRNA and RNA polymerase (RNAP) carry out the transcription of sRNA and the transcriptional process, respectively. Next, small-length (18–25 nucleotides) dsRNAs are synthesized, followed by the occurrence of sRNA methylation and incorporation into effector complexes [Budak and Akpinar, 2015]. As a result, respective target sites are identified and targeted [Voinnet, 2009].

This process is carried out in the presence of the mediator, which is a transcriptional coactivator. Transcription factor (TF) binding sites are identified in the promoter region of miRNA with the help of this coactivator, which then directs RNAP II [Chen, 2005; Budak and Akpinar, 2015]. Also, this transcription results in the synthesis of a primary single-stranded miRNA (pri-miRNA), which is processed by Dicer-like protein 1 (DCL1) in the nucleus of plant cells [Axtell and Meyers, 2018]. Cleavage reactions consist of (1) DCL1 forms precursor miRNA (pre-miRNA) of 70 nt while inducing the first cleavage from the 15 nt of the base of the stem, and (2) ATC 20–24 nt downstream of the first cleavage site where miRNA/miRNA (passenger strand/guide strand) duplex is released from the stem. SRNA-degrading nuclease (SDN) along with methylation by HEN1 inhibit degradation of miRNA/miRNA duplex [Voinnet, 2009] prior to the incorporation of miRNA into RISC (Figure 2.1) [Axtell et al., 2011].

When TAS transcripts with miRNA-guided cleavage, tasiRNAs are formed [Chen et al., 2010]. TAS1 and TAS2 transcripts are obtained and synthesized using miR173, whereas TAS3 and TAS4 are obtained from miR390 and miR828 [Howell et al., 2007].

This synthesis takes place in the nucleus in the presence of RDR2 [Vaucheret, 2006]. During the biogenesis, snRNAs move into the cytoplasm to and from the nucleus. The transcription process results in the attachment of several snRNA-associated proteins with snRNAs [Matera and Wang, 2014].

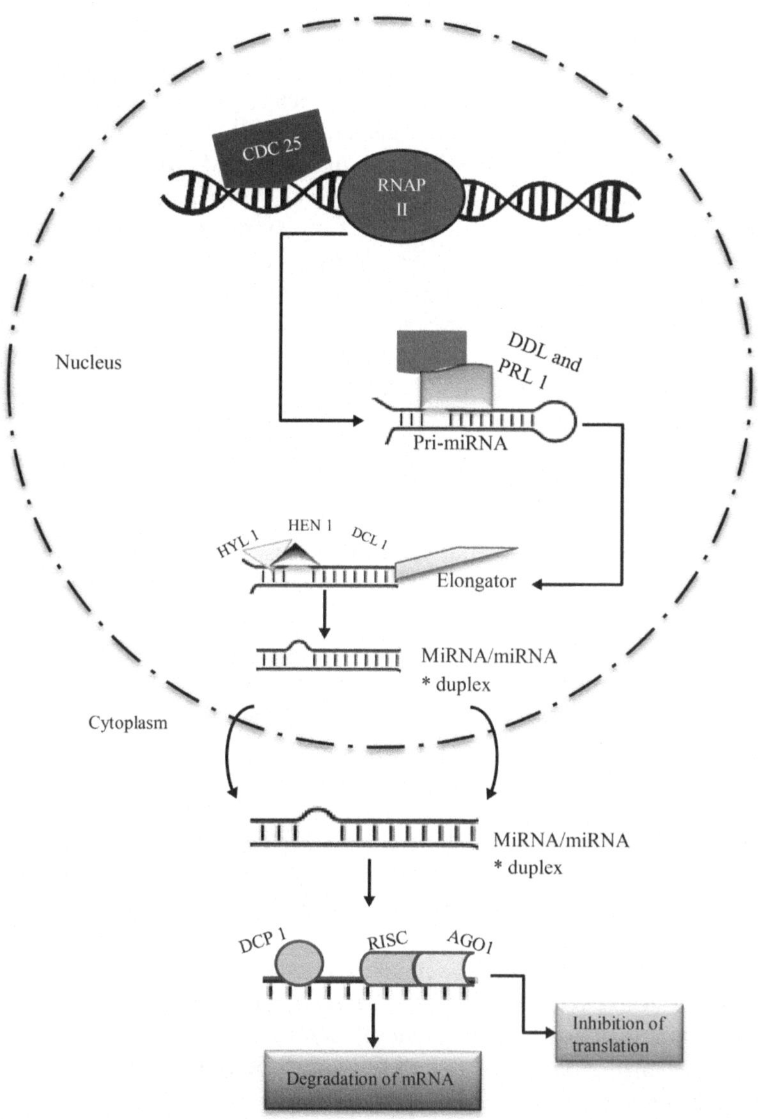

FIGURE 2.1 Schematic representation of miRNA biogenesis. In the nucleus, CDC 25 and RNA polymerase II binds to the miRNA gene and synthesize pri-miRNA, which is further associated and stabilized by Dawdle (DDL) and Pleiotropic Regulatory Locus 1 (PRL1). Further processing of transcripts into miRNA/miRNA* duplex is carried out by DICER-LIKE 1 (DCL), HYPONASTIC LEAVES1 (HYL1), and HUA ENHANCER1 (HEN1). miRNA/miRNA* duplex is translocated into the cytoplasm by HASTY (HST). In cytoplasm RNA-induced silencing complex (RISC), DCP 1 and Argonaute (AGO) binds to miRNA and helps in the inhibition of mRNA translation and mRNA degradation.

2.4 ROLE OF sRNA IN CROP PROTECTION AGAINST VIRAL PATHOGENS

Genetic engineering utilizing sRNA has made it possible for transgenic crops to develop virus-resistant genotypes. SRNA-based genetic engineering technology, especially SRGE, has been introduced to protect crops [Khalid et al., 2017]. Sense gene induced post-transcriptional gene silencing (S-PTGS), artificial miRNA induced PTGS (AMIR), hairpin RNA induced PTGS (hp-PTGS), and tasiRNA-induced PTGS (TAS) play a crucial role in the silencing of antiviral genes.

2.4.1 Maize, Rice, Wheat, and Cassava

These four genetically engineered crops are major staple food crops. Havoc destruction is, however, analyzed in the case of maize when exposed to viruses such as maize steak virus (MSV) and maize dwarf mosaic virus (MDMV). To prevent this, mechanisms of S-PTGS and hp-PTGS were carried out, resulting in the introduction of MSV-immune maize and MDMV-resistant maize. This was done by targeting the viral Rep gene and p1 along with CP genes [Shepherd et al., 2007; Zhang et al., 2011; Zhang et al., 2013]. In the case of rice, Tungroviruses, Phytoreoviruses, Tenuiviruses, and Waikaviruses are examples of viruses that destroy rice cultivation, thereby prioritizing the development of transgenic resistant rice strains using hp-PTGS [Roy et al., 2012; Sasaya et al., 2014]. Similarly, wheat that is immune to wheat steak mosaic virus (WSMV) was developed using three generations of SRGE [Fahim et al., 2012; Khalid et al., 2017]. Similarly, ACMV-immune Cassava was developed using hp-PTGS along with the S-PTGS mechanism via targeting the AC1 gene [Chellappan et al., 2004].

2.4.2 Watermelon, Squash, and Cucumber

Being members of the *Cucurbitaceae* family, various viruses hinder the successful production of these crops. To deal with this, virus-resistant watermelon and oriental melon were developed using the S-PTGS mechanism by targeting Chimeric CP sequences of viruses, the viral capsid protein gene, and the Rep gene, respectively. Squash mosaic virus-resistant squash and cucumber fruit mottle mosaic tobamovirus (CFMMV)-resistant cucumber are the two major examples of crop improvement using RNAs as regulators [Lin et al., 2012; Khalid et al., 2017].

2.4.3 Peanut, Soybean, and Common Bean

Being members of the *Fabaceae* family, these contain higher amounts of fatty acid, proteins, and other nutrients. Tobacco streak virus (TSV) resistant and peanut stripe virus (PStV) immune lines are the two improved peanut crops that were obtained using the S-PTGS mechanism [Mehta et al., 2013; Khalid et al., 2017]. In regards to soybean, soybean mosaic virus (SMV)-resistant transgenic soybean is a significant example, which is engineered using S-PTGS and hp-PTGS targeting CP and HcPro [Gao et al., 2015]. To protect common beans against bean golden mosaic virus (BGMV), BGMV-partial-resistant common beans were developed using the S-PTGS mechanism via targeting CP genes. This resulted in the generation of fully immune genotypes (Table 2.1) [Aragão et al., 2013; Faria et al., 2014].

2.5 MECHANISM OF ACTION OF sRNA IN PEARL MILLET AND WHEAT AGAINST SALINITY STRESS

Globally, salinity causes massive abiotic stress, which not only increases the amount of salt in the soil, but also reduces agricultural yield [Teh and Koh, 2016]. Between 20% and 30% of cultivated and irrigated agricultural lands have been affected due to salinity [Machado and Serralheiro, 2017]. In addition, Shinde et al. (2020) found that out of 130 million identified sRNA sequences in the salinity tolerant, 81 million and 14 million variants of pearl millets were identified using several sncRNA-derived conserved and novel miRNA, respectively. During salinity-stressed conditions, the miR166 family of miRNAs was downregulated, whereas the miR159 family was upregulated, playing crucial roles [Li et al., 2017a, 2017b; Abdelrahman et al., 2018]. In the latter case, promoter binding transcriptional factor genes, namely, SQUAMOSA (SBPL3 and SPL), are targeted in response to combating salinity stress [Shinde et al., 2020].

TABLE 2.1
Different Virus-Resistant Transgenic Lines Obtained Using sRNA Engineering

Sl No.	Crop	Modes of Mechanism to Develop Transgenic Lines	References
1.	**Watermelon**	Watermelon that is virus-free was developed using the S-PTGS mechanism via targeting Chimeric viral CP sequences.	Khalid et al., 2017
	Squash	S-PTGS was used to create Squash SqMV-immune squash by targeting the viral genome's CP gene.	Lin et al., 2012 Khalid et al., 2017
	Cucumber	Targeting the Rep gene via the action of S-PTGS helped in the production of CFMMV-immune cucumber.	Khalid et al., 2017
2.	**Maize**	Using the S-PTGS and hp-PTGS mechanisms, MSV-immune maize and MDMV-resistant maize were created.	Shepherd et al., 2007 Zhang et al., 2011 Zhang et al., 2013
	Rice	Using the hp-PTGS mechanism, transgenic resistant rice lines against viruses such as Phytoreoviruses, Tungroviruses, Tenuiviruses, and Waikaviruses were created.	Roy et al., 2012 Sasaya et al., 2014 Khalid et al., 2017
	Wheat	All three generations of SRGE were used to create WSMV-resistant wheat.	Fahim et al., 2012 Khalid et al., 2017
	Cassava	By combination of the hp-PTGS and the S-PTGS mechanism, ACMV-immune Cassava lines were created by targeting the AC1 gene.	Chellappan et al., 2004
3.	**Peanut**	The S-PTGS mechanism effectively developed TSV-resistant and PStV-immune strains.	Mehta et al., 2013 Khalid et al., 2017
	Common bean	S-PTGS targeting CP genes yielded BGMV-partial-resistant common beans, whereas hp-PTGS targeting the AC1 gene yielded totally immune genotypes.	Faria et al., 2006, 2014 Feng et al., 2018
	Soybean	S-PTGS and hp-PTGS were used to create SMV-resistant soybeans via targeting CP and HcPro, respectively. In addition, multi-resistant genotypes of BPMV, AMV, and SM were created by regulating the Rep gene.	Zhang et al., 2011 Gao et al., 2015

Additionally, detoxification of solutes from plant cells has been helping ion transporters, especially potassium transporters, to fight against sodium toxicity, i.e. abiotic salinity stress conditions [Brini and Masmoudi, 2012; Hamamoto et al., 2015]. In plants, salinity tolerance is modulated via the auxin signaling pathway. In pearl millet, miR162a and miR393a of the KEGG pathway analysis are specifically involved in the auxin signaling mechanism [Park et al., 2016; Bielach et al., 2017; Shinde et al., 2020]. Two positive modulators involved in the mechanism of auxin signaling are TIR1 and ARF TF genes, whereas these are targeted by miR393a and miR162a, respectively. Under the previously mentioned condition, miR393a is upregulated, which is often denoted as a negative modulator of the auxin signaling pathway [Shinde et al., 2020]. Contrarily, miR162a is downregulated, thereby assisting as a positive regulator of the same signaling pathway (Figure 2.2).

In the case of wheat, salinity resistance mechanisms like that of pearl millet are analyzed wherein 98 conserved and 219 novel known miRNAs have been identified via sRNA deep sequencing. Out of these, 26 conserved miRNAs target 58 genes, and 65 novel miRNAs target 549 genes [Han et al., 2018]. Specifically, miR300, a novel miRNA, has been regulating the auxin synthesis pathway. In response to salinity stress in wheat, DNA methylation is practiced. In this case, novel miR190 and miR223 are upregulated in contrast to the downregulation of miR284 and miR53 [Han et al., 2018].

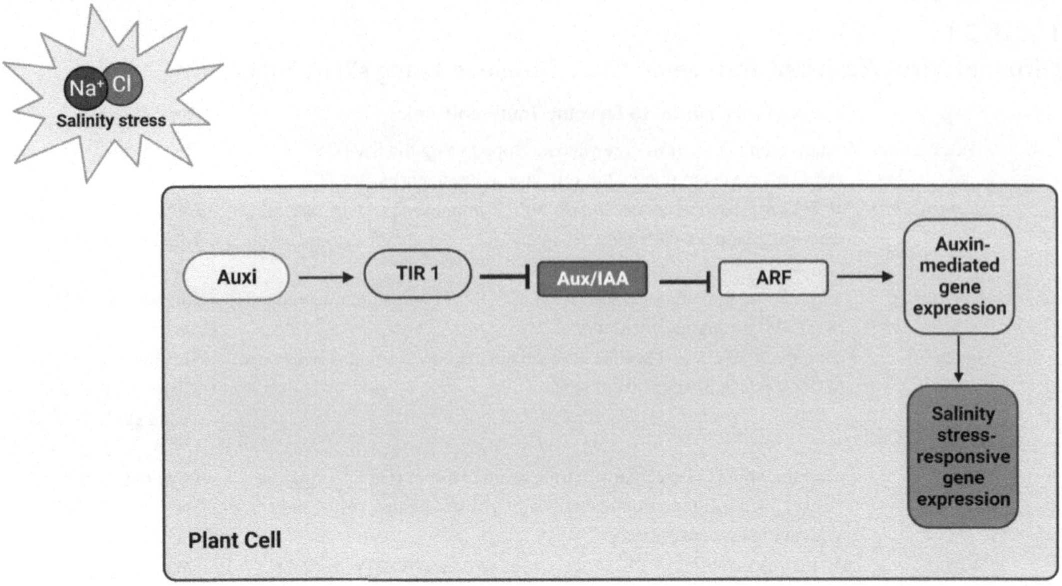

FIGURE 2.2 Schematic representation of auxin signaling pathway. During salinity stress, miR393a is upregulated and inhibits TIR 1 (inhibitor of Aux/IAA), whereas miR16a is downregulated, thus ARF binds to the ARF promoter and helps in the expression of salinity response genes. Figure created with Biorender.

2.6 ROLE OF sncRNAs IN REGULATING DROUGHT RESPONSE GENES IN DIFFERENT CROPS

Different types of miRNAs are analyzed to play crucial roles in the drought resistance responses in different crops, such as salinity resistance sRNA. The first miRNA that helped in the regulation of drought response was miR398 [Sunkar et al., 2006]. In this context, 13 drought-regulating miRNA families consisting of 65 miRNAs and 42 unique targets were found [Aravind et al., 2017]. Specifically in *Setaria italica* (foxtail millet), miR390 and miR167b target ubiquitin ARF and E2C binding protein genes, respectively, with two essential crops described next.

2.6.1 *Lycopersicon esculentum* (Tomato)

Three miRNAs were reported to be crucial to responding to drought, namely, miR9552b3p, miR845b5p, and miR398a5p [Candar-Cakir, Arican, and Zhang, 2016]. Out of these, the first two target UDP-glucosyltransferase (UGT) and the DELLA protein. Moreover, miR9552b3p is itself upregulated, which also assists in downregulating UGT expression [Candar-Cakir, Arican, and Zhang, 2016]. On the other hand, auxin proteins, HD-ZIP III, and NAC-domain genes are targeted by miR398a5p, 1.5 which, in turn, ensures drought resistance.

2.6.2 *Triticum aestivum* (Wheat)

Out of a variety of identified miRNAs, miRNA393 is an example that targets TIR1 and regulates the auxin signaling pathway, which lowers the rate of plant growth, thereby regulating the drought-induced response. On the other hand, miR5200 itself acts as guard cells, which then assist in regulating stomata opening. This is done by targeting the FT protein [Giusti et al., 2017]. Another miRNA, miR398, targets COX5b and Cu/Zn superoxide dismutase, which gradually decreases the chance of drought-induced damage. This is, however, reported to be possible in wheat via the regulation of ROS, CDS1/2, and COX5b [Akdogan et al., 2016]. Additionally, miR1029 and miR529a are observed to play key roles, as these target AP2 TF. In this process, miR1122c is further involved to regulate MADS-box TF, thereby modulating flowering in wheat [Bakhshi et al., 2017; Kaur et al., 2017].

2.7 ROLE OF tasiRNA IN ECONOMICALLY ESSENTIAL CROPS

2.7.1 *Cucumis Melo* (Melon)

Growth in semi-arid areas increases the chance of melons being exposed to adverse environmental changes [Bustamante et al., 2018; Wu et al., 2018; Sanz-Carbonell et al., 2019]. Two miR390 target sites that generate phased sRNAs, and reminiscent to TAS3 genes are reported in melon [Gonzalez-Ibeas et al., 2011]. TAS3-short (TAS3-S) and TAS3-long (TAS3-L) are the two families of TAS3 loci identified in *Cucumis melo* that transcribe short and long transcripts [Xia et al., 2017; Cervera-Seco et al., 2019]. In the first case, 5′ and 3′ target sites of miR390 are divided to generate one tasiARF [Xia et al., 2015; 2017]. However, in the latter case, 3′ end of miR390 is cleaved to generate two in-phase tasiARFs and 5′ end is non-cleavable [Allen et al., 2005; Axtell et al., 2006].

Auxin response factor (ARF) is a negative regulator (represented by miR390/TAS3-ARF) [Xia et al., 2017]. Downregulation of tasiARF, which is derived from TAS3-L and TAS3-S, assists in responding to stress conditions, especially cold temperature [Cervera-Seco et al., 2019]. This crop, when exposed to infection of Hop stunt viroid (HSVd), depicts a significant reduction in TAS3-S-derived tasiRNA due to the reduced accumulation level of mi390 [Sanz-Carbonell et al., 2019]. This, therefore, signifies the biogenesis of tasiRNA in melon to be a dynamic process. Under saline stress conditions, upregulation of TAS3-S-derived tasiRNAs is analyzed in polar plants, which contradicts with the downregulation of tasiRNAs in cold conditions. From this, it can, therefore, be stated that different stress conditions and species result in different accumulation levels of tasiRNA [Cervera-Seco et al., 2019].

2.7.2 *Oryza sativa* (Rice) and *Arabidopsis* (Rockcress)

They are the major staple foods across Asia. A melon-like mode of action of the auxin pathway via TAS3 was observed in this case. However, in *Oryza sativa*, modulation of ARF16 and ARF12 was the result of phosphate starvation, which further regulates phosphate homeostasis [Shen et al., 2013]. In *Arabidopsis*, tasiRNA-mutant derived from TAS3 was analyzed to be significant in floral development [Guilfoyle and Hagen, 2007] (Table 2.2).

2.8 FUNCTIONS OF sRNA IN REGULATING MALE STERILITY IN CROPS

2.8.1 *O. sativa*

Cheng et al. (2007) found that the development of hybrid varieties is an effective approach to increasing the productivity of *O. sativa* by 20%. One leading technology is the two-line system that is based on the application of environmentally sensitive genic male sterile (EGMS). Under varying environmental conditions, this system operates as male sterile as well as maintainer lines [Wang et al., 2006]. The next example is the three-line system, where maintainer lines, cytoplasmic male sterile lines, and restorer lines are utilized [Zhou et al., 2012]. In this context, photoperiod-sensitive genic male sterile (PGMS) and thermos-sensitive genic male sterile (TGMS) are the major components currently in use in agricultural fields [Mingsong and Jingyang, 1986].

Thermos- or photo-sensitive genic male sterility locus (also known as p/tms12-1) is, moreover, a major locus, which was cloned and mapped in chromosome 12 [Zhou et al., 2012]. Nongken 58S

TABLE 2.2
Different Roles Played by Small Non-Coding RNAs as Stress Responses

Stress Conditions	Crop	Role of the Stress Response by sRNAs	References
Salinity	Pearl millet	• This circumstance results in the upregulation of miR393a, which negatively modulates the auxin signaling pathway. • Contrarily, downregulation of miR162a leads to positive regulation of the auxin signaling. • Also, downregulation of miR159 occurs, which then results in the upregulation of SBPL3 and SPL, the two members SQUAMOSA. • Downregulation of miR172b is, moreover, responsible for the activation of three potassium transporters.	Bielach et al., 2017 Shinde et al., 2020
	Wheat	• Under this condition, modulation of miR300 is responsible for the regulation of auxin synthesis. • Novel miR190 and miR223 are upregulated in contrast to the downregulation of miR284 and miR53.	Han et al., 2018
Drought	Tomato	• The presence of miR398a5p results in regulating the auxin metabolism. • DELLA and UDP-glucosyltransferase (UGT) protein levels are modulated using miR845b5p and miR9552b3p.	Candar-Cakir et al., 2016
	Wheat	• Like the above, the auxin signaling pathway is regulated by miRNA393. This is possible when miRNA393 targets TIR1. • Stomatal opening is possible due to the presence of miR5200. • COX5b and Cu/Zn superoxide dismutase are targeted by miR398, which then reduces the chance of a cell becoming damaged during drought.	Akdogan et al., 2016 Giusti et al., 2017 Kaur et al., 2017
Adverse environmental changes	Melon	• Availability of tasiARF, which is derived from miR390/TAS3-L and miR390/TAS3-S, results in its downregulation, followed by obtaining tolerance to cold temperature. • Hop stunt viroid (HSVd) infection decreases the accumulation level of TAS3-S.	Cervera-Seco et al., 2019

(NK58S), which is the first spontaneous PGMS mutant, and a Japonica rice line, along with TGMS, which is an Indica rice line are conferred by PGMS using this locus. P/tms12-1, a wild-type allele located in a DNA, encodes sRNA osa-smR586w, which, in turn, assists in suppressing PGMS and TGMS [Zhou et al., 2012]. Being the first point mutation in locus p/tms12-1, loss-of-function of osa-smR586w was analyzed. This, however, restored PGMA and TGMS in Japonica and Indica, respectively. From a different perspective, synthesis of osa-smR5864w was reported to be possible in restoring pollen fertility in NK58S and PA64S plants because of overexpression of a sequence of 375-bp in p/tms12-1, which is not dependent on variable temperatures or day lengths [Zhou et al., 2012]. The ability to restore pollen fertility is, therefore, a key function of this sRNA.

1.8.2 *Brassica campestris* (Field Mustard)

This is cultivated across the globe as a vegetable as well as an oilseed crop. In this case, similar functions in regulating male sterility were observed. However, in *Brassica campestris*, the sncRNA, which is responsible for regulating male fertility as well as the development of pollen (microspore and tapetum), is known as BcMF11 of 828-bp. Researchers have succeeded in decreasing the efficiency of pollen germination with the inhibition of BcMF11 via antisense technology [Song et al., 2013].

2.9 ROLE OF sncRNAs IN BIOSYNTHESIS OF FATTY ACIDS IN OIL PALM

The most abundant oil-producing species on the planet is *Elaeis guineensis* Jacq. (oil palm), which is a monocot plant belonging in the *Arecaceae* family, producing roughly 36% of global oil output [US Department of Agriculture 2015]. Oil palm collects about 90% of its oil within its mesocarp [Loh, 2017]. Zheng et al. (2019) discovered the involvement of sRNA in fatty acid (FA) metabolism in oil palm. In addition, it was discovered using sRNA sequencing data that about 452 miRNAs were found in five developmental stages of mesocarp [Zheng et al., 2019]. These 452 miRNAs contain 282 novel-miRNAs (nov-miRNAs) and 170 known-miRNAs that have been conserved (kn-miRNAs). Nevertheless, after focusing these miRNAs to 37 genes for synthesis of FA, it was shown that 36 miRNAs (14 nov-miRNAs and 22 kn-miRNAs) were primarily responsible for controlling metabolism of FA in oil palm [Zheng et al., 2019].

A few of those 36 miRNAs have been identified to introduce double bonds, leading to the elongation of FA chains in plastids of plant cells. For instance, nov-miR129, eg-miR397, eg-miR156c, and eg-miR444b aid in transferring the FA-ACP to endoplasmic reticulum (ER) by regulating the genes enoyl-ACP reductase (ENR), long-chain acyl-CoA synthetase 9 (LACS 9), acetyl-CoA carboxylase 1 (ACC1), and LACS4, respectively [Pulsifer et al., 2012; Li-Beisson et al., 2013; Wang et al., 2016]. Moreover, nov-mi59 and nov-miR138, which target the glycerol-3-phosphate acyltransferase (GPAT) gene, have been revealed to be essential in the production of triacylglycerol (TAG) within the ER. The nov-miR274 gene regulates the phosphatidate phosphatase 1 (PAP1) gene and aids in the de-phosphorylation of glycerol 3-phosphate, which leads to the formation of diacylglycerol (DAG) [Li-Beisson et al., 2013]. Several miRNAs in the eg-miR156 family regulate the β-ketoacyl-CoA synthase 12 (KCS12) gene, which aids in the condensation of malonyl-CoA with acyl-CoA [Roudier et al., 2010]. Furthermore, nov-miR201 promotes the conversion of enoyl-CoA to acyl-CoA through regulating the genes for enoyl-CoA reductase (ECR) [DeBono et al., 2009; Lee et al., 2009]. We may deduce from this work that various short non-coding RNA-induced miRNAs are involved in controlling FA production and metabolism in the mesocarp of oil palm. Therefore, sRNA expression and its important function in the control of FA synthesis and retention in the mesocarp aid in the development of a better fruit in *E. guineensis* Jacq (oil palm) (Table 2.3).

TABLE 2.3
Summary of the Function of Different sRNAs in Crop Improvement

Roles of sRNA	Crops	Observations	References
Male fertility	Rice	The p/tms12-1 encoded small RNA osa-smR586w can act as a suppressor in photoperiod-sensitive genic male sterile (PGMS) and thermos-sensitive genic male sterile (TGMS), thereby conferring male fertility in rice.	Zhou et al., 2012
Male fertility	Cabbage	A small non-coding RNA called BcMF11 helps in the restoration of male fertility in cabbage.	Hua et al., 2012
Fatty acid synthesis	Oil palm	The nov-miR274 gene helps in synthesizing diacylglycerol production (DAG) via regulating the phosphatidate phosphatase 1 (PAP1) gene. Triacylglycerol (TAG) is produced in the ER via the nov-mi59 and nov-miR138 mechanisms, which target the glycerol-3-phosphate acyltransferase (GPAT) gene. The nov-miR129, eg-miR397, and other small non-coding RNA aid in the translocation of FA-ACP to the ER by regulating enoyl-ACP reductase (ENR) and long-chain acyl-CoA synthetase 9 (LACS9), respectively.	Debono et al., 2009 Roudier et al., 2010 Pulsifer et al., 2012 Li-Beisson et al., 2013 Wang et al., 2016

2.10 CONCLUSION

This chapter describes in depth the taxonomy, biosynthesis, and many uses of short non-coding RNA in crop improvement. As a result, sRNA engineering may be inferred to be beneficial in the production of transgenic lines of diverse crops belonging to the *Solanaceae*, *Cucurbitaceae*, *Fabaceae*, etc., and other families to combat against various types of viral infections. These transgenic lines assist to boost crop output and productivity, hence reducing crop failure. Many sRNA engineering techniques, such as S-PTGS, hg-PTGS, and others, aid in the development of numerous viral-immune crops. SRNAs also aided crops such as pearl millet and wheat in maintaining salt resistance by regulating numerous miRNA-induced saline responsive genes and suppressing inhibitory genes. In crops such as wheat, barley, and tomato, comparable sRNA activities were found in water deficit conditions, and numerous miRNAs played a role in regulating various drought-responsive genes, assisting these crops to survive drought conditions. In melon, sRNAs were particularly beneficial in resisting both abiotic and biotic stress circumstances caused by extreme temperature and HSV infection. Phosphate deficiency in rice can also be addressed through the activity of the sRNA-mediated auxin signaling pathway. Several sRNAs have been identified as aiding in the restoration of pollen fertility in rice and cabbage. In the case of oil palm, nevertheless, numerous new miRNAs participate in the production and FA metabolism. Considering these factors, it is possible to infer that short non-coding RNAs play a function in the regulation of numerous roles in crops, hence increasing crop survival and productivity. Furthermore, the introduction of several sRNA approaches has cleared the way for future study and improved agricultural productivity.

REFERENCES

Abdelrahman, M., Jogaiah, S., Burritt, D.J. and Tran, L.S., 2018. Legume genetic resources and transcriptome dynamics under abiotic stress conditions. *Plant, cell & environment*, 41(9), pp. 1972–1983.

Akdogan, G., Tufekci, E.D., Uranbey, S. and Unver, T., 2016. miRNA-based drought regulation in wheat. *Functional & integrative genomics*, 16(3), pp. 221–233.

Allen, E., Xie, Z., Gustafson, A.M. and Carrington, J.C., 2005. microRNA-directed phasing during trans-acting siRNA biogenesis in plants. *Cell*, 121(2), pp. 207–221.

Aragão, F.J., Nogueira, E.O., Tinoco, M.L.P. and Faria, J.C., 2013. Molecular characterization of the first commercial transgenic common bean immune to the bean golden mosaic virus. *Journal of biotechnology*, 166(1–2), pp. 42–50.

Aravind, J., Rinku, S., Pooja, B., Shikha, M., Kaliyugam, S., Mallikarjuna, M.G., Kumar, A., Rao, A.R. and Nepolean, T., 2017. Identification, characterization, and functional validation of drought-responsive microRNAs in subtropical maize inbreds. *Frontiers in plant science*, 8, p. 941.

Axtell, M.J., Jan, C., Rajagopalan, R. and Bartel, D.P., 2006. A two-hit trigger for siRNA biogenesis in plants. *Cell*, 127(3), pp. 565–577.

Axtell, M.J. and Meyers, B.C., 2018. Revisiting criteria for plant microRNA annotation in the era of big data. *The plant cell*, 30(2), pp. 272–284.

Axtell, M.J., Westholm, J.O. and Lai, E.C., 2011. Vive la différence: Biogenesis and evolution of microRNAs in plants and animals. *Genome biology*, 12(4), pp. 1–13.

Bakhshi, B., Fard, E.M., Gharechahi, J., Safarzadeh, M., Nikpay, N., Fotovat, R., Azimi, M.R. and Salekdeh, G.H., 2017. The contrasting microRNA content of a drought tolerant and a drought susceptible wheat cultivar. *Journal of plant physiology*, 216, pp. 35–43.

Bielach, A., Hrtyan, M. and Tognetti, V.B., 2017. Plants under stress: Involvement of auxin and cytokinin. *International journal of molecular sciences*, 18(7), p. 1427.

Brini, F. and Masmoudi, K., 2012. Ion transporters and abiotic stress tolerance in plants. *International scholarly research notices*, 2012.

Budak, H. and Akpinar, B.A., 2015. Plant miRNAs: Biogenesis, organization and origins. *Functional & integrative genomics*, 15(5), pp. 523–531.

Bustamante, A., Marques, M.C., Sanz-Carbonell, A., Mulet, J.M. and Gomez, G., 2018. Alternative processing of its precursor is related to miR319 decreasing in melon plants exposed to cold. *Scientific reports*, 8(1), pp. 1–13.

Candar-Cakir, B., Arican, E. and Zhang, B., 2016. Small RNA and degradome deep sequencing reveals drought-and tissue-specific micrornas and their important roles in drought-sensitive and drought-tolerant tomato genotypes. *Plant biotechnology journal*, 14(8), pp. 1727–1746.

Cervera-Seco, L., Marques, M.C., Sanz-Carbonell, A., Marquez-Molins, J., Carbonell, A., Darròs, J.A. and Gomez, G., 2019. Identification and characterization of stress-responsive TAS3-derived TasiRNAs in melon. *Plant and cell physiology*, 60(11), pp. 2382–2393.

Chellappan, P., Masona, M.V., Vanitharani, R., Taylor, N.J. and Fauquet, C.M., 2004. Broad spectrum resistance to ssDNA viruses associated with transgene-induced gene silencing in cassava. *Plant molecular biology*, 56(4), pp. 601–611.

Chen, X., 2005. MicroRNA biogenesis and function in plants. *FEBS letters*, 579(26), pp. 5923–5931.

Chen, H.M., Chen, L.T., Patel, K., Li, Y.H., Baulcombe, D.C. and Wu, S.H., 2010. 22-nucleotide RNAs trigger secondary siRNA biogenesis in plants. *Proceedings of the national academy of sciences*, 107(34), pp. 15269–15274.

Cheng, S.H., Zhuang, J.Y., Fan, Y.Y., Du, J.H. and Cao, L.Y., 2007. Progress in research and development on hybrid rice: A super-domesticate in China. *Annals of botany*, 100(5), pp. 959–966.

DeBono, A., Yeats, T.H., Rose, J.K., Bird, D., Jetter, R., Kunst, L. and Samuels, L., 2009. Arabidopsis LTPG is a glycosylphosphatidylinositol-anchored lipid transfer protein required for export of lipids to the plant surface. *The plant cell*, 21(4), pp. 1230–1238.

Fahim, M., Millar, A.A., Wood, C.C. and Larkin, P.J., 2012. Resistance to wheat streak mosaic virus generated by expression of an artificial polycistronic microRNA in wheat. *Plant biotechnology journal*, 10(2), pp. 150–163.

Faria, J.C., Valdisser, P.A., Nogueira, E.O. and Aragão, F.J., 2014. RNAi-based b ean golden mosaic virus-resistant common bean (E mbrapa 5.1) shows simple inheritance for both transgene and disease resistance. *Plant breeding*, 133(5), pp. 649–653.

Faria, A., Calhau, C., Freitas, V., Mateus, N. 2006. Procyanidins as antioxidants and tumor cell growth modulators. *Journal of agricultural and food chemistry*, 54(6), pp. 2392–2397.

Feng, X., Orellana, G. E., Myers, J. R., Karasev, A. V. 2018. Recessive resistance to bean common mosaic virus conferred by the bc-1 and bc-2 Genes in common bean (*Phaseolus vulgaris*) affects long-distance movement of the virus. *Phytopathology*, 108(8), pp. 1011–1018. doi: 10.1094/PHYTO-01-18-0021-R.

Gao, L., Ding, X., Li, K., Liao, W., Zhong, Y., Ren, R., Liu, Z., Adhimoolam, K. and Zhi, H., 2015. Characterization of soybean mosaic virus resistance derived from inverted repeat- SMV-HC-pro genes in multiple soybean cultivars. *Theoretical and applied genetics*, 128(8), pp. 1489–1505.

Giusti, L., Mica, E., Bertolini, E., De Leonardis, A.M., Faccioli, P., Cattivelli, L. and Crosatti, C., 2017. microRNAs differentially modulated in response to heat and drought stress in durum wheat cultivars with contrasting water use efficiency. *Functional & integrative genomics*, 17(2–3), pp. 293–309.

Gonzalez-Ibeas, D., Blanca, J., Donaire, L., Saladié, M., Mascarell-Creus, A., Cano-Delgado, A., Garcia-Mas, J., Llave, C. and Aranda, M.A., 2011. Analysis of the melon (Cucumis melo) small RNAome by high-throughput pyrosequencing. *BMC genomics*, 12(1), pp. 1–20.

Guilfoyle, T.J. and Hagen, G., 2007. Auxin response factors. *Current opinion in plant biology*, 10(5), pp. 453–460.

Hamamoto, S., Horie, T., Hauser, F., Deinlein, U., Schroeder, J.I. and Uozumi, N., 2015. HKT transporters mediate salt stress resistance in plants: From structure and function to the field. *Current opinion in biotechnology*, 32), pp. 113–120.

Han, H., Wang, Q., Wei, L., Liang, Y., Dai, J., Xia, G. and Liu, S., 2018. Small RNA and degradome sequencing used to elucidate the basis of tolerance to salinity and alkalinity in wheat. *BMC plant biology*, 18(1), pp. 1–17.

Howell, M.D., Fahlgren, N., Chapman, E.J., Cumbie, J.S., Sullivan, C.M., Givan, S.A., Kasschau, K.D. and Carrington, J.C., 2007. Genome-wide analysis of the RNA- DEPENDENT RNA POLYMERASE6/DICER-LIKE4 pathway in Arabidopsis reveals dependency on miRNA-and tasiRNA-directed targeting. *The plant cell*, 19(3), pp. 926–942.

Kaur, A., Gupta, O.P., Meena, N.L., Grewal, A. and Sharma, P., 2017. Comparative temporal expression analysis of microRNAs and their target genes in contrasting wheat genotypes during osmotic stress. *Applied biochemistry and biotechnology*, 181(2), pp. 613–626.

Khalid, A., Zhang, Q., Yasir, M. and Li, F., 2017. Small RNA based genetic engineering for plant viral resistance: Application in crop protection. *Frontiers in microbiology*, 8, p. 43.

Lee, R., Feinbaum, C. and Ambros, R.L., 1993. The c. elegans heterochronic Gene lin-4 encodes small RNAs with antisense complementarity to lin-14. *Cell*, 75(5), pp. 843–854.

Lee, S.B., Jung, S.J., Go, Y.S., Kim, H.U., Kim, J.K., Cho, H.J., Park, O.K. and Suh, M.C., 2009. Two Arabidopsis 3-ketoacyl CoA synthase genes, KCS20 and KCS2/DAISY, are functionally redundant in cuticular wax and root suberin biosynthesis, but differentially controlled by osmotic stress. *The plant journal*, 60(3), pp. 462–475.

Li-Beisson, Y., Shorrosh, B., Beisson, F., Andersson, M.X., Arondel, V., Bates, P.D., Baud, S., Bird, D., DeBono, A., Durrett, T.P. and Franke, R.B., 2013. Acyl-lipid metabolism. *The Arabidopsis book/ American society of plant biologists*, 11.

Li, Y., Wan, L., Bi, S., Wan, X., Li, Z., Cao, J., Tong, Z., Xu, H., He, F. and Li, X., 2017b. Identification of drought-responsive microRNAs from roots and leaves of alfalfa by high- throughput sequencing. *Genes*, 8(4), p. 119.

Li, X., Xie, X., Li, J., Cui, Y., Hou, Y., Zhai, L., Wang, X., Fu, Y., Liu, R. and Bian, S., 2017a. Conservation and diversification of the miR166 family in soybean and potential roles of newly identified miR166s. *BMC plant biology*, 17(1), pp. 1–18.

Loh, S.K., 2017. The potential of the Malaysian oil palm biomass as a renewable energy source. *Energy conversion and management*, 141, pp. 285–298.

Machado, R.M.A. and Serralheiro, R.P., 2017. Soil salinity: Effect on vegetable crop growth. Management practices to prevent and mitigate soil salinization. *Horticulturae*, 3(2), p. 30.

Matera, A.G. and Wang, Z., 2014. A day in the life of the spliceosome. *Nature reviews molecular cell biology*, 15(2), pp. 108–121.

Mehta, R., Radhakrishnan, T., Kumar, A., Yadav, R., Dobaria, J.R., Thirumalaisamy, P.P., Jain, R.K. and Chigurupati, P., 2013. Coat protein-mediated transgenic resistance of peanut (Arachis hypogaea l.) To peanut stem necrosis disease through agrobacterium-mediated genetic transformation. *Indian journal of virology*, 24(2), pp. 205–213.

Mingsong, S. and Jingyang, D., 1986. The discovery, determination and utilization of the hubei photosensitive genic Male4sterile rice (*Oryza sativa* subsp. Japonica). *Acta genetica sinica*, 13(2), p. 107.

Park, H.J., Kim, W.Y. and Yun, D.J., 2016. A new insight of salt stress signaling in plant. *Molecules and cells*, 39(6), p. 447.

Pulsifer, I.P., Kluge, S. and Rowland, O., 2012. Arabidopsis long-chain acyl-CoA synthetase 1 (LACS1), LACS2, and LACS3 facilitate fatty acid uptake in yeast. *Plant physiology and biochemistry*, 51, pp. 31–39.

Roudier, F., Gissot, L., Beaudoin, F., Haslam, R., Michaelson, L., Marion, J., Molino, D., Lima, A., Bach, L., Morin, H., Tellier, F., Palauqui, J.C., Bellec, Y., Renne, C., Miquel, M., Dacosta, M., Vignard, J., Rochat, C., Markham, J.E., Moreau, P., Napier, J., Faure, J.D., 2010. Very-long-chain fatty acids are involved in polar auxin transport and developmental patterning in Arabidopsis. *Plant cell*, 22(2), pp. 364–75. doi: 10.1105/tpc.109.071209.

Roy, S., Banerjee, A., Tarafdar, J., Senapati, B.K. and Dasgupta, I., 2012. Transfer of transgenes for resistance to rice tungro disease into high-yielding rice cultivars through gene- based marker-assisted selection. *The journal of agricultural science*, 150(5), p. 610.

Sanz-Carbonell, A., Marques, M.C., Bustamante, A., Fares, M.A., Rodrigo, G. and Gomez, G., 2019. Inferring the regulatory network of the miRNA-mediated response to biotic and abiotic stress in melon. *BMC plant biology*, 19(1), pp. 1–17.

Sasaya, T., Nakazono-Nagaoka, E., Saika, H., Aoki, H., Hiraguri, A., Netsu, O., Ichiki, T.U., Onuki, M., Toki, S., Saito, K. and Yatou, O., 2014. Transgenic strategies to confer resistance against viruses in rice plants. *Frontiers in microbiology*, 4, p. 409.

Shen, C., Wang, S., Zhang, S., Xu, Y., Qian, Q., Qi, Y. and Jiang, D.A., 2013. OsARF16, a transcription factor, is required for auxin and phosphate starvation response in rice (*Oryza sativa* l. *Plant, cell & environment*, 36(3), pp. 607–620.

Shepherd, D.N., Mangwende, T., Martin, D.P., Bezuidenhout, M., Thomson, J.A. and Rybicki, E.P., 2007. Inhibition of maize streak virus (MSV) replication by transient and transgenic expression of MSV replication-associated protein mutants. *Journal of general virology*, 88(1), pp. 325–336.

Shinde, H., Dudhate, A., Anand, L., Tsugama, D., Gupta, S.K., Liu, S. and Takano, T., 2020. Small RNA sequencing reveals the role of pearl millet miRNAs and their targets in salinity stress responses. *South African Journal of botany*, 132, pp. 395–402.

Song, J.H., Cao, J.S. and Wang, C.G. 2013. *BcMF11*, a novel non-coding RNA gene from *Brassica campestris*, is required for pollen development and male fertility. *Plant Cell Rep* 32, 21–30. https://doi.org/10.1007/s00299-012-1337-6

Sunkar, R., Kapoor, A. and Zhu, J.K., 2006. Posttranscriptional induction of two Cu/Zn superoxide dismutase genes in Arabidopsis is mediated by downregulation of miR398 and important for oxidative stress tolerance. *The plant cell*, 18(8), pp. 2051–2065.

Teh, S.Y. and Koh, H.L., 2016. Climate change and soil salinization: Impact on agriculture, water and food security. *International journal of agriculture, forestry and plantation*, 2, pp. 1–9.

USDA (US Department of Agriculture) (2015) Scientific Report of the 2015 Dietary Guidelines Advisory Committee. US Department of Agriculture, Washington, DC, 7.

Vaucheret, H., 2006. Post-transcriptional small RNA pathways in plants: Mechanisms and regulations. *Genes & development*, 20(7), pp. 759–771.

Voinnet, O., 2009. Origin, biogenesis, and activity of plant microRNAs. *Cell*, 136(4), pp. 669–687.

Wang, J., Jian, H., Wang, T., Wei, L., Li, J., Li, C. and Liu, L., 2016. Identification of microRNAs actively involved in fatty acid biosynthesis in developing *Brassica napus* seeds using high-throughput sequencing. *Frontiers in plant science*, 7, p. 1570.

Wang, Z., Zou, Y., Li, X., Zhang, Q., Chen, L., Wu, H., Su, D., Chen, Y., Guo, J., Luo, D. and Long, Y., 2006. Cytoplasmic male sterility of rice with boro II cytoplasm is caused by a cytotoxic peptide and is restored by two related PPR motif genes via distinct modes of mRNA silencing. *The plant cell*, 18(3), pp. 676–687.

Wassenegger, M. and Krczal, G., 2006. Nomenclature and functions of RNA-directed RNA polymerases. *Trends in plant science*, 11(3), pp. 142–151.

Wu, S., Zhang, B., Keyhaninejad, N., Rodríguez, G.R., Kim, H.J., Chakrabarti, M., Illa- Berenguer, E., Taitano, N.K., Gonzalo, M.J., Díaz, A. and Pan, Y., 2018. A common genetic mechanism underlies morphological diversity in fruits and other plant organs. *Nature communications*, 9(1), pp. 1–12.

Xia, R., Xu, J. and Meyers, B.C., 2017. The emergence, evolution, and diversification of the miR390-TAS3-ARF pathway in land plants. *The plant cell*, 29(6), pp. 1232–1247.

Xia, R., Ye, S., Liu, Z., Meyers, B.C. and Liu, Z., 2015. Novel and recently evolved microRNA clusters regulate expansive f-BOX Gene networks through phased small interfering RNAs in wild diploid strawberry. *Plant physiology*, 169(1), pp. 594–610.

Zhang, X., Sato, S., Ye, X., Dorrance, A.E., Morris, T.J., Clemente, T.E. and Qu, F., 2011. Robust RNAi-based resistance to mixed infection of three viruses in soybean plants expressing separate short hairpins from a single transgene. *Phytopathology*, 101(11), pp. 1264–1269.

Zhang, Z.Y., Wang, Y.G., Shen, X.J., Li, L., Zhou, S.F., Li, W.C. and Fu, F.L., 2013. RNA interference-mediated resistance to maize dwarf mosaic virus. *Plant cell, tissue and organ culture (PCTOC)*, 113(3), pp. 571–578.

Zhang, Z.Y., Yang, L., Zhou, S.F., Wang, H.G., Li, W.C. and Fu, F.L., 2011. Improvement of resistance to maize dwarf mosaic virus mediated by transgenic RNA interference. *Journal of biotechnology*, 153(3–4), pp. 181–187.

Zheng, Y., Chen, C., Liang, Y., Sun, R., Gao, L., Liu, T. and Li, D., 2019. Genome-wide association analysis of the lipid and fatty acid metabolism regulatory network in the mesocarp of oil palm (*Elaeis guineensis* jacq.) Based on small noncoding RNA sequencing. *Tree physiology*, 39(3), pp. 356–371.

Zhou, H., Liu, Q., Li, J., Jiang, D., Zhou, L., Wu, P., Lu, S., Li, F., Zhu, L., Liu, Z. and Chen, L., 2012. Photoperiod-and thermo-sensitive genic male sterility in rice are caused by a point mutation in a novel noncoding RNA that produces a small RNA. *Cell research*, 22(4), pp. 649–660.

3 Plant Non-Coding miRNA and Their Importance in Plant Physiology

Neelam Kumari, Ravi Kant Bhatia, Akanksha Sharma, and Nitesh Kumar
Department of Bio-sciences, Himachal Pradesh University, Shimla, India

3.1 INTRODUCTION

MicroRNAs (miRNAs) are a group of endogenous small RNAs (sRNAs) 21–22 nucleotides (nt) in size (Bartel, 2004). miRNAs have been recognized as key regulators of a variety of plant developmental and physiological processes (Chen, 2005; Jones-Rhoades et al., 2006; Zhang et al., 2006; Willmann & Poethig, 2007;Sunkar et al., 2012). Many genes targeted by miRNAs encode transcription factors; this places miRNA in a pivotal hub within the regulation of protein–DNA and protein–protein interactions in plant development. Furthermore, these transcription factors control important developmental processes such as cell division/expansion/differentiation, specification of organ identities, phase transition, and nutrition homeostasis, thus, acting as effectors of miRNAs in plant growth (Bartel & Bartel, 2003; Jones-Rhoades et al., 2006b). In plants, biogenesis of miRNAs initiates from the transcription of the primary miRNAs (pri-miRNAs) that are recognized by the Dicer-like 1 (DCL1) enzyme (Kurihara & Watanabe, 2004; Fahlgren et al., 2007). Dicer enzymes are characterized by an amino-terminal DExH-box RNA helicase domain, a PAZ domain that is also found in Argonaute (AGO) proteins, two RNase III domains, and carboxy-terminal dsRNA-binding domains (Schauer et al., 2002). Arabidopsis has four Dicer-like (DCL) proteins: DCL3 is needed for endogenous siRNA accumulation, DCL2 is required for the efficient accumulation of certain virus-derived siRNAs (Xie et al., 2004), and DCL1 (originally known as CARPEL FACTORY, SHORT INTEGUMENTS1, or SUSPENSOR1) is needed for miRNA accumulation. After two steps of the DCL1 cleaving processes, pri-miRNAs produce miRNA duplex, which is exported by HASTY from the nucleus to cytoplasm (Bollman et al., 2003; Yeon Park et al., 2005). In cytoplasm, the mature miRNA interacts with AGO proteins to form the RNA-induced silencing complex (RISC), while the miRNA strand is often degraded during this process (Kurihara & Watanabe 2004). The RISC complex directs miRNA to its targets and represses the target genes by messenger RNA cleavage, translational inhibition, or chromatin modifications (Baumberger & Baulcombe, 2005; Pratt & MacRae, 2009). In 2001, it was discovered that these two RNAs are members of a large class of 21–24 nt non-coding RNAs, called miRNAs, found in nematodes, fruitflies (*Drosophila melanogaster*), and humans (Lagos-Quintana et al., 2001; Lee et al., 1993). Also, among the miRNAs known to be involved in plant development, some are highly conserved in the plant kingdom, while others are more unique in certain plant species (Axtell & Bowman, 2008; Cuperus et al., 2011; Djami-Tchatchou et al., 2017). The degree of the conservation of a miRNA indicates its "age" in evolution time and is closely related to its expression patterns and functions.

Although most genes use RNA in the form of mRNA as a coding intermediate for protein production, there are many genes whose final products are RNA. These non-coding RNAs range from the familiar transfer and ribosomal RNAs to the more recently discovered regulatory RNAs. One

DOI: 10.1201/9781003369288-3

type of regulatory RNA was first discovered during the study of nematode larval development. Two approximately 22 nt RNAs (the *lin-4* and *let-7* RNAs) control developmental timing by binding to their respective mRNA targets and preventing productive use of these messages, perhaps by attenuating translation (Lee et al., 1993). The *let-7* RNA was found broadly throughout bilateral animals, including humans, suggesting that these two ribo-regulators were more than oddities of worm larval development. Recent computational and molecular analyses indicate that humans have more than 200 miRNA genes, nearly all of which are conserved in mice, and at least 80% of which are conserved in fish (Lim et al., 2003). Although the functions of nearly all metazoan miRNAs are unknown, their abundance and evolutionary conservation, together with the analogy to the two founding nematode miRNAs, which are post-transcriptional gene regulators, suggest that an important mode of metazoan gene regulation had gone virtually undetected until just recently. In animals, formation of miRNAs occurs from introns, while in plants, formation of miRNAs occurs from the intergenic regions (Figure 3.1). A gene that is coding for miRNA is called MIR. Transcription of MIR with the action of RNA polymerase II generates long-stem loop primary transcript, known as pri-miRNA. The size of pri-miRNA may vary from 100 nucleotides to several kilo bases. In the plant nucleus, Dicer (DCL1) processes pri-miRNA into the duplex miRNA via the intermediate hairpin-stem-like structure known as pre-miRNA. DCL1 Through the interaction with HYL1, undergo splicing within the miRNA precursor to liberate miRNA/miRNA duplex.

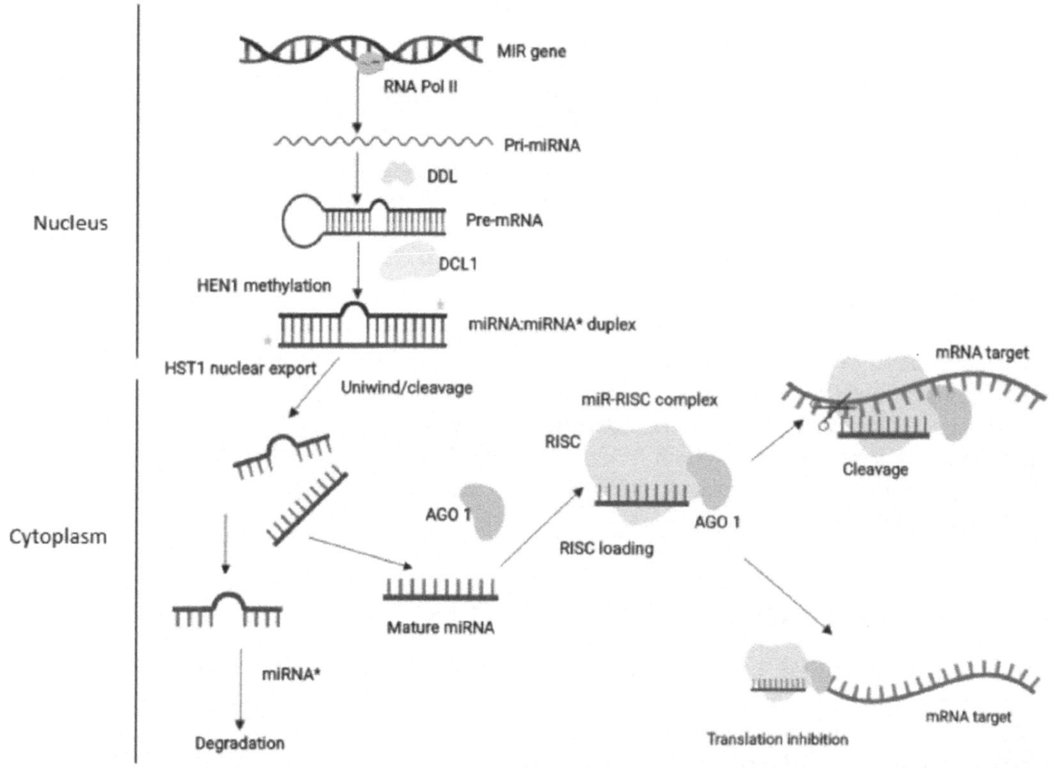

FIGURE 3.1 Plant miRNA is transcribed by RNA polymerase II from the transcripts located in intergenic regions. Pri-miRNA was stabilized by DDL, then binds with hyponasty leaves (HYL1), and is processed by DCL1 into pre-miRNA. The pre-miRNA duplex is methylated by Hua enhancer 1 (HEN1) and transported from the nucleus to the cytoplasm by HASTY. The guide miRNA strand binds AGO1 and is loaded into the RISC complex to carry out the silencing process (Chaudhary et al., 2021).

miRNAs play an important role in the regulation of a wide range of biological processes such as growth, development, reproduction, and various stresses (Gupta et al., 2012; Feng et al., 2014; Chen et al., 2015). Polycistronic pri-miRNA transcripts can generate multiple miRNAs in animals. These polycistronic miRNAs are rare in plants (Baldrich et al., 2016). Transcription of pri-miRNA to miRNA duplex occurs in the plant nucleus, whereas pre-miRNAs are transported to the cytoplasm for further processing in animals.

Enzymes transform double stranded RNA into 21–23 bp miRNA. Small interfering RNAs generated in the initiator step are loaded into the RISC or ribonucleoprotein complex that mediates unwinding of the siRNA duplexes into two strands. Only one strand, known as a guide strand, binds to the AGO protein, which results into transcript degradation or translational repression, while another strand, known as a passenger strand, is degraded during RISC activation. From the nucleus, the mature single-stranded miRNA or duplex miRNA is exported in the cytoplasm through the HASTY (plant exportin-5 protein) for incorporation into the ribonucleoprotein complex (miRNP). This miRNP-miRNA duplex avoids the expression of target genes through homologous mRNA cleavage or by the suppression of translation after binding to the untranslated region at 3′ end of homologous mRNA.

3.2 INFLUENCE OF miRNA IN GENE EXPRESSION

miRNAs can regulate other genes' expression via target cleavage or translational inhibition (Xie et al., 2010; Wang & Guo, 2015; Huang et al., 2016). The mechanism of gene regulation depends on the complementarity between the miRNA sequence and its target site. It has been reported that a perfect match between miRNAs and their targets resulted in cleavage, while mismatch in some bases led to translational inhibition. In plants, specific cleavage is the predominant mode of target gene regulation, as the plant miRNAs bind to their targets with high specificity (Fahlgren & Carrington, 2010). When miRNA and the target site are perfectly complementary, then target cleavage occurs. The sRNAs direct the AGO1, a component of RISC to the complementary site of the target transcript, and then cleavage is started. Cleavage occurs generally in between the 10th and 11th nucleotide pairing residues relative to the 5′ end of miRNA (Park & Shin, 2014). The majority of the plant miRNAs' target sites are present in the coding region of the target transcript. In plants, some miRNAs can guide sequence-specific DNA methylation, which is called miRNA-directed DNA methylation (Wu et al., 2010; Zhang et al., 2013). This DNA methylation is associated with transcriptional gene 25 silencing (TGS). In Arabidopsis, AGO1 and AGO10 were proposed to be involved in translational inhibition (Brodersen et al., 2008), as mutant protein levels of several miRNA targets were increased in both AGO1 and AGO10. Although miRNAs have been implicated in many normal and disease-related biological processes, the potential regulatory activity of the miRNA species is less well-known.

3.3 miRNA REGULATING PLANT GROWTH AND DEVELOPMENT

MiRNAs play a crucial role in controlling different developmental phenomena of plants. Transcription factor encoding genes like no apical meristem (NAM) and cup-shaped cotyledon (CUC) are specifically involved in the development of root and shoot, and are regulated by the miR164 family in *Arabidopsis thaliana*. The short root (SHR) and scarecrow-like (SCR) proteins up-regulate the expression of the miR165a and miR166b genes, which in turn down-regulate the HD-ZIP III transcription factor and are, thus, involved in root development. miR156 regulates the expression of Squamosa Promoter binding-Like (SPL) genes involved in development and phase transition. Plants over-expressing miR156 showed a semi-dwarf phenotype and the number of leaves also changed. miR172 is also involved in flower development by regulating floral patterning genes like APETALA2 (Zhu & Helliwell, 2011).

3.4 DURING SEED GERMINATION AND SEED DEVELOPMENT, miRNA REGULATION ON DIFFERENT PHYTOHORMONES IS ESSENTIAL

Several atmospheric conditions like oxygen, light, soil, temperature, moisture, stress, etc., and some physiological factors such as dormancy period, viability of seeds, thickness of seed coat, etc., also play significant roles in seed germination stages (Weitbrecht et al., 2011). Various recent studies have shown that the interactions between different phytohormones like auxin, abscisic acid (ABA), gibberellins (GAs), cytokinins (CKs), ethylene, and brassinosteroids (BRs) conduct molecular actions that control the discharging from dormancy and activation stages of seed germination (Finkelstein et al., 2008). As some phytohormones exert contrasting influence on the process, the activity of plant hormones, therefore, needs to be precisely regulated. It is reported that ABA acts as a negative regulator of seed germination, but it acts as an up-regulator of dormancy and its maintenance (Finkelstein et al., 2008). The insensitivity or absence of ABA during later stages of seed development gives rise to viviparous or precociously germinating seeds, such as Arabidopsis ABA-deficient, ABA-insensitive, and maize viviparous. GA helps to release from dormancy and its role is quite similar to ethylene and BR in promoting germination by counteracting ABA effects. Studies have shown that in both plants and animals, GAs lay multiple significant roles in cases of growth, development, morphogenesis, and various stress responses (Chen, 2012). From double-stranded RNA precursors with the help of RNA-dependent RNA polymerase, DCL, and AGO proteins, mature and functional miRNAs are produced. miRNAs bind to the complementary sequences of their target genes and, thus, negatively regulate them.

3.5 HORMONE BIOSYNTHESIS AND SIGNALING

In Arabidopsis, MYB33, MYB65, and MYB101 are the targets of miR159. These target genes are linked to the MYB gene GAMYB, which triggers the response of gibberellin-responsive (GA) genes in the 31 aleurone layer. MYB genes are required in developmental processes that involve GA, like flowering in short-day conditions, etc., although these genes don't mediate the GA signaling process. Over-expression of miR159 causes a decrease in transcripts of MYB65 and MYB33 and a reduction in male sterility, leading to delayed flowering under short days (Millar & Gubler, 2005). The ARF genes are targeted by miR160, miR167, and tasiRARFs. The regulation of miR160, in many developmental concerns of root and shoot by ARF10, ARF16 and ARF17, is regarded as the most significant. ARF6 and ARF8 are the target genes of miR167. Various studies have shown that miR167-resistant ARF6 expression led to detained anthers and ovule development. Therefore, it could be concluded that both ARF6 and ARF8 play significant roles in ovule and anther development.

3.6 BOUNDARY FORMATION/ORGAN SEPARATION

miR164 targets the Cup- Shaped Cotyledons (CUC)1, 2, and 3 and NAC gene family members. These two gene families share some common functions in the initiation of shoot apical meristem (SAM) and in organ boundary formation processes. These three target genes (CUC)1,2 and 3 are mainly expressed in the terminal regions of cotyledons within embryos and are further expressed in the terminal regions of floral organs. Also, a fusion of cotyledons (embryonic organ), sepals, and stamens was observed in the phenotype of this double mutant. Though obvious, but also *in vivo,* it was observed that over-expression of miR164 in wild type background with the 35S promoter causes both floral organ and cotyledon fusions (Chen, 2009). Experimentally, it has been observed that miR164-resistant CUC2 can rescue sepal separation in miRNA164 over-expressing lines. The CUC2 and CUC1 transcript amounts were higher in mutant miRNA164c, which is not applicable in the case of the other four targets of miR164 (Baker et al., 2005). Additionally, the phenotype of CUC2, CUC1 miRNA164 mutant flowers are identical with CUC2, CUC1 flowers, indicating that the reason behind the improved number of petals in miR164c is the over-expression of CUC2 or CUC1 (Baker et al., 2005).

3.7 ORGAN POLARITY

The portion of an organ that is in front of the meristem in the primordium is termed the adaxial side. The portion of the organ that appears distant from the meristem is known as the abaxial side. These two sides differ from each other in lateral organs like floral and leafy organs. They are also called polar organs. The distinctions of these two sides are clearly visible through their epidermis structure and also through the mesophyll tissues on both sides. Earlier, it was studied that at the adaxial region the restriction of PHB expression is controlled by miRNA-mediated regulation. However, the mechanism behind this is not clear to date. miR165/166 mediates the transcript level cleavage of PHB, PHV, and REV, therefore, it could be inferred that HD-ZIP transcripts receive restrictions via miR165/166 in the abaxial domain that clears the mRNAs from this region. miR165/166 is also responsible for the DNA methylation of the two genes of the HD-ZIP gene family, PHB and PHV. Therefore, it seems that there is a substitute mechanism through which the mRNAs of transcription factors in the abaxial region receive resistance by the miRNAs. Simultaneously, the miR390–TAS3 node is the determinant of abaxialization (Rubio-Somoza & Weigel, 2011).

3.8 miRNA REGULATOR OF REPRODUCTION

In plants, flowering plays an important role in evolution and reproduction. To ensure successful reproduction, plants have undergo particular phase changes in their lifespan, from juvenile to adult and adult to the reproductive stage. This is achieved by controlling the precise expression of important flowering-time genes at both the transcriptional and post-transcriptional levels. miRNAs related to flowering-time act as key positive and negative factors in plant development from vegetative to reproductive phase transition. The involvement of miRNAs and their targets affect and cross-talk with other miRNA-pathways by interacting with the biochemical and environmental factors coordinating flowering time.

3.9 DEVELOPMENTAL TRANSITIONS

Leaves and flowers are produced from SAM during the vegetative and reproductive phases of development in plants. In the vegetative phase, the juvenile leaves are normally distinct from the adult leaves (that develop later). Recent findings suggest that various developmental transitions of SAM are regulated by sRNAs (Chen, 2009; Rubio-Somoza & Weigel, 2011). Apart from AP2, miR172 governs a number of AP2-like genes, TOE1, TOE2, TOE3, SNZ, and SMZ. The mutant TOE1-1 results in early flowering whereas the TOE2-1 mutant has no actual effect on flowering time. On the other hand, the double mutant TOE1-1 TOE2-1 shows more early flowering compared to wild type, indicating that TOE2 and TOE1 are not essential repressors of the vegetative-to-reproductive phase transition in plants. These studies suggest that the miR172-TOE node significantly controls the vegetative-to-reproductive transition (Aukerman & Sakai, 2003). The lack of a change in TOE1 transcript level due to miR172 over-expression could result in translational inhibition of the TOE1 node. Oddly, over-expression of miR172 results in a depletion of the TOE2 transcript amount, which showed that miR172 regulates various targets in a number of ways (In Figure 3.2). Developmental transitions in plants are regulated by the antagonistic activities of two miRNAs, miR156 and miR172. While the miR156 expression level reduces with leaf age, the miR172 level of expression increases. miR172 and miRNA156target Apetala 2 and SPL, respectively, and their expression patterns are complementary. Lower-level activity of miR172 or higher activity of miR156 results in delay of flowering, extended juvenile characteristics, and pin-down developmental transitions. The SPL can be divided into two categories: small SPLs like SPL5, SPL4, and SPL3 and large SPLs like SPL15, SPL10, and SPL9.

The larger SPLs, particularly SPL9, bind to the promoter region of miR172b and play a significant role in juvenile-to-reproductive-phase transition. A third set of sRNA nodes includes trans-acting

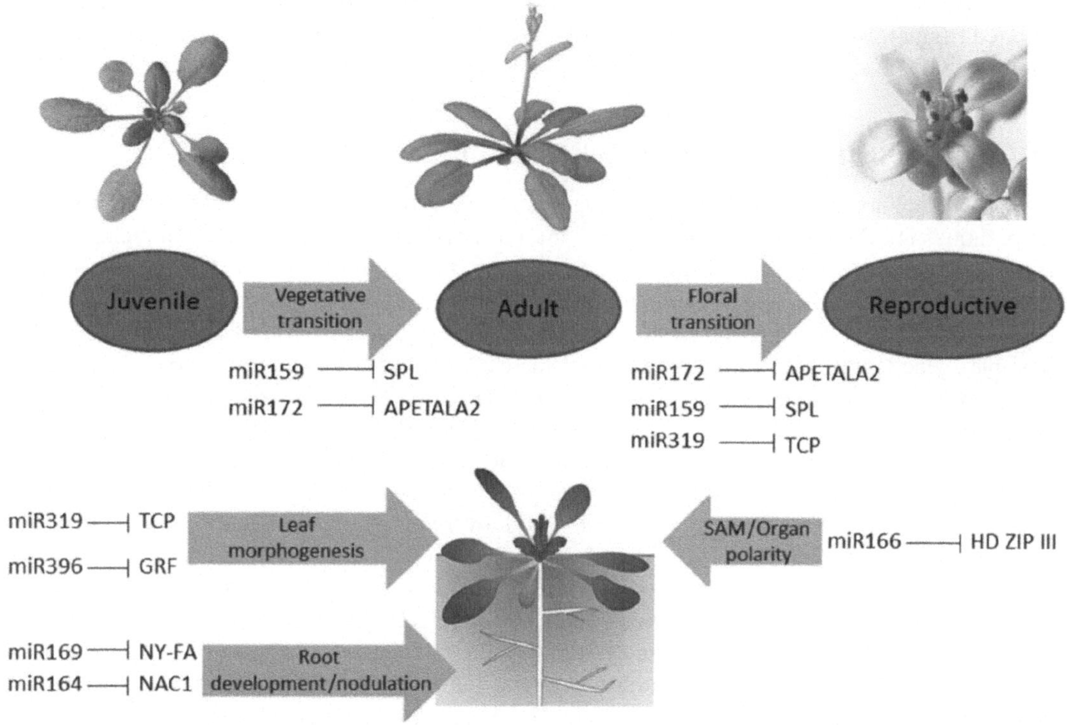

FIGURE 3.2 The involvement of various miRNAs in the different flowering-related phases of plant development, from the transition from juvenility to maturity, to floral induction and development.(Hong & Jackson, 2015)

ta-siRNAs from the TAS3 locus, which are involved in regulating the vegetative-to-reproductive-phase transition. ARF3 and AGO7 are elevated in mutants that express a Ta-siRNA biogenesis process, which is not possible with RNA-dependent ARN 6 (RDR6), DCL4, or AGG7 (Yant et al., 2010). Interestingly, apart from targeting ARF3 and ARF4, TAS3-tasiRNAs target members of the AP2-like family of transcription factors (Rubio-Somoza & Weigel, 2011).

3.10 FLORAL ORGAN AND REPRODUCTIVE DEVELOPMENT

The origin of floral organs is the floral meristem in plants. Floral homeotic genes, named as A, B, and C genes, specify the floral organ identities. The genes A and C perform antagonistically to confine each other within the floral meristem. One example of this phenomenon is the class C gene AGAMOUS (AG) mutations, which give rise to perianth organs in replacement of the reproductive organs, whereas class A gene APETALA2 (AP2) mutations give rise to the reverse effect. Essentially, the transcript of AG was observed in the floral meristem (inner two whorls) only, which will later convert into reproductive organs, whereas the transcript of APETALA1 (AP1), a class A gene, was only observed in the outer two whorls of the meristem, which will soon be the perianth organs; the AP2 mRNAs are present throughout the floral meristem.

The over-expression of miR172 produces lower levels of the AP2 protein, but this transcript level alone cannot suggest that miR172 has a translational inhibition mode of action (Schwab et al., 2005). miR172 may cause decreased amounts of AP2 production in the inner two whorls even though the uniform production of AP2 RNA was observed in all the floral meristem. In reproductive development, miR159 also plays a considerable role in the regulation of two MYB genes, MYB65 and MYB33. The function of miR159 is critical in restricting the MYB65 and MYB33 expressions to anthers. Transgenic plants consisting of the miR159-resistant version of MYB33 that is blocked or inhibited from growth at different stages, clearly indicates that the limitation of the MYB33 expression by miR159 is significant for plant growth and development (Millar & Gubler, 2005).

3.11 miRNA IN REGULATING PHENOTYPIC PLASTICITY

The ability of an organism to alter its physiology/morphology in response to changes in environmental conditions is called phenotypic plasticity. This ability is particularly necessary in plants, whose sessile lifestyle requires them to withstand variable environmental conditions. This review focuses on changes related to the miRNA-regulating phenotypic plasticity in plants. The development of shoot-borne roots, or adventitious roots, is necessary for mass propagation of elite genotypes. It is a complex genetic trait with a high phenotypic plasticity due to various internal and environmental regulatory factors. A subtle balance of activator and repressor Auxin Response Factor (*ARF*) transcripts controls adventitious root initiation. Moreover, miRNA activity appears to be required for fine regulation of this process. Thus, *ARF17* is a negative regulator target of miR160 and *ARF6* and *ARF8*, targets of miR167 are positive regulators of adventitious rooting. The three *ARF*s interact genetically and regulate each other's expression with overlapping expression at both the transcriptional and post-transcriptional levels by modulating miR160 and miR167 availability. This regulatory domain includes feedback regulation of miRNA homeostasis by direct and non-direct target transcription factors. These results provide evidence of miRNA control of phenotypic variability and are an inevitable step forward in understanding the molecular mechanism. The regulation of gene expression through the miRNA pathway is a relatively new discovery. In the past 10 years, rapid progress has led to the identification of genes involved in the processing and maturation of miRNAs (Voinnet, 2009). However, mechanisms for regulation of their production are still unclear in both plants and animals. miRNAs are required for fine activity of gene expression and for adaptation to subtle endogenous (e.g., hormonal) or environmental fluctuations. Phenotypic variation linked with complex genetic traits is one of the manifestations of these variations in gene expression.

3.12 LEAF GROWTH

The undifferentiated association of cells in SAM led to the origin of leaves, while leaf primordia will be decided by auxin maxima. miR164 targets the boundary region between the undifferentiated cells is established by CUP-SHAPED COTYLEDON1 37 (CUC1) and CUC2. Plants have no regulation in CUC2 and CUC1 action has fused cotyledons, nearly similar to mutants affected in the miR319–TCP4 node of regulation (Palatnik et al., 2003).The regulation of cell proliferation in Arabidopsis by miR319–TCP nodes is moderately intervened by the conserved miR396–GRF (growth regulator factor) node (Rodriguez et al., 2010). Any kind of issue in the regulation of the miR319–TCP4 node gave rise to a decrease in GRF expression and also lowered the mitotic activity. In Arabidopsis, the miR390–TAS3–ARF node is crucial in the regulation of leaf margin growth.

3.13 MODULATION OF ROOT ARCHITECTURE PLASTICITY

The root system shows equal plasticity like the aerial part of the plant, as it interacts with highly crucial physical environmental responses. Root system plasticity primarily depends on the secondary growth of lateral roots and root hairs, which empower the plant to forage efficiently for nutrients. The periodicity with which lateral roots emerge along the main root is set by two oscillating opposite waves of gene expression that correlate with pulses of auxin-induced transcription (Moreno Risueno, 2010). Lateral root development starts when single pericycle cells turn into founder cells, which proliferate to form a lateral root primordium. Depending on

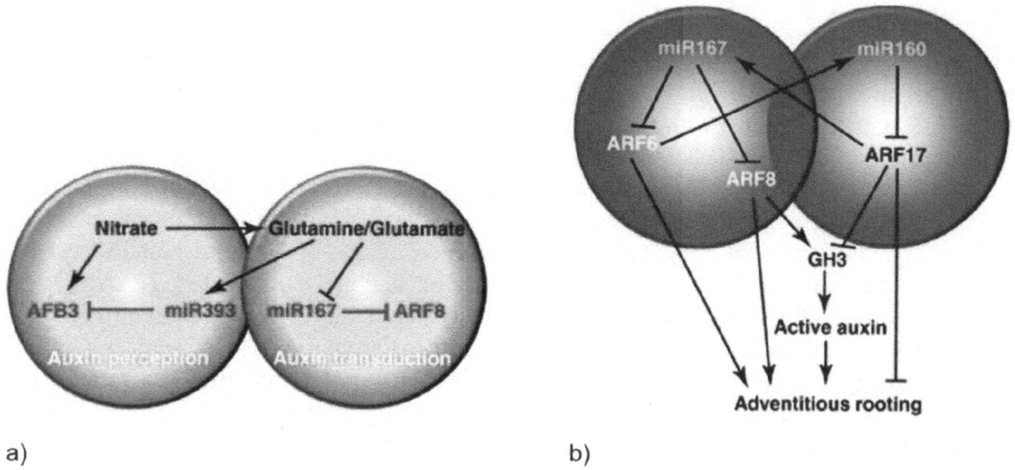

FIGURE 3.3 miRNA-targeting nodes in root plasticity. (a) miRNA nodes coordinating nitrate- and auxin-dependent formation of lateral roots. Nitrogen metabolic pathway and effect on the expression of node components are shown in the figure. Elements from the miRNA nodes and their regulatory relation are also depicted as AFB3, miR393, miR167, and ARF8. (b) miRNA nodes in the regulation of auxin homeostasis and adventitious rooting by inhibiting and activating expression of GH3. (Rubio-Somoza & Weigel, 2011).

environmental cues, the primordium continues to develop, breaking through the surface of the primary root.

The miR164–NAC1 and miR160– ARF10 nodes affect initiation of adventitious rooting (In Figure 3.3). TAS3–ARFs together with two nitrate-regulated miRNA nodes, miR167–ARF8 and miR393–AFB3, act subsequently during emergence and elongation of the lateral root (Yoon, 2010). The miR167 and miR393 nodes are an example of how miRNAs can coordinate the integration of metabolic state and soil nutrient availability and results in dose-dependent, auxin-based plasticity of the root system. Nitrate is a key source of nitrogen, which is often limiting for plant growth and agricultural productivity.

3.14 miRNA RESPONSE TO EXTERNAL STIMULI

3.14.1 Abiotic Stimulation

Plants are sessile in nature and are thus prone to several abiotic stimulations in nature such as high temperature, chilling, salinity, drought, etc. To cope with these conditions, plants can modulate their stimuli-response gene expression. Stimuli-responsive transcription factors play key role in withstanding harsh environmental conditions. miRNAs are found to target these transcription factors and, thus, regulate their expression under stress conditions. miR398 can target two Cu/Zn superoxide dismutases (SOD) involved in the reduction of superoxide anions to hydrogen peroxide, which is, in turn, converted into water and oxygen by the action of catalase, ascorbate peroxidase (APX), or glutathione peroxidase (GPX) (Bouché, 2010). miR159, which targets MYB transcription factors, was found to be up-regulated by the stress hormone ABA under drought conditions (Buxdorf et al., 2010). During stress conditions, miR169 is induced. miR395 is another abiotic stress-related miRNA involved in sulfur metabolism and transport. It has been found that phosphate deficiency induces miR399 expression with a corresponding decrease in the target gene transcript, ubiquitin-conjugating enzyme (UBC), in Arabidopsis (Fujii et al., 2005). miR168, miR171, and miR396 are commonly found to be up-regulated during salt, osmotic, and cold stresses, respectively (Khaksefidi et al., 2015)

3.14.2 miRNAs in Biotic Stimuli Response

Plants in nature coexist with various microorganisms. Some of them are beneficial for plants and some are harmful pathogens. For their food and habitat, pathogens such as fungi, bacteria, viruses, etc., can infect plants. To deal with these potentially fatal pressures, plants have developed a number of defense systems. One such reaction by which plants can fend off a pathogen onslaught is the activation of defense genes. In the plasma membrane of plants, there are specialized binding sites for elicitors of pathogen origin that serve as a detection mechanism for pathogen invasion and signal transduction activation of the immune system.. Different fungal or bacterial proteins can act as elicitors that the plant can recognize as a pathogen-associated molecular pattern (PAMP) specific for a particular type of pathogen. This kind of immunity is called PAMP-triggered immunity (PTI) (Rasmussen et al., 2012). The Arabidopsis miRNA effector protein AGO1 is required for a number of PTI responses, including gene expression and seedling growth inhibition. Bacterial flagellin flg22 induces the transcription of miRNA and flg22 treatment regulates miR393 expression as well as in the infection of bacterial pathogens. In wheat, 24 miRNAs responsive to powdery mildew infection have been identified (e.g., miR156, miR164, miR393, etc.) (Gupta et al., 2012). During other biotic stresses like fungi, bacteria, virus, nematode infections, etc., several miRNAs are found to be differentially regulated.

REFERENCES

Aukerman, M. J. & Sakai, H. (2003). Regulation of flowering time and floral organ identity by a MicroRNA and its *APETALA2*-like target genes. *Plant Cell* **15**: 2730–2741.

Axtell, M. J. & Bowman, J. L. (2008). Evolution of plant microRNAs and their targets. *Trends Plant Sci.* **13**(7): 343–349. https://doi.org/10.1016/j.tplants.2008.03.009

Baker, C. C., Sieber, P., Wellmer, F. & Meyerowitz, E. M. (2005). The early extra petals1 mutant uncovers a role for microRNA miR164c in regulating petal number in *Arabidopsis. Curr. Biol.* **15**: 303–315.

Baldrich, P., Hsing, Y. I. & San Segundo, B. (2016). Genome-wide analysis of polycistronic MicroRNAs in cultivated and wild rice. *Genome Biol. Evol.* **8**: 1104–1114.

Bartel, D. P. (2004). Review MicroRNAs: Genomics, biogenesis, mechanism, and function. *Cell* **116**: 281–297.

Bartel, B. & Bartel, D. P. (2003). MicroRNAs: At the root of plant development? *Plant Physiol.* **132**(2): 709–717. https://doi.org/10.1104/pp.103.023630

Baumberger, N. & Baulcombe, D. C. (2005). *Arabidopsis* ARGONAUTE1 is an RNA Slicer that selectively recruits microRNAs and short interfering RNAs. PNAS **102**: 11928–11933.

Bollman, K. M., Aukerman, M. J., Park, M. Y., Hunter, C., Berardini, T. Z. & Scott Poethig, R. (2003). HASTY, the Arabidopsis ortholog of exportin 5/MSN5, regulates phase change and morphogenesis. *Development* **130**(8): 1493–1504. https://doi.org/10.1242/dev.00362

Bouché, N. (2010). New insights into miR398 functions in *Arabidopsis. Plant Signal Behav.* **5**: 684–686.

Brodersen, P., Sakvarelidze-Achard, L., Bruun-Rasmussen, M., Dunoyer, P., Yamamoto, Y. Y., Sieburth, L. & Voinnet, O. (2008). Widespread translational inhibition by plant miRNAs and siRNAs. *Science* **320**: 1185–1190.

Buxdorf, K. A., Hendelman, R., Stav, M., Lapidot, N. & Arazi, T. (2010). Identification and characterization of a novel miR159 target not related to MYB in tomato. *Planta* **232**: 1009–1022.

Chaudhary, S., Grover, A. & Sharma, P. C. (2021). MicroRNAs: Potential targets for developing stress-tolerant crops. *Life* **11**(4): 289. https://doi.org/10.3390/life11040289

Chen, X. (2005). microRNA biogenesis and function in plants. *FEBS Letters* **579**(26): 5923–5931. https://doi.org/10.1016/j.febslet.2005.07.071

Chen, X. (2009). Small RNAs and their roles in plant development. *Annu. Rev. Cell Dev. Biol.* **25**: 21–44.

Chen, X. (2012). Small RNAs in development—Insights from plants. *Curr. Opin. Genet. Dev.* **22**: 361–367.

Chen, L., Luan, Y. & Zhai, J. (2015). Sp-miR396a-5p acts as a stress-responsive genes regulator by conferring tolerance to abiotic stresses and susceptibility to *Phytophthora nicotianae* infection in transgenic tobacco. *Plant Cell Rep.* **34**: 2013–2025.

Cuperus, J. T., Fahlgren, N. & Carrington, J. C. (2011). Evolution and functional diversification of MIRNA genes. *Plant Cell* **23**(2): 431–442. https://doi.org/10.1105/tpc.110.082784

Djami-Tchatchou, A. T., Sanan-Mishra, N., Ntushelo, K. & Dubery, I. A. (2017). Functional roles of microRNAs in agronomically important plants-potential as targets for crop improvement and protection. In *Front. Plant Sci.* **8**: 378. https://doi.org/10.3389/fpls.2017.00378

Fahlgren, N. & Carrington, J. C. (2010). MiRNA target prediction in plants. *Methods Mol. Biol.* **592**: 51–57.

Fahlgren, N., Howell, M. D., Kasschau, K. D., Chapman, E. J., Sullivan, C. M., Cumbie, J. S., Givan, S. A., Law, T. F., Grant, S. R., Dangl, J. L. & Carrington, J. C. (2007). High-throughput sequencing of Arabidopsis microRNAs: Evidence for frequent birth and death of MIRNA genes. *PLoS ONE* **2**(2). https://doi.org/10.1371/journal.pone.0000219

Feng, H., Duan, X., Zhang, Q., Wang, X., Li, B., Huang, L., Wang, X. & Kang, Z. (2014). The target gene of tae-miR164, a novel NAC transcription factor from the NAM subfamily, negatively regulates resistance of wheat to stripe rust. *Mol. Plant Pathol.* **15**: 284–296.

Finkelstein, R., Reeves, W., Ariizumi, T. & Steber, C. (2008). Molecular aspects of seed dormancy. *Annu. Rev. Plant. Biol.* **59**: 387–415.

Fujii, H., Chiou, T., Lin, J. S. I., Aung, K. & Zhu, J. K. (2005). A miRNA involved in phosphatestarvation response in *Arabidopsis. Curr. Biol.* **15**: 2038–2043.

Gupta, O. P., Permar, V., Koundal, V., Singh, U. D. & Praveen, S. (2012). MicroRNA regulated defense responses in *Triticum aestivum* L. During *Puccinia graminis f.sp. tritici* infection. *Mol. Biol. Rep.* **39**: 817–824.

Hong, Y. & Jackson, S. (2015). Floral induction and flower formation-the role and potential applications of miRNAs. *Plant Biotechnol. J.* **13**(3): 282–292. https://doi.org/10.1111/pbi.12340

Huang, J., Yang, M. & Zhang, X. (2016). The function of small RNAs in plant biotic stress response. *J. Integr. Plant Biol.* **58**: 312–327.

Jones-Rhoades, M. W., Bartel, D. P. & Bartel, B. (2006a). MicroRNAs and their regulatory roles in plants. *Annu. Rev. Plant Biol.* **57**: 19–53. https://doi.org/10.1146/annurev.arplant.57.032905.105218

Jones-Rhoades, M. W., Bartel, D. P. & Bartel, B. (2006b). MicroRNAs and their regulatory roles in plants. *Annu. Rev. Plant Biol.* 57: 19–53. https://doi.org/10.1146/annurev.arplant.57.032905.105218

Khaksefidi, R. E., Mirlohi, S., Khalajin, F., Fakharin, Z., Shiran, B., Fallahi, H., Rafiei, F., Budak, H. & Ebrahimie, E. (2015). Differential expression of seven conserved microRNAs in response to abiotic stress and their regulatory network in *Helianthus annuus. Front. Plant Sci.* **6**. https://doi.org/10.3389/fpls.2015.00741

Kurihara, Y. & Watanabe, Y. (2004). *Arabidopsis* micro-RNA biogenesis through Dicer-like 1 protein functions. *PNAS* **101**: 12753–12758.

Lagos-Quintana, M., Rauhut, R., Lendeckel, W. & Tuschl, T. (2001). Identification of novel genes coding for small expressed RNAs. *Science* **294**(5543): 853–858. https://doi.org/10.1126/science.1064921

Lee, R. C., Feinbaum, R. L. & Ambrost, V. (1993). The C. elegans heterochronic gene lin-4 encodes small RNAs with antisense complementarity to lin-14. *Cell* **75**: 843–854.

Lim, L. P., Glasner, M. E., Yekta, S., Burge, C. B. & Bartel, D. P. (2003). Vertebrate MicroRNA genes. *Science* **299**: 1540.

Millar, A. A. & Gubler, F. (2005). The *Arabidopsis* GAMYB-like genes, MYB33 and MYB65, are microRNA regulated genes that redundantly facilitate anther development. *Plant Cell* **17**: 705–721.

Moreno-Risueno, M. A. (2010). Oscillating gene expression determines competence for periodic *Arabidopsis* root branching. *Science* **329**: 1306–1311.

Palatnik, J. F., Allen, E., Wu, X., Schommer, C., Schwab, R., Carrington, J. C. & Weigel, D. (2003). Control of leaf morphogenesis by microRNAs. *Nature* **425**: 257–263.

Park, J. H. & Shin, C. (2014). MicroRNA-directed cleavage of targets: Mechanism and experimental approaches. *BMB Rep.* **47**: 417–423.

Pratt, A. J. & MacRae, I. J. (2009). The RNA-induced silencing complex: A versatile gene-silencing machine. *J. Biol. Chem.* **284**(27): 17897–17901. https://doi.org/10.1074/jbc.R900012200

Rasmussen, M. W., Roux, M., Petersen, M. & Mundy, J. (2012). MAP kinase cascades in *Arabidopsis* innate immunity. *Front. Plant Sci.* **3**.

Rodriguez, R. E., Mecchia, M. A., Debernardi, J. M., Schommer, C., Weigel, D. & Palatnik, J. F. (2010). Control of cell proliferation in *Arabidopsis thaliana* by microRNA miR396. *Development* **137**: 103–112.

Rubio-Somoza, I. & Weigel, D. (2011). MicroRNA networks and developmental plasticity in plants. *Trends in Plant Sci.* **16**(5): 258–264. https://doi.org/10.1016/j.tplants.2011.03.001

Schauer, S. E., Jacobsen, S. E., Meinke, D. W. & Ray, A. (2002). DICER-LIKE1: Blind men and elephants in Arabidopsis development. *Trends Plant Sci.* **7**: 487–491.

Schwab, R., Palatnik, J. F., Riester, M., Schommer, C., Schmid, M. & Weigel, D. (2005). Specific effects of microRNAs on the plant transcriptome. *Dev. Cell* **8**: 517–527.

Sunkar, R., Li, Y. F. & Jagadeeswaran, G. (2012). Functions of microRNAs in plant stress responses. *Trends in Plant Sci.* **17**(4): 196–203. https://doi.org/10.1016/j.tplants.2012.01.010

Voinnet, O. (2009). Origin, biogenesis, and activity of plant microRNAs. *Cell* **136**: 669–687.

Weitbrecht, K., Muller, K. & Leubner-Metzger, G. (2011). First off the mark early seed germination. *J. Exp. Bot.* **62**: 3289–3309.

Willmann, M. R. & Poethig, R. S. (2007). Conservation and evolution of miRNA regulatory programs in plant development. *Curr. Opin. Plant Biol.* **10**(5) 503–511. https://doi.org/10.1016/j.pbi.2007.07.004

Wu, L., Zhou, H., Zhang, Q., Zhang, J., Ni, F., Liu, C. & Qi, Y. (2010). DNA methylation mediated by a microRNA pathway. *Mol. Cell.* **38**: 465–475.

Xie, Z., Johansen, L. K., Gustafson, A. M., Kasschau, K. D., Lellis, A. D., Zilberman, D., Jacobsen, S. E. & Carrington, J. C. (2004). Genetic and functional diversification of small RNA pathways in plants. *PLoS Biol.* **2**: E104.

Xie, Z., Khanna, K. & Ruan, S. (2010). Expression of microRNAs and its regulation in plants. *Semin. Cell Dev. Biol.* **21**: 790–797.

Yant, L., Mathieu, J., Dinh, T. T., Ott, F., Lanz, C., Wollmann, H., Chen, X. & Schmid, M. (2010). Orchestration of the floral transition and floral development in *Arabidopsis* by the bifunctional transcription factor APETALA2. *Plant Cell* **22**: 2156–2170.

Yeon Park, M., Wu, G., Gonzalez-Sulser, A., Vaucheret, H. & Scott Poethig, R. (2005). Nuclear processing and export of microRNAs in *Arabidopsis*. *PNAS* **102**: 3691–3696. https://doi.org/10.1073/pnas.0405570102

Yoon, E. K. (2010). Auxin regulation of the microRNA390- dependent transacting small interfering RNA pathway in *Arabidopsis* lateral root development. *Nucleic Acids Res.* **38**: 1382–1391.

Zhang, B., Pan, X., Cobb, G. P. & Anderson, T. A. (2006). Plant microRNA: A small regulatory molecule with big impact. *Dev. Biol.* **289**(1): 3–16. https://doi.org/10.1016/j.ydbio.2005.10.036

Zhang, S., Xie, M., Ren, G. & Yu, B. (2013). CDC5, a DNA binding protein, positively regulates posttranscriptional processing and/or transcription of primary microRNA transcripts. *Proc. Natl. Acad. Sci. USA* **110**: 17588–17593.

Zhu, Q. H. & Helliwell, C. A. (2011). Regulation of flowering time and floral patterning by miR172. *J. Exp. Bot.* **62**: 487–495.

4 miRNA-Regulated Transcription Factor Networks in Response to Osmotic Stress

Sudhir Kumar, Yuniet Hernandez, and Neeti Sanan-Mishra
Plant RNAi Biology Group, International Centre for Genetic Engineering and Biotechnology, New Delhi, India

4.1 INTRODUCTION

The global changes in climate conditions and the decrease in the availability of arable land are imposing great challenges for sustainable agriculture and food security by negatively impacting plant growth and productivity. Changes in climate have resulted in an increase in temperature and imbalances in the water cycle. These factors accompanied by poor drainage facilities of the irrigated land are leading to accumulation of salts in the soil (Serrano et al., 1999). As a result, crops are being challenged by a combination of high temperatures, water deficiency and soil salinity. In cultivated crops, 10–90% of losses can be attributed to instances of drought and salt.

Several policies have been implemented to improve irrigation and soil management, but these are restricted to sustainable farm operations. This has put forth an urgent need to design effective strategies to enhance the tolerance and adaptability of plants to the environment cues. The success of scientific methodologies will be ensured when the genetic basis of stress tolerance is known and the responses to abiotic stress are better understood. Therefore, it is necessary to identify the master molecules regulating the plant stress response and understand the genetic mechanisms of their regulation.

The emergence of high-throughput genomics has provided plant biologists the ability to identify the global changes in gene regulation. This has helped in designing systematic research strategies for the functional analysis of complex gene regulatory networks that are coordinated at the transcriptional, post-transcriptional and translation levels in a spatio-temporal fashion (Hasegawa et al., 2000; Seki et al., 2002; Barrera-Figueroa et al., 2012; Yu et al., 2012). A new dimension was added to the studies with the unearthing of small RNA molecules as gene-silencing regulatory components. The microRNAs (miRNAs) comprise major stakeholders in regulating the growth, development and response to stress. Their discovery has necessitated the requirement to revisit the molecular mechanisms underlying the plant response to stress. In this chapter, we discuss the crucial role of miRNAs as regulators of this response.

4.2 CONSEQUENCE OF STRESS IN PLANTS

Water constitutes approximately 90% of the weight of the plant and acts in practically all the physiological and biochemical processes. It performs the function of a solvent, through which gases, minerals and other solutes enter the cells and move through the plant. It also plays an important role in the thermal regulation of the plant, acting both on cooling and on the maintenance and distribution of heat (Fuganti-Pagliarini et al., 2017). The adverse effects of deficiency of water and increased soil salinity are primarily due to osmotic stress caused by a reduction in water uptake (Syvertsen and Garcia-Sanchez, 2014). Drought responses in plants result in stomatal closure and reduced transpiration to conserve water. At the cellular level, division and cell expansion

DOI: 10.1201/9781003369288-4

are regulated to maintain turgor (Valliyodan and Nguyen, 2006). It is generally accepted that a more robust root system with increased root length or root biomass is a positive feature under water deficit conditions.

Increased soil salt limits the ability of plants to extract water, causing hyperosmotic stress. It also leads to the accumulation of high concentrations of salts within the cells, causing hyperionic stress (Hasanuzzaman et al., 2013). This results in reduced osmotic potential and ion toxicity, which in turn, disrupt the normal functioning of enzymatic and metabolic activities. Salt stress results in damage to the cell wall, plasmolysis, accumulation of electron-dense protein bodies and damage to endoplasmic reticulum (Garcia et al., 1997; Khan et al., 1997; Pareek et al., 1997; Sivakumar et al., 1998; Munns and Tester, 2008). High Na^+ levels also lead to a reduction in photosynthesis and the production of reactive oxygen species (ROS) (Greenway and Munns, 1980; Flowers et al., 1990; Yeo, 1998). The injury can manifest as a reversible inhibition of metabolism and growth, or an irreversible damage involving cell death.

Response to stress is a complex mechanism involving numerous adaptations such as osmoregulation and osmotic adjustment, hormonal regulation, activation of the antioxidant defence system and ion homeostasis. This can be achieved by induction and/or repression of a large array of genes that govern the adaptability/susceptibility response of each plant. A series of mechanisms involved in tolerating soil salt and drought stresses have been identified. Several excellent reviews are available on the different aspects of stress signalling and response mechanisms in plants (Zhu, 2002; Flowers et al., 2010; Deinlein et al., 2014; Farooq et al., 2015; Parihar et al., 2015). A brief overview of the various genetic components involved in this process is presented in this section.

4.2.1 Receptors and Kinases

The plasma membrane serves as a physical barrier to the environment and also helps to perceive and transmit external information to the genetic module. The response to ionic (salt) and non-ionic (drought) osmotic stresses is perceived by membrane receptors to trigger signalling cascades as a first step towards the activation of plant's response machinery.

The receptor-like kinases (RLKs) have been implicated in the perception of osmotic stresses. A novel transmembrane protein, NtC7, which resembles the receptor domain of RLK, was identified in *Nicotiana tabacum* (Nt). Over-expression of NtC7 in transgenic tobacco activated the production of cellular metabolites, which confer tolerance to osmotic stress (Tamura et al., 2003). *Arabidopsis* histidine kinase1 (AtHK1) also performs a similar role by sensing the stress signal (Wohlbach et al., 2008). This results in the activation of a variety of kinases that lead to increased synthesis and accumulation of osmolytes (Maeda et al., 1994).

The mitogen-activated protein kinase (MAPK) cascades are known to control stress responses in plants by activating the antioxidative genes and transcription factors (TFs). The MAPK pathways are activated in response to osmotic stress and accumulation of ROS. The ROS-responsive MAPKKK, MEKK1, MPK4 and MPK6 pathway is the best understood so far (Xing et al., 2008; Jammes et al., 2009).

Phosphorylation of cellular lipids can generate intracellular secondary messenger molecules such as inositol 1,4,5-trisphosphate (IP_3) and diacylglycerol (DAG) to activate protein kinase C and trigger calcium ion (Ca^{2+}) release, respectively (Kopka et al., 1998; Mikami et al., 1998; DeWald et al., 2001; Hunt et al., 2003; Peters et al., 2010). The perception of drought and salt stress activates specific phosphatidylinositol (PI) phospholipase C (PI-PLC) isoforms (Drobak and Watkins, 2000), which cleave PI-4, 5-biphosphate (PI-4,5-P_2 or PIP_2) to produce IP_3 and DAG (Gaude et al., 2008; Munnik and Testerink, 2009; Peters et al., 2010; Pokotylo et al., 2014), thereby activating a downstream cascade that results in the manipulation of gene expression (Sanders et al., 2002).

4.2.2 Channels and Transporters

Cellular homeostasis also involves interplay of different ions through the coordinated regulation of their channels and transporters. The role of ion transporters in maintaining osmotic balance, membrane organization, enzyme functions, cell division and expansion in plant cells has received much attention recently. Great progress has been made in the characterization of ionic transporters and their functional evaluation.

Salt stress increases the Na^+ influx through non-selective cation channels. This results in membrane depolarization that activates the K^+ efflux through the outward rectifying of K^+ channels. Most cells normally maintain relatively high K^+ and low Na^+ concentrations in the cytosol. This is facilitated by the Na^+/K^+ symporters that have a high affinity for Na^+ and favour the intracellular influx of K^+ (Niu et al., 1995; Blumwald and Grover, 2006). K^+ is an essential co-factor of important enzymes like pyruvate kinase (Mahajan and Tuteja, 2005). An increase in salt around roots dissipates the membrane potential by affecting the Na^+/K^+ symporters, and Na^+ competes with K^+ for entry (Niu et al., 1995; Blumwald and Grover, 2006). It has been shown that externally supplied Ca^{2+} reduces the toxic effects of NaCl, by facilitating higher K^+/Na^+ selectivity (Cramer et al., 1987; Lauchli and Schubert, 1989; Liu and Zhu, 1998).

Many other transporters are also activated to counter salt stress by reducing K^+ efflux. The plasma membrane Na^+/H^+ antiporter (SOS1) and the vacuolar Na^+/H^+ exchanger are up-regulated in response to the osmotic stress-induced ABA signalling pathway and work to oppose NaCl-induced membrane depolarization. At the same time, the K^+–Na^+ co-transporter (HKT1) is negatively regulated by the SOS pathway so that it reduces Na^+ influx in cytoplasm. Sucrose non-fermenting 1-related protein kinase2 (SnRK2) is another enzyme that regulates expression of genes involved in ion homeostasis and oxidative stress response. Over-expression of the *sapk4* in rice conferred increased tolerance to salt (Diédhiou et al., 2008).

4.2.3 Calcium Signals

Calcium ions (Ca^{2+}) play a central role in regulating and specifying the cellular responses to disparate environmental stresses (Sanders et al., 2002; White and Broadley, 2003; Xiong et al., 2006; Dodd et al., 2010). Each stress stimulus triggers a unique Ca^{2+} transient or Ca^{2+} signature with specific amplitude, frequency, oscillation duration and spatio-temporal pattern (McAinsh and Hetherington, 1998). The Ca^{2+} concentrations are partitioned in the apoplast as well as the intracellular compartments, and stress increases its influx in the cytoplasm (Knight et al., 1997). These activate the Ca^{2+}-binding proteins to initiate signal transduction, leading to the activation of adaptation responses (Knight and Knight, 2001; Mahajan et al., 2006; Kudla et al., 2010).

The salt (Na^+)-induced increase in intercellular Ca^{2+} is perceived by the primary sensor SOS3. It is a myristoylated Ca^{2+}-binding protein (Kim et al., 2007; Quan et al., 2007), which physically interacts with a Ca^{2+}-dependent serine-threonine protein kinase, SOS2 (Hrabak et al., 2003), to activate the transmembrane Na^+/H^+ antiporter, SOS1, which helps to remove Na^+ from cytoplasm (Shi et al., 2000; Qiu et al., 2002; Quintero et al., 2002; Qiu et al., 2003; Lin et al., 2009). There are three main families of Ca^{2+} sensor proteins viz calmodulin (CaM), calcineurin B-like (CBL) and Ca^{2+}-dependent protein kinases (CDPKs).

CBL participates in the salt stress signal transduction pathway via the CBL-interacting protein kinases (CIPKs) to control the efflux of Na^+. Expression of the *cbl* gene from maize and soybean in *Arabidopsis* enhanced salt tolerance in plants (Li et al., 2012). The CDPKs are serine/threonine protein kinases that directly bind Ca^{2+} and get auto-phosphorylated (Chaudhuri et al., 1999; Rutschmann et al., 2002; Franz et al., 2011) to activate the downstream stress signal transduction. It has been reported that CDPKs are involved in salt, cold and drought stress signalling (Urao et al., 1994; Tähtiharju et al., 1997). Over-expression of *cdpk7* resulted in enhanced osmotic stress tolerance in rice (Saijo et al., 2000).

4.2.4 ROS Scavengers

ROS such as superoxide (O_2^-), hydrogen peroxide (H_2O_2) and nitric oxide (NO) cause extensive damage to the cellular membranes and extensively impact ion homeostasis (Baier et al., 2005). Moreover, increased ROS levels can cause salicylic acid accumulation, contributing to cell death and stomatal closure (Khokon et al., 2011). The regulation of ROS levels is one of the key mechanisms for restoring the oxidative balance and minimizing the cellular damage for increasing the adaptation to adverse environmental conditions in plants (Pang and Wang, 2008; Gill and Tuteja, 2010). The ROS are scavenged by antioxidant metabolites such as ascorbate, glutathione, tocopherols and by ROS-detoxifying enzymes such as glutathione reductase, superoxide dismutase (SOD), peroxidase, ascorbate peroxidase (APX) and catalase (CAT). It was shown that expression of *Triticum turgidum* MnSOD in *Arabidopsis thaliana* enhanced tolerance to multiple abiotic stresses by lowering H_2O_2 content and promoting proline accumulation (Kaouthar et al., 2016).

4.2.5 Osmoprotectants

The cellular metabolites that help to maintain osmotic adjustments are known as compatible solutes or osmolytes (Bressan et al., 1998). Accumulation of osmolytes lowers the water potential inside the cell to prevent loss of intracellular water (Garg et al., 2002; Nelson et al., 2004; Carillo et al., 2011). These include sugars, mainly fructose and sucrose, sugar alcohols such as mannitol, sorbitol, ononitol, pinitol and complex sugars such as trehalose and fructans. Additionally, charged metabolites such as glycine, glycine betaine, betaine, proline, ectoine and polyamines are also accumulated. They play an important role in maintaining stomatal conductance and relative leaf water content to improve tolerance to stress (Lopez et al., 2002; Nayyar and Walia, 2003). Osmolytes such as mannitol and polyamines also act as antioxidants by quenching ROS that are generated as a consequence of osmotic stress and protecting the membranes from oxidative damage (Crowe et al., 1992; Smirnoff, 1998; Fariduddin et al., 2013). Their accumulation, storage and utilization are intricately balanced in response to salt and drought stress (Garg et al., 2002).

4.3 REGULATION OF STRESS-INDUCED GENE EXPRESSION BY TRANSCRIPTION FACTORS

Signals from osmotic stresses as well as phytohormones converge on various TFs to bring about rapid changes in regulation of gene expression (Golldack et al., 2011). The resulting changes affect primary metabolism of the cell to influence development (Dey and Harbone, 1997). Various families of TFs are known (Tripathi et al., 2014).

Dehydration Responsive Element Binding (DREB) proteins have gained great interest in the study of plant tolerance to water stress. The *DREB/CBF* sub-family of the *AP2/ERF* genes was first identified in *Arabidopsis* and since then has been studied by several groups (Kudo et al., 2017). They have been shown to play a vital role in managing the adaptations to different stresses through ABA-dependent and independent pathways (Yamaguchi-Shinozaki and Shinozaki, 2006; Khatib et al., 2011; Yang et al., 2012). At-DREB2A, AtB7 and AtBF3 also modulate salt tolerance by accumulating salt-responsive solutes such as proline and glycine betaine (Pruthvi et al., 2014). Over-expression of *RD29 (AtDREB1A)* in a Brazilian soybean genotype that is sensitive to drought enhanced drought-stress tolerance by regulating stomatal conductance, rate of photosynthesis and induction of a series of stress-responsive genes (Polizel et al., 2011; de Paiva Rolla et al., 2014; Fuganti-Pagliarini et al., 2017).

The NAC (NAM, ATAF1 2, CUC2) TF family also plays an important role in regulating the developmental networks operating in plants in response to various abiotic and biotic stresses (Ma et al., 2013; Hernandez and Sanan-Mishra, 2017). The ABA-responsive NAC-TFs, VND-Interacting2 (VNI2) and H_2O_2 responsive Jungbrunnen1 (JUB1) play important roles in leaf maturing and aging in response to hyperosmotic and salt stress (Yang et al., 2011; Wu et al., 2012). In a recent study, a stress-related NAC1 from *Phyllostachys edulis* was characterized. Its ectopic expression in *Arabidopsis* indicated that *PeSNAC1* together with *ped-miR164b* participated in tolerance to salinity and drought stresses through regulation of root development (Wang et al., 2016).

Basic leucine zipper (bZIP) TFs play a crucial role in the regulation of Na^+ and Cl^- homeostasis during salt stress. Transgenic lines over-expressing *GmFDL19*, a bZIP-TF, exhibited significantly lower Na^+ ion content when compared to WT plants, indicating that it might function in regulating absorption of Na^+ (Li et al., 2017). It seems likely that this TF acts by increasing the chlorophyll content and activities of several antioxidative enzymes.

Likewise, WRKY-TF and MYB-TF confer tolerance to salt and drought stress in plants by suppressing negative effectors of stress tolerance (Wang et al., 2011). Functional analysis reveals that WRKY27 interacts with MYB174 to synergistically inhibit transcription of *NAC29* under drought and salt stress. The *GmNAC29* transcript may be involved in ROS production by directly increasing the expression of *GmSPOD1*, so its repression leads to reduction in intracellular ROS levels. GmWRKY27 is also involved in proline accumulation via indirect suppression of proline dehydrogenase. ABA and salt responsive At-WRKY33 and Th-WRKY4 induce genes involved in the detoxification of ROS such as glutathione S-transferase (GST) U11, peroxidases and lipoxygenase LOX1 (Jiang and Deyholos, 2009; Zheng et al., 2013).

4.4 MECHANISM OF REGULATION BY miRNAs

Plants utilize several mechanisms for regulating and reprogramming gene expression during abiotic stress response. The miRNA regulatory pathways play a crucial role in controlling the expression of a variety of different stress-regulated genes. The miRNAs constitute the main class of small RNAs involved in the regulation of gene expression, acting in a series of biological processes, controlling growth, development and adaptation to diverse stresses (Chen, 2005; Mallory and Vaucheret, 2006; Shukla et al., 2008).

The first miRNA, lin-4, was discovered in 1993 and identified as a post-transcriptional regulator of the lin-14 gene that is involved in controlling the larval development of *Caenorhabditis elegans* (Lee and Ambros, 2001). In plants, miRNAs were first reported in *Arabidopsis* in 2002 (Park et al., 2002). Ever since, they have been recognized as the most important among gene regulators.

4.4.1 Origin of Functional miRNAs

In plants, miRNAs are encoded by the *MIR* gene located mostly in the intergenic regions throughout the genome (Naqvi et al., 2012) and transcribed as independent transcription units (Tong et al., 2013). The biogenesis of miRNAs is a complex process that involves several distinct steps. This process is compartmentalized among the nucleus and cytoplasm (Naqvi et al., 2012). In the nucleus, miRNA loci are transcribed into primary miRNAs (pri-miRNAs) mainly by DNA-dependent RNA polymerase II (RNA Pol II) (Jones-Rhoades et al., 2006; Lee et al., 2004) and in some cases by RNA Pol III (Faller and Guo, 2008).

During transcription, the miRNA processing complex is recruited to the pri-miRNAs. The Elongator complex plays an important role in this process as it is able to interact with both Pol II and DCL1 (Yang et al., 2009; Fang et al., 2015). This complex also contains MOS4, a conserved protein associated with the spliceosome (Jia et al., 2017; Li et al., 2018).

The pri-miRNA transcript is associated with the nuclear processing centre known as a D-body or SmD3-body. Multiple RNA binding proteins associate with distinct regions of the pri-miRNA to maintain the structural determinants and to recruit DCL1. After transcription, the imperfect hairpin structure is protected by the binding of Dawdle (DDL), Serrate (SE), a C_2H_2-zinc finger protein (Ori et al., 2000; Yang et al., 2006; Yu et al., 2008) and Hyponastic Leaves 1 (HYL1), a nuclear dsRNA binding phosphoprotein (Han et al., 2004; Vazquez et al., 2004). It is hypothesized that SE binds at ssRNA/dsRNA junctions and recruits C-terminal domain Phosphatase Like 1 (CPL1) and HYL1 to the DCL1 complex. HYL1 is a phospho-protein that binds to dsRNA region and it requires the activity of CPL1 to maintain its hypo-phosphorylated state.

Here, the imperfect hairpin structure is cleaved out of the main transcript by RNaseIII activity of the Dicer-Like 1 (DCL1) enzyme (Kurihara et al., 2006) to form the precursor miRNA (pre-miRNA). In exceptional cases, the processing is mediated by DCL4 (Rajagopalan et al., 2006).

The same proteins process the cleavage of pre-miRNA to release the mature miRNA duplex (Zhang et al., 2006; Manavella et al., 2012). The other major proteins involved in miRNA processing include TOUGH (TGH), an RNA binding protein with a G-patch and a SWAP domain (Ren et al., 2012a), and Cap binding complex (CBC), a heterodimeric nuclear cap-binding riboprotein (Raczynska et al., 2014).

The newly processed miRNA duplex has two unpaired nucleotides at the 3′ terminal end of each strand so they need to be protected from the exonucleases (Ramachandran and Chen, 2008). HUA Enhancer 1 (HEN1) plays an important role at this step by methylating the nucleotides at the 3′ end of each strand (Yu et al., 2005; Ren et al., 2012b). The miRNA duplex is then transported from the nucleus to the cytoplasm through the HASTY membrane protein (Bollman et al., 2003). In the cytoplasm, the miRNA duplex is incorporated into Argonaute 1 (AGO1) containing RNA-induced silencing complex (RISC) with the help of HFSP90 (Bologna et al., 2018). Here, the two strands are separated and only one is retained to guide the RISC to the target mRNA for cleavage by AGO1 (Chen, 2005) or inhibition of translation.

Post-translational modifications of core miRNA-biogenesis machinery have indicated the modulation of miRNA biogenesis in response to environmental stresses and ABA. The phosphorylation of DCL1 is essential for its interaction with DDL (Machida and Yuan, 2013). Likewise, HYL1 and SE activity could be modulated by phosphorylation under stress (Yan et al., 2017). The HYL1 phosphorylation by MPK3 and SnRK2 makes it inactive and restricts its localization to the nucleus, thereby protecting it from proteolytic degradation triggered during stress. In the microprocessor complex, HYL1 is activated by de-phosphorylation by CPL1/2 and PP4 (Su et al., 2017). Regulator of CBF Gene Expression 3 (RCF3)/HOS5/SHINY assists CPL1 during this process. The expression of RCF3 is reduced by salt, hyperosmotic stress and ABA (Karlsson et al., 2015). The rice nucleotide transferase OsNTP3 and *Arabidopsis* HESO1 are implicated in the uridylation and degradation of unmethylated miRNAs. The expression of these proteins is enhanced by ABA, indicating that the environment cues may control miRNA stability (Ren et al., 2010; Tu et al., 2015; Yang et al., 2017).

4.4.2 Stress-Induced miRNAs

Several studies have reported the stress induction of miRNAs (Sunkar and Zhu, 2004; Fujii et al., 2005; Phillips et al., 2007; Ruiz-Ferrer and Voinnet, 2009; Pérez-Quintero et al., 2010; Zhou et al., 2010; Meng et al., 2011; Yu et al., 2012; Zheng et al., 2012; Mittal et al., 2016; Sharma et al., 2017), indicating that gene networks are finely regulated to help plants manage and cope with osmotic stresses.

The first direct evidence for the involvement of miRNAs in plant stress responses came from the work of Jones-Rhoades and Bartel (2004) in *Arabidopsis*, where they identified several novel miRNAs, including miR395 to be up-regulated upon sulphate starvation. MiR395 targets a low-affinity *sulphate transporter* (*AST68*) and three *ATP sulfurylases* (*APS1, APS3* and *APS4*) involved in sulphate assimilation. This was followed by the identification of phosphate deficiency-induced expression of miR399 with a corresponding decrease in the target *ubiquitin-conjugating enzyme* (*UBC*) in *Arabidopsis* (Fujii et al., 2005; Aung et al., 2006; Bari et al., 2006). The identification of miR398, which targets two *Cu/Zn superoxide dismutases* (*CSDs*), linked miRNAs to the ROS pathway (Sunkar et al., 2006). The down-regulation of miR398 under stresses such as drought, salt and bacterial infection, resulted in the up-regulation of its target gene *CSD1* and *CSD2*, which resulted in sequestration of ROS to protect from cellular damage (Jagadeeswaran et al., 2009).

Several other studies have unravelled the overlapping regulation of miRNAs by multiple stresses and phytohormones as exemplified by the induction of miR159 by the stress hormone ABA under drought condition (Reyes and Chua, 2007). Ath-miR398 was down-regulated under salt stress leading to up-regulation of *CSD1* and *CSD2*, which probably helps in maintaining cellular oxygen metabolism (Jagadeeswaran et al., 2009). Likewise, expression of Osa-miR393 was down-regulated under salt stress and levels of this miRNA are altered in response to many heavy metals like aluminium (Lima et al., 2011), arsenite (Liu and Zhang, 2012; Yu et al., 2012), cadmium (Huang et al., 2009; Ding et al., 2011), etc. Its over-expression in rice negatively regulated the salt-alkali stress tolerance of the plant (Gao et al., 2011). There are reports that salt- and drought-responsive miRNAs are also involved in adaptive response mechanisms to boron deficiency (Lu et al., 2015). For example, miR159 was down-regulated in salt-stressed sugarcane leaves (Patade and Suprasanna, 2010) and in B-deficient *Citrus sinensis* leaves (Lu et al., 2015). Identification of miRNAs that participate in regulating the combined effect of drought and B-deficit stress might improve the plant tolerance to both environmental stresses.

The variations in expression patterns of more than 250 miRNAs have been reported in response to drought, salt, cold or ABA treatments and have been mapped in several plant species including *Arabidopsis, Glycine max, Glycine soja, Gossypium hirsutum, Medicago truncatula, Nicotiana tabacum, Oryza sativa, Panicum virgatum, Phaseolus vulgaris, Populus euphratica, Saccharum officinarum, Triticum aestivum* and *Zea mays.* Conserved miRNAs show contrasting behaviour in different plant species. For example, the salt-induced deregulation of miR167 is known in *Arabidopsis* but not in rice (Sunkar and Zhu, 2004; Lv et al., 2010).

High-throughput sequencing and bioinformatics analysis have allowed the generation of global miRNA profiles (Liu et al., 2008; Barrera-Figueroa et al., 2012). The comparative and integrated analysis of miRNAs and their targets have identified the regulatory networks operative in response to osmotic stress (Goswami et al., 2017). The differences in expression profiles between drought tolerant N22 and drought sensitive PB1 identified miRNAs with variety-specific expression patterns during phase transition (miR164, miR396, miR812 and miR1881) as well as drought stress (Kansal et al., 2015). Our group has reported that salt stress induced the expression of a large number of miRNAs in the salt-tolerant PK, while it repressed the expression of several miRNAs in salt-sensitive PB. The up-regulation of the miRNAs seemed to play an important role in enhancing the tolerance of the plants to salt stress. By comparing the data within the genetically similar backgrounds using the Gly-transgenics in which salt tolerance was artificially engineered, the changes in the regulatory networks were apparent (Tripathi et al., 2017). It also indicated that manipulating one pathway alone may not be sufficient to completely alter the physiology of the plants (Goswami et al., 2017).

A crucial role for miRNAs in maintaining the target gene expression under stress is also indicated by over-expression studies. The drought and salinity stress induced miR393 targets an auxin transporter gene (Os*UX1*) and a rice tiller inhibitor gene (Os*TIR1*). Transgenic rice plants over-expressing miR393 showed an increase in tillers and early flowering, together with decreased tolerance to salt and hypersensitiveness to auxin (Xia et al., 2012). Under salt-stressed conditions, expression of Osa-miR396c also decreased in an ABA-dependent manner and over-expression of the miRNA resulted in reduced salt-stress tolerance (Gao et al., 2011). Plants with elevated levels of miR394 that display hypersensitivity to salinity stress in an ABA-dependent manner provide another example, whereas plants harbouring an miR394-resistant LCR have increased salinity and ABA tolerance (Song et al., 2013). MiRNVL5 from cotton and miR417 from *Arabidopsis* also negatively influence plant responses to salinity stress (Jung and Kang, 2007; Gao et al., 2016). Conversely, two salinity-inducible miRNAs, miR319 and miR528, can positively affect plant response to salinity stress through the down-regulation of their targets (Zhou et al., 2013; Yuan et al., 2015). Over-expression of miR408 using CaMV35S promoter in *Arabidopsis* resulted in enhanced salt tolerance and better root development as compared to the wild type by down-regulating copper-containing phytocyanin and laccase proteins (Ma et al., 2015). Expression of miR408 was induced in response to salt stress and deficiency of copper (Ma et al., 2015). Likewise, it was observed that miR319a over-expression improves salt and drought stress tolerance in transgenic plants by significantly down-regulating its target genes, including As*NAC60* (Zhou et al., 2013).

4.5 REGULATION OF TRANSCRIPTION FACTORS BY miRNAs

Several stress-responsive miRNAs have been shown to negatively or positively affect plant responses to stress by modulating the hormone-signalling pathways (Meng et al., 2010; Zhang, 2015; Gao et al., 2016). The miRNAs may also act by targeting transcripts of stress-responsive proteins that may be associated with diverse metabolic pathways or physiological processes (Yan et al., 2005; Cheng and Long, 2007). Most of the miRNAs are conserved across plant species, so it is likely that they may regulate similar targets in all plants. Many of these proteins have also been reported as stress-responsive TFs in plants (Fang and Grzymala-Busse, 2006; Xu et al., 2008). The targets for conserved miRNAs mainly include the TFs that regulate plant development and organ formation (Jones-Rhoades and Bartel, 2004; Kebrom et al., 2013; Fan et al., 2015; Xie et al., 2015). The miRNA and TF modules may serve as versatile regulatory mechanisms in plant stress responses. Some of the well-studied examples include:

miR156::*SPL*-TF (*Squamosa promoter binding protein-like* TF) and **miR172::*AP2*-TF** (*Apetala2*) are evolutionarily conserved among all land plants. It was shown that miR156 regulated *SPL*-TF promotes transcription of miR172 (Wu et al., 2009) which targets *AP2*, a crucial regulator of flower development (Mehrpooyan et al., 2012; Liu et al., 2013). The AP2-like TFs such as *Target Of EAT 1* (*TOE1*), *TOE2*, *TOE3*, *Schlafmütze* (*SMZ*) and *Schnarchzapfen* (*SNZ*) are also targeted by miR172. These modules operate during juvenile-to-adult phase transitions to regulate vegetative development, floral induction and fruiting (Wu and Poethig, 2006; Martin et al., 2010; Bhogale et al., 2014; Stief et al., 2014; Wang et al., 2014). MiR156 is abundantly expressed during early shoot development and decreases with time, while miR172 is highly down-regulated during early shoot development but its levels increase with time (Aukerman and Sakai, 2003; Wu and Poethig, 2006; Chuck et al., 2007; Wu et al., 2009). Over-expression of miR156 results in a delay in flowering of *Arabidopsis* plants, while over-expression of miR172 accelerates flowering response in *Arabidopsis* (Wu et al., 2009; Jung et al., 2011; Rubio-Somoza and Weigel 2011; Jung et al., 2014). The miR156 is differentially expressed under various abiotic stresses (Preston and Hileman, 2013; Wang et al., 2016). Members of the miR156 family were down-regulated in maize but up-regulated in *Arabidopsis* and cotton under salt stress (Ding et al., 2009; Xie et al., 2014; Mittal et al., 2015). Alfalfa plants over-expressing miR156 exhibited robust salinity tolerance (Arshad et al., 2017).

miR159::*MYB*-TF is directly associated with hormone signalling in plants (Zhao et al., 2016). MiR159 acts as a negative regulator for the gibberellic acid-regulated *MYB* (*GA-MYB*) TF family that plays a crucial role in juvenile-to-adult transition, flower development and seed germination (Millar and Gubler, 2005; Reyes and Chua, 2007; Guo et al., 2017). Salt treatment to wheat and tobacco plants resulted in down-regulation of miR159, whereas the same showed up-regulation in *Arabidopsis* (Liu et al., 2008; Frazier et al., 2011; Gupta et al., 2014). MiR159 expression was differentially regulated in response to heat stress in wheat, rice and switchgrass (Xin et al., 2010; Hivrale et al., 2016; Mangrauthia et al., 2017). *Arabidopsis* and rice plants over-expressing miR159 showed increased sensitivity to heat stress (Wang et al., 2012; Liu et al., 2015).

miR160/miR167::*ARF*-TF (*auxin response factor* TF) are involved in phytohormone signalling and play important roles in development of roots, shoots and floral organs (Gutierrez et al., 2009; Xin et al., 2010; Luo et al., 2013; Kruszka et al., 2014; Ebrahimi-Khaksefidi et al., 2015; Zhao et al., 2016). MiR160 and miR167 were also identified as salt-responsive in radish, wheat, chickpea, maize, sorghum, etc. (Ding et al., 2009; Wang et al., 2014; Xie et al., 2014; Kohli et al., 2014; Sun et al., 2015; El Sanousi et al., 2016). Both the miRNAs are regulated by heat and drought stress as well (Xin et al., 2010; Kruszka et al., 2014; Hivrale et al., 2016). Transgenic *Arabidopsis* plants over-expressing miR160 showed improved seed germination and seedling survival due to altered expression of heat shock protein (HSPs) (Lin et al., 2018).

miR164::*NAC*-TF module plays a crucial role in regulating plant development and stress responses (Aida et al., 1997; Rhoades et al., 2002; Mallory et al., 2004; Guo et al., 2005). It was shown that auxin-responsive miR164 expression provided a homeostatic mechanism to cleave At*NAC1* transcripts, which in turn down-regulate auxin signals (Guo et al., 2005). An earlier study demonstrated that miR164 may also regulate age-dependent cell death and senescence in *Arabidopsis* leaves. This occurs through a trifurcate feed-forward pathway involving NAC-TF ORE1/AtNAC2, miR164 and *ethylene insensitive 2* (*EIN2*, positive regulator of the ethylene response). In this pathway, miR164 expression gradually decreases with aging through negative regulation by *EIN2*, which leads to up-regulation of *ORE1* expression, resulting in cell death (Kim et al., 2009). Similarly, it was shown that ethylene regulates cell expansion by fine-tuning the miRNA164/Rh*NAC100* module (Pei et al., 2013). Deep sequencing revealed the role of miR164 in response to different abiotic stresses. Salinity stress response down-regulated miR164 expression in rice, maize, cotton and wheat, whereas its expression was increased in *Arabidopsis* (Amor et al., 2009; Ding et al., 2009; Macovei et al., 2012; Xie et al., 2014; Gupta et al., 2014; Mittal et al., 2016). Kumar et al. (2015) reported that miR164 directly targeted *heat shock protein-17* (*HSP17*) in wheat, which may help in better adaptation to stress.

miR165/166::*HD-ZIP*-TF is involved in SAM maintenance, lateral organ polarity, ABA homeostasis and vascular development (Willmann and Poethig, 2007; Liu et al., 2009; Zhang et al., 2012; Sakaguchi and Watanabe, 2012; Zhou et al., 2015; Du et al., 2015; Yan et al., 2016). The two families seem to have played important roles in adaptation of plants to salt stress (Zhu et al., 2013; Kohli et al., 2014; Sun et al., 2015). For instance, up-regulation of miR165 was observed in *Arabidopsis* under high salinity, whereas a decrease in its expression was seen in wheat (Liu et al., 2008; Wang et al., 2014; Gao et al., 2016). Similarly, reports on miR166 suggest its abundance during salinity stress in maize, soybean and *Arabidopsis*, while a decrease was observed in cotton (Amor et al., 2009; Kong et al., 2010; Li et al., 2011; Xie et al., 2014; Mittal et al., 2015). Both miRNA families were studied in switchgrass where high salt concentrations caused their overall down-regulation (Xie et al., 2014). Jia and coworkers performed over-expression of Small Tandem Target Mimics of miR165 and miR166 (STTM165/166-31) in tomato and *Arabidopsis* to show that these families play a role in auxin signalling. The degradation of mature miR165/166 in *Arabidopsis* and tomato transgenic lines resulted in an up-regulation of *HD-ZIP III* genes and improved the plant tolerance to salinity stress (Jia et al., 2015).

miR169::*NF-Y*-TF is associated with nodule development, response to ABA, control of flowering time and stress responses (Rhoades et al., 2002; Warpeha et al., 2003; Combier et al., 2006; Kumimoto et al., 2010; Petroni et al., 2012; Sorin et al., 2014; Luan et al., 2015). The over-expression studies of miR169 in *Arabidopsis* resulted in early flowering response (Xu et al., 2014). Another report stated that the over-expression of the rice *heme activator protein* gene (Os*HAP2E*), which is a target of miR169, resulted in resistance to biotic (pathogens) and abiotic stresses (salinity, drought). It also increased the photosynthetic efficiency and number of tillers in transgenic rice plants (Alam et al., 2015). MiR169 is widely known to be high salinity-responsive miRNA in various plants like rice, tobacco, maize, soybean, *Arabidopsis*, aspen, cowpea, poplar, etc. (Liu et al., 2008; Zhao et al., 2009; Jia et al., 2009; Paul et al., 2011; Frazier et al., 2011; Li et al., 2011; Luan et al., 2015). Members of the miR169 family were also found to be heat responsive (Guan et al., 2013; Sailaja et al., 2014; Wang et al., 2015; Hivrale et al., 2016). Analysis of individual family members showed that Tas-miR169g and Tas-miR169n demonstrated improved expression under salt response by down-regulating the expression of its target transcript, *Nuclear Factor Y Subunit A* (*NF-YA*). The authors reported a cis-acting ABA-responsive element (ABRE) in the up-stream region of Tas-miR169n, which suggested that its expression may be ABA regulated (Zhou et al., 2010). Similarly, drought stress down-regulates miR169, to up-regulate its target *NF-YA5*. Transgenic plants over-expressing *NF-YA5* showed enhanced drought tolerance while plants over-expressing miR169a were more sensitive to drought stress (Li et al., 2008). A similar observation was found in case of soybean, where over-production of the GmNFYA3 gene, a newly identified target gene for miR169a, led to improved drought tolerance (Ni et al., 2013), though the salinity tolerance was reduced (Ni et al., 2013).

miR170/miR171::*SCL*-TF (*ScareCrow-like* TF) are known to regulate stress responses, radial patterning in roots and light and phytohormone signalling (Rhoades et al., 2002; Bej and Basak, 2014; Zhu et al., 2015). MiR171 is produced in leaves and flowers, but not in roots, and tightly regulates *SCL* expression in these tissues for normal development (Cho and Paszkowski, 2017), whereas miR170 promotes radial patterning in roots, floral development and shoot branching by regulating *SCL* genes (Zhou et al., 2007; Zhang, 2015). Both miRNAs are differentially regulated in response to salt and other abiotic stresses.

miR319::*TCP*-TF module has a primary role in different aspects of plant development including seedling development and embryo patterning, leaf morphogenesis, complexity and senescence as well as petal and stamen development (Schommer et al., 2012). Over-expression of miR319 reduces the expression of *TCPs*, resulting in serrated leaf margins and crinkled leaf shape, which can be partially rescued by the over-expression of *TCPs* (Palatnik et al., 2003). It has also been shown to regulate root growth in response to the presence of ethylene by suppressing the *MYB33*-TF (Zhang et al., 2016). Expression of miR319 has shown to be affected in response to salt stress miR319 (Liu et al., 2008). A differential down-regulation of miR319 expression was observed between salt-tolerant (NC286) and salt-sensitive (Huangzao4) maize inbred lines (Ding et al., 2009). Over-expression of *OsamiR319a*, resulted in enhanced salt tolerance, possibly via a salt exclusion mechanism, in creeping bentgrass (*Agrostis stolonifera*) (Zhou and Luo, 2014).

miR393::*bHLH*-TF plays a crucial role in regulating plant architecture and growth. MiR393 also targets *F-box* genes that encode auxin receptors to control the root system architecture. *AFB3*, an auxin receptor, targeted by miR393 is involved in plants' response to nitrate in the soil, resulting in decreased primary root growth, increased lateral root branching and the development of flag leaves (Vidal et al., 2010). The miRNA is up-regulated in rice, wheat and *Arabidopsis* under salt stress (Sunkar and Zhu, 2004; Gupta et al., 2014; Chen et al., 2014). Over-expression of *osa-miR393* in rice and *Arabidopsis* resulted in increased sensitivity to salt and alkalinity stress (Gao et al., 2011). Over-expression of the miR393-resistant auxin receptor *Transport Inhibitor Response 1* (*TIR1*) in *Arabidopsis* resulted in enhanced salt tolerance, possibly via osmoregulation and Na^+ exclusion (Chen et al., 2014).

miR396::*GRF* (growth-regulating factor) regulates leaf development (Gupta and Nath, 2015). This module is involved in regulating cell division and differentiation during leaf development (Jones-Rhoades and Bartel, 2004; Wang et al., 2011). In *Arabidopsis* plants over-expressing miR396, there was repression of the expression of *GRF* genes, so transgenic plants showed narrow-leaf phenotypes. This resulted in the reduction of cell numbers and stomatal density, which made the transgenic plants more tolerant to drought (Liu et al., 2009). Over-expression of miR396 in tobacco confirmed its role in the leaf development and drought tolerance of plants (Yang and Yu, 2009). Intriguingly, miR396 has been reported to mediate the cleavage of Ha*WRKY6* in sunflowers and it is unique to *Asteraceae* plants (Giacomelli et al., 2012). Transgenic plants expressing the miR396-resistant form of Ha*WRKY6* showed increased sensitivity to the heat stress response (Giacomelli et al., 2012).

miR398::*HsfA1* (heat shock TF A1) forms a positive regulatory feedback loop by regulating each other's expression during heat stress response in *Arabidopsis* (Ohama et al., 2017). MiR398 is found to be heat-inhibited in *Brassica rapa, O. sativa, Raphanus sativus* and *P. virgatum* while showing accumulation in *A. thaliana* and *T. aestivum* during heat stress (Guan et al., 2013; Sailaja et al., 2014; Kumar et al., 2015; Wang et al., 2015; Hivrale et al., 2016). The dynamic regulation of miR398 in response to salinity stress has also been well studied. It was found to be up-regulated in poplar and sugarcane but down-regulated in *A. thaliana, Populus tomentosa, R. sativus, P. vulgaris, Caragana intermedia; Populus cathayana* and *G. hirsutum* (Jia et al., 2009; Jagadeeswaran et al., 2009; Chen et al., 2012; Zhou et al., 2012; Wang et al., 2013; Carnavale Bottino et al., 2013; Zhu et al., 2013; Nageshbabu et al., 2013; Sun et al., 2015).

miR1514:: NAC-TF triggers phased small interfering RNA (phasiRNA) formation in response to drought in *P. vulgaris.* In water deficit conditions, miR1514a is induced and it targets *NAC-700* transcript for cleavage. The cleaved product is amplified by RNA-dependent RNA polymerases to initiate the production of phasiRNAs, which are again recruited into RISC for mediating post-transcriptional gene silencing (Sosa-Valencia et al., 2017). Similarly, in *Glycine max,* miR1514a targets an *NTL 9* transcript and triggers the production of phasiRNA (Arikit et al., 2014). NTL9 has been shown to be an important mediator of osmotic stress responses that affect leaf senescence in *Arabidopsis* (Yoon et al., 2008).

4.6 CONCLUDING REMARKS AND PERSPECTIVES

Salinity and drought exert ionic and non-ionic osmotic stress, respectively, to severely retard plant growth development. This field has been actively researched and several aspects of the molecular mechanisms controlling stress responses in plants have been studied. The discovery of miRNAs has unravelled a new layer in the efficient regulation of gene expression. MiRNA-based gene manipulation strategies are being extensively applied in functional genomics to achieve various qualitative or quantitative effects. Some excellent studies have provided proof of concept for the manipulation of miRNAs to meet agriculture demands. MiR397 is one such molecule, which increases grain yield by down-regulating the *OsLAC* gene. In another case, miR444a was shown to interact with *OsMADS57, OsTB1* (*TEOSINTE BRANCHED* 1) and *D14* (*Dwarf14*), which regulated tillering in rice, thereby directly affecting the grain yield (Guo et al., 2013).

However, additional knowledge of the miRNA function is a prerequisite to these strategies as it may affect plant morphology. It was recently shown that plants over-expressing Os-miR393 showed high tillering and early flowering (Jiao et al., 2010), but these plants were more susceptible to drought as well as salt stress (Xia et al., 2012). Likewise, the *Arabidopsis* lines over-expressing *miR399f* displayed tolerance to salt stress and ABA treatment, but were hypersensitive to drought (Baek et al., 2016). In the case of stress down-regulated miR319, the targets *OsPCF5, OsPCF6, OsPCF7, OsPCF8* and *OsTCP21* were up-regulated under stress (Sunkar and Zhu, 2004; Liu et al., 2008; Lv et al., 2010). The over-expression of Osa-miR319 led to increased cold stress tolerance, but the plants had developmental defects. To avoid this effect, two RNA silenced lines for *OsPCF5* and *OsTCP21* were generated, which showed better cold tolerance than wild type controls. The miRNA biogenesis and stability of the mature miRNAs also show tissue-specific variations in response to environmental cues. Thus, it will be necessary to study the regulatory pathways in a tissue-specific or cell-dependent manner. Such studies will shed light on the genetic measures adopted by plants to respond/adapt to stresses. This information will provide acceptable alternatives to classical plant breeding and transgenic methods for improving plant responses to stress.

ACKNOWLEDGEMENTS

We apologize to colleagues whose work could not be included owing to space constraints. The study on miRNAs in our lab was supported by financial grants from the Department of Biotechnology.

REFERENCES

Aida, M., Ishida, T., Fukaki, H., Fujisawa, H., Tasaka, M. 1997. Genes involved in organ separation in Arabidopsis: An analysis of the cupshaped cotyledon mutant. *Plant Cell* 9:841–57.

Alam, M.M., Nakamura, H., Ichikawa, H., Kobayashi, K., Yaeno, T., Yamaoka, N., Nishiguchi, M. 2015. Overexpression of OsHAP2E for a CCAAT-binding factor confers resistance to *Cucumber mosaic virus* and *Rice necrosis mosaic virus*. *Journal of General Plant Pathology* 81(1):32–41.

Arikit, S., Xia, R., Kakrana, A., Huang, K., Zhai, J., Yan, Z., Valdés-López, O., Prince, S., Musket, T.A., Nguyen, H.T., Stacey, G. 2014. An atlas of soybean small RNAs identifies phased siRNAs from hundreds of coding genes. *Plant Cell* 26(12):4584–601.

Arshad, M., Gruber, M.Y., Wall, K., Hannoufa, A. 2017. An insight into microRNA156 role in salinity stress responses of alfalfa. *Frontiers in Plant Science* 8:356.

Aukerman, M.J., Sakai, H. 2003. Regulation of flowering time and floral organ identity by a microRNA and its APETALA2-like target genes. *Plant Cell* 15:2730–41.

Aung, K., Lin, S.-I., Wu, C.-C., Huang, Y.-T., Su, C.-l, Chiou, T.-J. 2006. pho2, a phosphate over accumulator, is caused by a nonsense mutation in a microRNA399 target gene. *Plant Physiology* 141(3):1000–1011.

Baek, D., Chun, H.J., Kang, S., Shin, G., Park, S.J., Hong, H., Kim, C., Kim, D.H., Lee, S.Y., Kim, M.C. 2016. A role for *Arabidopsis* miR399f in salt, drought and ABA signaling. *Molecular Cells* 39(2):111.

Baier, M., Kandlbinder, A., Golldack, D., Dietz, K.J. 2005. Oxidative stress and ozone: Perception, signalling and response. *Plant Cell Environment* 28(8):1012–20.

Bari, R., Datt Pant, B., Stitt, M., Scheible, W.R. 2006. PHO2, microRNA399 and PHR1 define a phosphate-signaling pathway in plants. *Plant Physiology* 141:988–99.

Barrera-Figueroa, B.E., Gao, L., Wu, Z., Zhou, X., Zhu, J., Jin, H., Liu, R., Zhu, J.-K. 2012. High throughput sequencing reveals novel and abiotic stress-regulated microRNAs in the inflorescences of rice. *BMC Plant Biology* 12(1):132.

Bej, S., Basak, J. 2014. MicroRNAs: The potential biomarkers in plant stress response. *American Journal of Plant Science* 5(5):748–59.

Amor, B., Wirth, S., Merchan, F., Laporte, P., d'Aubenton-Carafa, Y., Hirsch, J., Maizel, A., Mallory, A., Lucas, A., Deragon, J.M., Vaucheret, H., Thermes, C., Crespi, M. 2009. Novel long non-protein coding RNAs involved in Arabidopsis differentiation and stress responses. *Genome Research* 19(1):57–69.

Bhogale, S., Mahajan, A.S., Natarajan, B., Rajabhoj, M., Thulasiram, H.V., Banerjee, A.K. 2014. MicroRNA156: A potential graft-transmissible microRNA that modulates plant architecture and tuberization in *Solanum tuberosum ssp. Andigena*. *Plant Physiology* 164(2):1011–27.

Blumwald, E., Grover, A. 2006. Salt tolerance. In: *Plant Biotechnology: Current and Future uses of Genetically Modified Crops*, ed. N.G. Halford, 206–224. John Wiley and Sons Ltd.

Bollman, K.M., Aukerman, M.J., Park, M.Y., Hunter, C., Berardini, T.Z., Poethig, R.S. 2003. HASTY, the Arabidopsis ortholog of exportin 5/MSN5, regulates phase change and morphogenesis. *Development* 130(8):1493–1504.

Bologna, N.G., Iselin, R., Abriata, L.A., Sarazin, A., Pumplin, N., Jay, F., Grentzinger, T., Dal Peraro, M., Voinnet, O. 2018. Nucleo-cytosolic shuttling of ARGONAUTE1 prompts a revised model of the plant microRNA pathway. *Molecular Cells* 69(4):709–19.

Bressan, R.A., Hasegawa, P.M., Pardo, J.M. 1998. Plants use calcium to resolve salt stress. *Trends in Plant Science* 3(11):411–12.

Carillo, P., Annunziata, M.G., Pontecorvo, G., Fuggi, A., Woodrow, P. 2011. Salinity stress and salt tolerance. In: *Abiotic Stress in Plants-Mechanisms and Adaptations*, ed. A. Shanker and B. Venkateswarlu, 21–38. InTech Open.

Carnavale Bottino, M., Rosario, S., Grativol, C., Thiebaut, F., Rojas, C.A., Farrineli, L., Hemerly, A.S., Ferreira, P. 2013. High-throughput sequencing of small RNA transcriptome reveals salt stress regulated microRNAs in sugarcane. *PLoS One* 8(3):e59423.

Chaudhuri, S., Seal, A., Gupta, M.D. 1999. Autophosphorylation-dependent activation of a calcium-dependent protein kinase from groundnut. *Plant Physiology* 120(3):859–66.

Chen, C., Ridzon, D.A., Broomer, A.J., Zhou, Z., Lee, D.H., Nguyen, J.T., Barbisin, M., Xu, N.L., Mahuvakar, V.R., Andersen, M.R. and Lao, K.Q., 2005. Real-time quantification of microRNAs by stem–loop RT–PCR. *Nucleic acids research*, 33(20), pp.e179–e179.

Cheng, Y., Long, M. 2007. A cytosolic NADP-malic enzyme gene from rice (*Oryza sativa* l.) confers salt tolerance in transgenic Arabidopsis. *Biotechnology Letters* 29:1129–34.

Chen, S., Hawighorst, P., Sun, J., Polle, A. 2014. Salt tolerance in Populus: Significance of stress signaling networks, mycorrhization, and soil amendments for cellular and whole-plant nutrition. *Environmental and Experimental Botany* 107:113–24.

Chen, L., Zhang, Y., Ren, Y., Xu, J., Zhang, Z., Wang, Y. 2012. Genome-wide identification of cold-responsive and new microRNAs in *Populus tomentosa* by high-throughput sequencing. *Biochemistry Biophysics Research Communication* 417(2):892–6.

Cho, J., Paszkowski, J. 2017. Regulation of rice root development by a retrotransposon acting as a microRNA sponge. *eLife* 6:e30038.

Chuck, G., Cigan, A.M., Saeteurn, K., Hake, S. 2007. The heterochronic maize mutant Corngrassl results from overexpression of a tandem microRNA. *Nature Genetics* 39:544–49.

Combier, J.P., Frugier, F., de Billy, F., Boualem, A., El-Yahyaoui, F., Moreau, S., Vernié, T., Ott, T., Gamas, P., Crespi, M., Niebel, A. 2006. MtHAP2-1 is a key transcriptional regulator of symbiotic nodule development regulated by microRNA169 in *Medicago truncatula*. *Genes and Development* 20:3084–88.

Cramer, G.R., Lynch, J., Läuchli, A., Epstein, E. 1987. Influx of Na^+, K^+ and Ca^{2+} into roots of salt-stressed cotton seedlings effects of supplemental Ca^2. *Plant Physiology* 83(3):510–16.

Crowe, J.H., Hoekstra, F.A., Crowe, L.M. 1992. Anhydrobiosis. *Annual Review of Physiology* 54(1):579–99.

de Paiva Rolla, A.A., de Fátima Corrêa Carvalho, J., Fuganti-Pagliarini, R., Engels, C., do Rio, A., Marin, S.R., de Oliveira, M.C., Beneventi, M.A., Marcelino-Guimarães, F.C., Farias, J.R., Neumaier, N., Nakashima, K., Yamaguchi-Shinozaki, K., Nepomuceno, A.L. 2014. Phenotyping soybean plants transformed with rd29A: AtDREB1A for drought tolerance in the greenhouse and field. *Transgenic Research* 23(1):75–87.

Deinlein, U., Stephan, A.B., Horie, T., Luo, W., Xu, G., Schroeder, J.I. 2014. Plant salt-tolerance mechanisms. *Trends in Plant Science* 19(6):371–79.

DeWald, D.B., Torabinejad, J., Jones, C.A., Shope, J.C., Cangelosi, A.R., Thompson, J.E., Prestwich, G.D., Hama, H. 2001. Rapid accumulation of phosphatidylinositol 4, 5-bisphosphate and inositol 1, 4, 5-trisphosphate correlates with calcium mobilization in salt-stressed *Arabidopsis*. *Plant Physiology* 126(2):759–69.

Diédhiou, C.J., Golldack, D., Dietz, K.-J., Popova, O.V. 2008. The SNF1-type serine-threonine protein kinase SAPK4 regulates stress-responsive gene expression in rice. *BMC Plant Biology* 8(1):49.

Ding, Y., Chen, Z., Zhu, C. 2011. Microarray-based analysis of cadmium-responsive microRNAs in rice (*Oryza sativa*). *Journal of Experimental Biology* 62:3563–73.

Ding, D., Zhang, L., Wang, H., Liu, Z., Zhang, Z., Zheng, Y. 2009. Differential expression of miRNAs in response to salt stress in maize roots. *Annals of Botany* 103:29–38.

Dodd, A.N., Kudla, J., Sanders, D. 2010. The language of calcium signaling. *Annual Review in Plant Biology* 61:593–620.

Drobak B. K., Watkins P. A. 2000. Inositol(1,4,5)trisphosphate production in plant cells: an early response to salinity and hyperosmotic stress. *FEBS Letters* 481(3):240–244. doi:10.1016/s0014-5793(00)01941-4

Du, Q., Avci, U., Li, S., Gallego-Giraldo, L., Pattathil, S., Qi, L., Hahn, M.G., Wang, H. 2015. Activation of miR165b represses at HB 15 expression and induces pith secondary wall development in Arabidopsis. *Plant Journal* 83(3):388–400.

Ebrahimi-Khaksefidi, R., Mirlohi, S., Khalaji, F., Fakhari, Z., Shiran, B., Fallahi, H., Rafiei, F., Budak, H., Ebrahimie, E. 2015. Differential expression of seven conserved microRNAs in response to abiotic stress and their regulatory network in *Helianthus annuus*. *Frontiers in Plant Science* 6:741.

El Sanousi, R.S., Hamza, N.B., Abdelmula, A.A., Mohammed, I.A., Gasim, S.M., Sanan-Mishra, N. 2016. Differential expression of miRNAs in *Sorghum bicolor* under drought and salt stress. *American Journal of Plant Science* 7(06):870.

Faller, M., Guo, F. 2008. MicroRNA biogenesis: there's more than one way to skin a cat. *Biochimica et Biophysica Acta* 1779(11):663–67.

Fang, X., Cui, Y., Li, Y., Qi, Y. 2015. Transcription and processing of primary microRNAs are coupled by elongator complex in Arabidopsis. *Nature Plants* 1(6):1–9.

Fang, J., Grzymala-Busse, J.W. 2006. Mining of microRNA expression data—a rough set approach. In: *International Conference on Rough Sets and Knowledge Technology*, ed. G.Y. Wang, J.F. Peters, A. Skowron, Y. Yao, 758–765. Springer, Berlin, Heidelberg.

Fan, W., Lou, H.Q., Gong, Y.L., Liu, M.Y., Cao, M.J., Liu, Y., Yang, J.L., Zheng, S.J. 2015. Characterization of an inducible C2H2-type zinc finger transcription factor Vu STOP 1 in rice bean (*Vigna umbellata*) reveals differential regulation between low pH and aluminum tolerance mechanisms. *New Phytologist* 208(2):456–68.

Fariduddin, Q., Varshney, P., Yusuf, M., Ahmad, A. 2013. Polyamines: Potent modulators of plant responses to stress. *Journal of Plant Interactions* 8(1):1–16.

Farooq, M., Hussain, M., Wakeel, A., Siddique, K.H. 2015. Salt stress in maize: Effects, resistance mechanisms and management. A review. *Agronomy for Sustainable Development* 35(2):461–81.

Flowers, T.J., Flowers, S.A., Hajibagheri, M.A., Yeo, A.R. 1990. Salt tolerance in the halophytic wild rice, *Porteresia coarctata Tateoka. New Phytologist* 114(4):675–84.

Flowers, T.J., Galal, H.K., Bromham, L. 2010. Evolution of halophytes: Multiple origins of salt tolerance in land plants. *Functional Plant Biology* 37(7):604–12.

Franz, S., Ehlert, B., Liese, A., Kurth, J., Cazalé, A.-C., Romeis, T. 2011. Calcium-dependent protein kinase CPK21 functions in abiotic stress response in *Arabidopsis thaliana. Molecular Plant* 4(1):83–96.

Frazier, T.P., Sun, G., Burklew, C.E., Zhang, B. 2011. Salt and drought stresses induce the aberrant expression of microRNA genes in tobacco. *Molecular Biotechnol*ogy 49(2):159–65.

Fuganti-Pagliarini, R., Ferreira, L.C., Rodrigues, F.A., Molinari, H.B.C., Marin, S.R.R., Molinari, M.D.C., Marcolino-Gomes, J., Mertz-Henning, L.M., Farias, J.R.B., de Oliveira, M.C.N., Neumaier, N., Kanamori, N., Fujita, Y., Mizoi, J., Nakashima, K., Yamaguchi-Shinozaki, K., Nepomuceno, A.L. 2017. Characterization of soybean genetically modified for drought tolerance in field conditions. *Frontiers in Plant Science* 8:448.

Fujii, H., Chiou, T.J., Lin, S.I., Aung, K., Zhu, J.K. 2005. A miRNA involved in phosphate starvation response in arabidopsis. *Current Biology* 15:2038–43.

Gao, P., Bai, X., Yang, L., Lv, D., Pan, X., Li, Y., Cai, H., Ji, W., Chen, Q., Zhu, Y. 2011. Osa-miR393: A salinity and alkaline stress-related microRNA gene. *Molecular Biology Reports* 38:237–42.

Gao, S., Guo, C., Zhang, Y., Zhang, F., Du, X., Gu, J., Xiao, K. 2016. Wheat microRNA member tamir444a is nitrogen deprivation-responsive and involves plant adaptation to the nitrogen-starvation stress. *Plant Molecular Biology Reporter* 34(5):931–46.

Garcia, A.B., Engler, J.D.A., Iyer, S., Gerats, T., Van Montagu, M., Caplan, A. 1997. Effects of osmoprotectants upon NaCl stress in rice. *Plant Physiology* 115(1):159–69.

Garg, A.K., Kim, J.-K., Owens, T.G., Ranwala, A.P., Do Choi, Y., Kochian, L.V., Wu, R.J. 2002. Trehalose accumulation in rice plants confers high tolerance levels to different abiotic stresses. *Proceedings of the National Academy of Science USA* 99(25):15898–903.

Gaude, N., Nakamura, Y., Scheible, W.R., Ohta, H., Dörmann, P. 2008. Phospholipase C5 (NPC5) is involved in galactolipid accumulation during phosphate limitation in leaves of *Arabidopsis. Plant Journal* 56(1):28–39.

Giacomelli, J.I., Weigel, D., Chan, R.L., Manavella, P.A. 2012. Role of recently evolved miRNA regulation of sunflower HaWRKY6 in response to temperature damage. *New Phytologist* 195(4):766–73.

Gill, S.S., Tuteja, N. 2010. Polyamines and abiotic stress tolerance in plants. *Plant Signaling and Behaviour* 5(1):26–33.

Golldack, D., Lüking, I., Yang, O. 2011. Plant tolerance to drought and salinity: Stress regulating transcription factors and their functional significance in the cellular transcriptional network. *Plant Cell Reports* 30(8):1383–91.

Goswami, K., Tripathi, A., Sanan-Mishra, N. 2017. Comparative miRomics of salt-tolerant and salt-sensitive rice. *Journal of Integrative Bioinformatics* 14(1):20170002.

Greenway, H., Munns, R. 1980. Mechanisms of salt tolerance in nonhalophytes. *Annual Review in Plant Physiology* 31(1):149–90.

Guan, Q., Lu, X., Zeng, H., Zhang, Y., Zhu, J. 2013. Heat stress induction of miR398 triggers a regulatory loop that is critical for thermotolerance in *Arabidopsis. Plant Journal* 74(5):840–51.

Guo, H., Wang, Y., Wang, L., Hu, P., Wang, Y., Jia, Y., Zhang, C., Zhang, Y., Zhang, Y., Wang, C., Yang, C. 2017. Expression of the MYB transcription factor gene Bpl MYB 46 affects abiotic stress tolerance and secondary cell wall deposition in *Betula platyphylla. Plant Biotechnology Journal* 15(1):107–12.

Guo, H.S., Xie, Q., Fie, J.F., Chua, N.H. 2005. MicroRNA directs mRNA cleavage of the transcription factor NAC1 to downregulate auxin signals for Arabidopsis lateral root development. *Plant Cell* 17:1376–86.

Guo, S., Xu, Y., Liu, H., Mao, Z., Zhang, C., Ma, Y., Zhang, Q., Meng, Z., Chong, K. 2013. The interaction between OsMADS57 and OsTB1 modulates rice tillering via DWARF14. *Nature Communications* 4:1566.

Gupta, M.D., Nath, U. 2015. Divergence in patterns of leaf growth polarity is associated with the expression divergence of miR396. *Plant Cell* 27(10):2785–99.

Gupta, O., Sharma, P., Gupta, R., Sharma, I. 2014. MicroRNA mediated regulation of metal toxicity in plants: Present status and future perspectives. *Plant Molecular Biology* 84(1–2):1–18.

Gutierrez, L., Bussell, J.D., Păcurar, D.I., Schwambach, J., Păcurar, M., Bellini, C. 2009. Phenotypic plasticity of adventitious rooting in Arabidopsis is controlled by complex regulation of auxin response factor transcripts and microRNA abundance. *Plant Cell* 21(10):3119–32.

Han, M.H., Goud, S., Song, L., Fedoroff, N. 2004. The Arabidopsis double stranded RNA-binding protein HYL1 plays a role in microRNA mediated gene regulation. *Proceedings of the National Academy of Science USA* 101:1093–98.

Hasanuzzaman, M., Nahar, K., Fujita, M. 2013. Plant response to salt stress and role of exogenous protectants to mitigate salt-induced damages. In: *Ecophysiology and Responses of Plants Under salt Stress*, ed. P. Ahmad, M. Azooz, M. Prasad, 25–87. Springer, New York.

Hasegawa, P.M., Bressan, R.A., Zhu, J.-K., Bohnert, H.J. 2000. Plant cellular and molecular responses to high salinity. *Annual Review in Plant Biology* 51(1):463–99.

Hernandez, Y., Sanan-Mishra, N. 2017. miRNA mediated regulation of NAC transcription factors in plant development and environment stress response. *Plant Gene* 11:190–8.

Hivrale, V., Zheng, Y., Puli, C.O.R., Jagadeeswaran, G., Gowdu, K., Kakani, V.G., Barakat, A., Sunkar, R. 2016. Characterization of drought-and heat-responsive microRNAs in switchgrass. *Plant Science* 242:214–23.

Hrabak, E.M., Chan, C.W., Gribskov, M., Harper, J.F., Choi, J.H., Halford, N., Kudla, J., Luan, S., Nimmo, H.G., Sussman, M.R. 2003. The Arabidopsis CDPK-SnRK superfamily of protein kinases. *Plant Physiology* 132(2):666–80.

Huang, S.Q., Peng, J., Qiu, C.X., Yang, Z.M. 2009. Heavy metal-regulated new microRNAs from rice. *Journal of Inorganic Biochemistry* 103:282–87.

Hunt, L., Mills, L.N., Pical, C., Leckie, C.P., Aitken, F.L., Kopka, J., Mueller-Roeber, B., McAinsh, M.R., Hetherington, A.M., Gray, J.E. 2003. Phospholipase C is required for the control of stomatal aperture by ABA. *Plant Journal* 34(1):47–55.

Jagadeeswaran, G., Saini, A., Sunkar, R. 2009. Biotic and abiotic stress down-regulate miR398 expression in Arabidopsis. *Planta* 229:1009–14.

Jammes, F., Song, C., Shin, D., Munemasa, S., Takeda, K., Gu, D., Cho, D., Lee, S., Giordo, R., Sritubtim, S. 2009. MAP kinases MPK9 and MPK12 are preferentially expressed in guard cells and positively regulate ROS-mediated ABA signaling. *Proceedings of the National Academy of Science USA* 106(48):20520–25.

Jia, X., Ding, N., Fan, W., Yan, J., Gu, Y., Tang, X., Li, R., Tang, G. 2015. Functional plasticity of miR165/166 in plant development revealed by small tandem target mimic. *Plant Science* 233:11–21.

Jiang, Y., Deyholos, M.K. 2009. Functional characterization of *Arabidopsis* NaCl-inducible WRKY25 and WRKY33 transcription factors in abiotic stresses. *Plant Molecular Biology* 69(1–2):91–105.

Jiao, Y., Wang, Y., Xue, D., Wang, J., Yan, M., Liu, G., Dong, G., Zeng, D., Lu, Z., Zhu, X., Qian, Q., Li, J. 2010. Regulation of *OsSPL14* by *OsmiR156* defines ideal plant architecture in rice. *Nature Genetics* 42:541–44.

Jia, X., Wang, W.-X., Ren, L., Chen, Q.-J., Mendu, V., Willcut, B., Dinkins, R., Tang, X., Tang, G. 2009. Differential and dynamic regulation of miR398 in response to ABA and salt stress in *Populus tremula* and *Arabidopsis thaliana*. *Plant Molecular Biology* 71(1–2):51–9.

Jia, T., Zhang, B., You, C., Zhang, Y., Zeng, L., Li, S., Johnson, K., Yu, C.M., Li, B., Chen, X. 2017. The Arabidopsis MOS4-associated complex promotes microRNA biogenesis and precursor messenger RNA splicing. *Plant Cell* 29(10):2626–43.

Jones-Rhoades, M.W., Bartel, D.P. 2004. Computational identification of plant microRNAs and their targets, including a stress-induced miRNA. *Molecular Cell* 14:787–99.

Jones-Rhoades, M.W., Bartel, D.P., Bartel, B. 2006. MicroRNAs and their regulatory roles in plants. *Annual Review in Plant Biology* 57:19–53.

Jung, H.J., Kang, H. 2007. Expression and functional analyses of microRNA417 in Arabidopsis thaliana under stress conditions. *Plant Physiology and Biochemistry* 45(10–11):805–11.

Jung, J.H., Lee, S., Yun, J., Lee, M., Park, C.M. 2014. The miR172 target TOE3 represses AGAMOUS expression during Arabidopsis floral patterning. *Plant Science* 215:29–38.

Jung, J.H., Seo, P.J., Kang, S.K., Park, C.M. 2011. miR172 signals are incorporated into the miR156 signaling pathway at the SPL3/4/5 genes in Arabidopsis developmental transitions. *Plant Molecular Biology* 76(1–2):35–45.

Kansal, S., Devi, R.M., Balyan, S.C., Arora, M.K., Singh, A.K., Mathur, S., Raghuvanshi, S. 2015. Unique miRNome during anthesis in drought-tolerant indica rice var. Nagina 22. *Planta* 241(6):1543–59.

Kaouthar, F., Ameny, F.-K., Yosra, K., Walid, S., Ali, G., Faical, B. 2016. Responses of transgenic *Arabidopsis* plants and recombinant yeast cells expressing a novel durum wheat manganese superoxide dismutase TdMnSOD to various abiotic stresses. *Journal of Plant Physiology* 198:56–68.

Karlsson, P., Christie, M.D., Seymour, D.K., Wang, H., Wang, X., Hagmann, J., Kulcheski, F., Manavella, P.A. 2015. KH domain protein RCF3 is a tissue-biased regulator of the plant miRNA biogenesis cofactor HYL1. *Proceedings of the National Academy of Science USA* 112(45):14096–101.

Kebrom, T.H., Spielmeyer, W., Finnegan, E.J. 2013. Grasses provide new insights into regulation of shoot branching. *Trends in Plant Science* 18(1):41–8.

Khan, M., Hamid, A., Salahuddin, A., Quasem, A., Karim, M. 1997. Effect of sodium chloride on growth, photosynthesis and mineral ions accumulation of different types of rice (*Oryza sativa* L.). *Journal of Agronomy and Crop Science* 179(3):149–61.

Khatib, F., Makris, A., Yamaguchi-Shinozaki, K., Kumar, S., Sarker, A., Erskine, W., Baum, M. 2011. Expression of the DREB1A gene in lentil (*Lens culinaris Medik.* subsp. *culinaris*) transformed with the *Agrobacterium* system. *Crop Pasture Science* 62(6):488–95.

Khokon, M., Okuma, E., Hossain, M.A., Munemasa, S., Uraji, M., Nakamura, Y., Mori, I.C., Murata, Y. 2011. Involvement of extracellular oxidative burst in salicylic acid-induced stomatal closure in *Arabidopsis*. *Plant Cell and Environment* 34(3):434–43.

Kim, B.G., Waadt, R., Cheong, Y.H., Pandey, G.K., Dominguez-Solis, J.R., Schültke, S., Lee, S.C., Kudla, J., Luan, S. 2007. The calcium sensor CBL10 mediates salt tolerance by regulating ion homeostasis in *Arabidopsis*. *Plant Journal* 52(3):473–84.

Kim, J.H., Woo, H.R., Kim, J., Lim, P.O., Lee, I.C., Choi, S.H., Hwang, D., Nam, H.G. 2009. Trifurcate feed-forward regulation of age-dependent cell death involving miR164 in Arabidopsis. *Science* 323(5917):1053–57.

Knight, H., Knight, M.R. 2001. Abiotic stress signalling pathways: Specificity and cross-talk. *Trends in Plant Science* 6(6):262–67.

Knight, H., Trewavas, A.J., Knight, M.R. 1997. Calcium signalling in *Arabidopsis thaliana* responding to drought and salinity. *Plant Journal* 12(5):1067–78.

Kohli, D., Joshi, G., Deokar, A.A., Bhardwaj, A.R., Agarwal, M., Katiyar-Agarwal, S., Srinivasan, R., Jain, P.K. 2014. Identification and characterization of wilt and salt stress-responsive microRNAs in chickpea through high-throughput sequencing. *PLoS One* 9(10):e108851.

Kong, Y., Elling, A.A., Chen, B., Deng, X. 2010. Differential expression of microRNAs in maize inbred and hybrid lines during salt and drought stress. *American Journal of Plant Science* 1(02):69.

Kopka, J., Pical, C., Hetherington, A.M., Müller-Röber, B. 1998. Ca2+/phospholipid-binding (C2) domain in multiple Plant Proteins: Novel Components of the Calcium-Sensing Apparatus. *Plant Molecular Biology* 36(5):627–37.

Kruszka, K., Pacak, A., Swida-Barteczka, A., Nuc, P., Alaba, S., Wroblewska, Z., Karlowski, W., Jarmolowski, A., Szweykowska-Kulinska, Z. 2014. Transcriptionally and post-transcriptionally regulated microRNAs in heat stress response in barley. *Journal of Experimental Botany* 65(20):6123–35.

Kudla, J., Batistič, O., Hashimoto, K. 2010. Calcium signals: The lead currency of plant information processing. *Plant Cell* 22(3):541–63.

Kudo, M., Kidokoro, S., Yoshida, T., Mizoi, J., Todaka, D., Fernie, A.R., Shinozaki, K., Yamaguchi-Shinozaki, K. 2017. Double overexpression of DREB and PIF transcription factors improves drought stress tolerance and cell elongation in transgenic plants. *Plant Biotechnology Journal* 15(4):458–471.

Kumar, R.R., Pathak, H., Sharma, S.K., Kala, Y.K., Nirjal, M.K., Singh, G.P., Goswami, S., Rai, R.D. 2015. Novel and conserved heat-responsive microRNAs in wheat (*Triticum aestivum* l). *Functional and Integrative Genomics* 15(3):323–48.

Kumimoto, R.W., Zhang, Y., Siefers, N., Holt, B.F. 2010. NF–YC3, NF–YC4 and NF–YC9 are required for CONSTANS-mediated, photoperiod-dependent flowering in *Arabidopsis thaliana*. *Plant Journal* 63(3):379–91.

Kurihara, Y., Takashi, Y., Watanabe, Y. 2006. The interaction between DCL1 and HYL1 is important for efficient and precise processing of pri-miRNA in plant microRNA biogenesis. *RNA* 12:206–12.

Lauchli, A., Schubert, S. 1989. The role of calcium in the regulation of membrane and cellular growth processes under salt stress. In: *Environmental Stress in Plants*, ed. J.H. Cherry, 131–138. NATO ASI Series (Series G: Ecological Sciences), Springer.

Lee, R.C., Ambros, V. 2001. An extensive class of small RNAs in *Caenorhabditis elegans*. *Science* 294(5543):862–64.

Li, Y., Chen, Q., Nan, H., Li, X., Lu, S., Zhao, X., Liu, B., Guo, C., Kong, F., Cao, D. 2017. Overexpression of GmFDL19 enhances tolerance to drought and salt stresses in soybean. *PLoS One* 12(6):e0179554.

Li, W.-X., Oono, Y., Zhu, J., He, X.-J., Wu, J.-M., Iida, K., Lu, X.-Y., Cui, X., Jin, H., Zhu, J.-K. 2008. The *Arabidopsis* NFYA5 transcription factor is regulated transcriptionally and post transcriptionally to promote drought resistance. *Plant Cell* 20(8):2238–51.

Li, F., Pignatta, D., Bendix, C., Brunkard, J.O., Cohn, M.M., Tung, J., Sun, H., Kumar, P., Baker, B. 2012. MicroRNA regulation of plant innate immune receptors. *Proceedings of the National Academy of Science USA* 109:1790–95.

Li, T., Li, H., Zhang, Y.X., Liu, J.Y. 2011. Identification and analysis of seven H_2O_2-responsive miRNAs and 32 new miRNAs in the seedlings of rice (*Oryza sativa* l. Ssp. indica). *Nucleic Acids Research* 39:2821–33.

Li, W., Xu, G., Alli, A., Yu, L. 2018. Plant HAK/KUP/KT K^+ transporters: Function and regulation. *Seminars in Cell and Developmental Biology* 74:133–141.

Lima, J.C., Arenhart, R.A., Margis, M. 2011. Aluminum triggers broad changes in microRNA expression in rice roots. *Genetics and Molecular Research* 10(4):2817–32.

Lin, J.S., Kuo, C.C., Yang, I.C., Tsai, W.A., Shen, Y.H., Lin, C.C., Liang, Y.C., Li, Y.C., Kuo, Y.W., King, Y.C., Lai, H.M., Jeng, S.T. 2018. MicroRNA160 modulates plant development and heat shock protein gene expression to mediate heat tolerance in *Arabidopsis*. *Frontiers in Plant Science* 9:68.

Lin, H., Yang, Y., Quan, R., Mendoza, I., Wu, Y., Du, W., Zhao, S., Schumaker, K.S., Pardo, J.M., Guo, Y. 2009. Phosphorylation of SOS3-like calcium binding protein8 by SOS2 protein kinase stabilizes their protein complex and regulates salt tolerance in *Arabidopsis*. *Plant Cell* 21(5):1607–19.

Liu, J., Feng, L., Li, J., He, Z. 2015. Genetic and epigenetic control of plant heat responses. *Frontiers in Plant Science* 6:267.

Liu, D., Song, Y., Chen, Z., Yu, D. 2009. Ectopic expression of miR396 suppresses GRF target gene expression and alters leaf growth in *Arabidopsis*. *Physiologia Plantarum* 136(2):223–36.

Liu, H.-H., Tian, X., Li, Y.-J., Wu, C.-A., Zheng, C.-C. 2008. Microarray-based analysis of stress-regulated microRNAs in *Arabidopsis thaliana*. *RNA* 14(5):836–43.

Liu, N., Yang, J., Guo, S., Xu, Y., Zhang, M. 2013. Genome-wide identification and comparative analysis of conserved and novel microRNAs in grafted watermelon by high-throughput sequencing. *PLoS One* 8(2):e57359.

Liu, Q., Zhang, H. 2012. Molecular identification and analysis of arsenite stress-responsive miRNAs in rice. *Journal of Agricultural and Food Chemistry* 60(26):6524–36.

Liu, Q., Zhang, Y.C., Wang, C.Y., Luo, Y.C., Huang, Q.J., Chen, S.Y., Zhou, H., Qu, L.H., Chen, Y.Q. 2009. Expression analysis of phytohormone-regulated microRNAs in rice, implying their regulation roles in plant hormone signaling. *FEBS Letters* 583(4):723–28.

Liu, J., Zhu, J.-K. 1998. A calcium sensor homolog required for plant salt tolerance. *Science* 280(5371):1943–45.

Lopez, C., Takahashi, H., Yamazaki, S. 2002. Plant–water relations of kidney bean plants treated with NaCl and foliarly applied glycinebetaine. *Journal of Agronomy and Crop Science* 188(2):73–80.

Luan, M., Xu, M., Lu, Y., Zhang, L., Fan, Y., Wang, L. 2015. Expression of zma-miR169 miRNAs and their target ZmNF-YA genes in response to abiotic stress in maize leaves. *Gene* 555(2):178–85.

Luo, Y., Guo, Z., Li, L. 2013. Evolutionary conservation of microRNA regulatory programs in plant flower development. *Development Biology* 380(2):133–144.

Lu, Y.B., Qi, Y.P., Yang, L.T., Guo, P., Li, Y., Chen, L.S. 2015. Boron-deficiency-responsive microRNAs and their targets in *Citrus sinensis* leaves. *BMC Plant Biology* 15(1):271.

Lv, D.-K., Bai, X., Li, Y., Ding, X.-D., Ge, Y., Cai, H., Ji, W., Wu, N., Zhu, Y.-M. 2010. Profiling of cold-stress-responsive miRNAs in rice by microarrays. *Gene* 459(1):39–47.

Ma, X., Xin, Z., Wang, Z., Yang, Q., Guo, S., Guo, X., Cao, L., Lin, T. 2015. Identification and comparative analysis of differentially expressed miRNAs in leaves of two wheat (*Triticum aestivum* L.) genotypes during dehydration stress. *BMC Plant Biology* 15(1):21.

Ma, N.N., Zuo, Y.Q., Liang, X.Q., Yin, B., Wang, G.D., Meng, Q.W. 2013. The multiple stress-responsive transcription factor SlNAC1 improves the chilling tolerance of tomato. *Physiologia Plantarum* 149(4):474–86.

Machida, S., Yuan, Y.A. 2013. Crystal structure of *Arabidopsis thaliana* Dawdle forkhead-associated domain reveals a conserved phospho-threonine recognition cleft for dicer-like 1 binding. *Molecular Plant* 6(4):1290–1300.

Macovei, A., Gill, S.S., Tuteja, N. 2012. microRNAs as promising tools for improving stress tolerance in rice. *Plant Signaling and Behaviour* 7(10):1296–1301.

Maeda, T., Wurgler-Murphy, S.M., Saito, H. 1994. A two-component system that regulates an osmosensing MAP kinase cascade in yeast. *Nature* 369(6477):242–45.

Mahajan, S., Sopory, S.K., Tuteja, N. 2006. CBL-CIPK paradigm: Role in calcium and stress signaling in plants. *Proceedings of the Indian National Science Academy* 72(2):63–78.

Mahajan, S., Tuteja, N. 2005. Cold, salinity and drought stresses: An overview. *Archives of Biochemistry Biophysics* 444(2):139–58.

Mallory, A.C., Reinhart, B.J., Jones-Rhoades, M.W., Tang, G., Zamore, P.D., Barton, M.K., Bartel, D.P. 2004. MicroRNA control of PHABULOSA in leaf development: Importance of pairing to the microRNA 5′ region. *EMBO Journal* 23:3356–64.

Mallory, A.C., Vaucheret, H. 2006. Functions of microRNAs and related small RNAs in plants. *Nature Genetics* 38:S31–S36.

Manavella, P.A., Hagmann, J., Ott, F., Laubinger, S., Franz, M., Macek, B., Weigel, D. 2012. Fast-forward genetics identifies plant CPL phosphatases as regulators of miRNA processing factor HYL1. *Cell* 151(4):859–870.

Mangrauthia, S.K., Bhogireddy, S., Agarwal, S., Prasanth, V.V., Voleti, S.R., Neelamraju, S., Subrahmanyam, D. 2017. Genome-wide changes in microRNA expression during short and prolonged heat stress and recovery in contrasting rice cultivars. *Journal of Experimental Botany* 68(9):2399–2412.

Martin, R.C., Asahina, M., Liu, P.P., Kristof, J.R., Coppersmith, J.L., Pluskota, W.E., Bassel, G.W., Goloviznina, N.A., Nguyen, T.T., Martínez-Andújar, C., Arun Kumar, M.B., Pupel, P., Nonogaki, H. 2010. The microRNA156 and microRNA172 gene regulation cascades at post-germinative stages in Arabidopsis. *Seed Science and Research* 20(2):79–87.

McAinsh, M.R., Hetherington, A.M. 1998. Encoding specificity in Ca2+ signalling systems. *Trends in Plant Science* 3(1):32–6.

Mehrpooyan, F., Othman, R., Harikrishna, J. 2012. Tissue and temporal expression of miR172 paralogs and the AP2-like target in oil palm (*Elaeis guineensis Jacq.*). *Tree Genetics and Genomes* 8(6):1331–43.

Meng, Y., Gou, L., Chen, D., Wu, P., Chen, M. 2010. High-throughput degradome sequencing can be used to gain insights into microRNA precursor metabolism. *Journal of Experimental Botany* 61:3833–37.

Meng, Y., Shao, C., Gou, L., Jin, Y., Chen, M. 2011. Construction of microRNA and microRNA*-mediated regulatory networks in plants. *RNA Biology* 8:1–25.

Mikami, K., Katagiri, T., Iuchi, S., Yamaguchi-Shinozaki, K., Shinozaki, K. 1998. A gene encoding phosphatidylinositol-4-phosphate 5-kinase is induced by water stress and abscisic acid in *Arabidopsis thaliana*. *Plant Journal* 15(4):563–8.

Millar, A.A., Gubler, F. 2005. The arabidopsis GAMYB-like genes, MYB33 and MYB65 are microRNA-regulated genes that redundantly facilitate anther development. *Plant Cell* 17:705–21.

Mittal, A., Jiang, Y., Ritchie, G.L., Burke, J.J., Rock, C.D. 2015. AtRAV1 and AtRAV2 overexpression in cotton increases fiber length differentially under drought stress and delays flowering. *Plant Science* 241:78–95.

Mittal, D., Sharma, N., Sharma, V., Sopory, S.K., Sanan-Mishra, N. 2016. Role of microRNAs in rice plant under salt stress. *Annals of Applied Biology* 168(1):2–18.

Munnik, T., Testerink, C. 2009. Plant phospholipid signaling: "in a nutshell". *Journal of Lipid Research* 50:S260–5.

Munns, R., Tester, M. 2008. Mechanisms of salinity tolerance. *Annual Review of Plant Biology* 59:651–81.

Nageshbabu, R., Jyothi, M.N., Sharadamma, N. 2013. Expression of miRNAs regulates growth and development of French bean (*Phaseolus vulgaris*) under salt and drought stress conditions. *International Research Journal of Biological Sciences* 2(1):52–56.

Naqvi, A.R., Sarwat, M., Hasan, S., Roychodhury, N. 2012. Biogenesis, functions and fate of plant microRNAs. *Journal of Cellular Physiology* 227(9):3163–68.

Nayyar, H., Walia, D. 2003. Water stress induced proline accumulation in contrasting wheat genotypes as affected by calcium and abscisic acid. *Biologia Plantarum* 46(2):275–79.

Nelson, P.T., Baldwin, D.A., Scearce, L.M., Oberholtzer, J.C., Tobias, J.W., Mourelatos, Z. 2004. Microarray-based, high-throughput gene expression profiling of microRNAs. *Nature Methods* 1(2):155–61.

Ni, Z., Hu, Z., Jiang, Q., Zhang, H. 2013. GmNFYA3, a target gene of miR169, is a positive regulator of plant tolerance to drought stress. *Plant Molecular Biology* 82(1–2):113–29.

Niu, X., Bressan, R.A., Hasegawa, P.M., Pardo, J.M. 1995. Ion homeostasis in NaCl stress environments. *Plant Physiology* 109(3):735.

Ohama, N., Sato, H., Shinozaki, K., Yamaguchi-Shinozaki, K. 2017. Transcriptional regulatory network of plant heat stress response. *Trends in Plant Science* 22(1):53–65.

Ori, N., Eshed, Y., Chuck, G., Bowman, J.L., Hake, S. 2000. Mechanisms that control Knox gene expression in the Arabidopsis shoot. *Development* 127(24):5523–32.

Palatnik, J.F., Allen, E., Wu, X.L., Schommer, C., Schwab, R., Carrington, J.C., Weigel, D. 2003. Control of leaf morphogenesis by microRNAs. *Nature* 425:257–63.

Pang, C.H., Wang, B.S. 2008. Oxidative stress and salt tolerance in plants. In *Progress in Botany*, ed. U. Lüttge, W. Beyschlag, J. Murata, 231–245. Heidelberg Springer, Berlin.

Pareek, A., Singla, S., Grover, A. 1997. Salt responsive proteins/genes in crop plants. In *Strategies for Improving Salt in Higher Plants*. P.K. Singh, A. Gulati, Science Pub, Jaiwal, R.P.

Parihar, P., Singh, S., Singh, R., Singh, V.P., Prasad, S.M. 2015. Effect of salinity stress on plants and its tolerance strategies: a review. *Environmental Science and Pollution Research International* 22(6):4056–75.

Park, W., Li, J., Song, R., Messing, J., Chen, X. 2002. CARPEL FACTORY, a dicer homolog, and HEN1, a novel protein, act in microRNA metabolism in *Arabidopsis thaliana*. *Current Biology* 12:1484.

Patade, V.Y., Suprasanna, P. 2010. Short-term salt and PEG stresses regulate expression of microRNA, miR159 in sugarcane leaves. *Journal of Crop Science and Biotechnology* 13(3):177–82.

Paul, S., Kundu, A., Pal, A. 2011. Identification and validation of conserved microRNAs along with their differential expression in roots of *Vigna unguiculata* grown under salt stress. *Plant Cell, Tissue and Organ Culture* 105(2):233–42.

Pei, H., Ma, N., Tian, J., Luo, J., Chen, J., Li, J., Zheng, Y., Chen, X., Fei, Z., Gao, J. 2013. An NAC transcription factor controls ethylene-regulated cell expansion in flower petals. *Plant Physiology* 163(2):775–91.

Pérez-Quintero, A.L., Neme, R., Zapata, A., López, C. 2010. Plant microRNAs and their role in defense against viruses: A bioinformatics approach. *BMC Plant Biology* 10:138.

Peters, C., Li, M., Narasimhan, R., Roth, M., Welti, R., Wang, X. 2010. Nonspecific phospholipase C NPC4 promotes responses to abscisic acid and tolerance to hyperosmotic stress in *Arabidopsis*. *Plant Cell* 22(8):2642–59.

Petroni, K., Kumimoto, R.W., Gnesutta, N., Calvenzani, V., Fornari, M., Tonelli, C., Holt, B.F., Mantovani, R. 2012. The promiscuous life of plant NUCLEAR FACTOR Y transcription factors. *Plant Cell* 24(12):4777–92.

Phillips, J.R., Dalmay, T., Bartels, D. 2007. The role of small RNAs in abiotic stress. *FEBS Letters* 581:3592–97.

Pokotylo, I., Kolesnikov, Y., Kravets, V., Zachowski, A., Ruelland, E. 2014. Plant phosphoinositide-dependent phospholipases C: Variations around a canonical theme. *Biochimie* 96:144–57.

Polizel, A.M., Medri, M.E., Nakashima, K., Yamanaka, N., Farias, J.R.B., de Oliveira, M.C.N., Marin, S.R.R. 2011. Molecular, anatomical and physiological properties of a genetically modified soybean line transformed with rd29A: AtDREB1A for the improvement of drought tolerance. *Genetics and Molecular Research* 10(4):3641–56.

Preston, J.C., Hileman, L. 2013. Functional evolution in the plant SQUAMOSA-PROMOTER BINDING PROTEIN-LIKE (SPL) gene family. *Frontiers in Plant Science* 4:80.

Pruthvi, V., Narasimhan, R., Nataraja, K.N. 2014. Simultaneous expression of abiotic stress responsive transcription factors, AtDREB2A., AtHB7 and AtABF3 improves salinity and drought tolerance in peanut (*Arachis hypogaea* L). *PLoS One* 9(12):e111152.

Qiu, Q.-S., Barkla, B.J., Vera-Estrella, R., Zhu, J.-K., Schumaker, K.S. 2003. Na^+/H^+ exchange activity in the plasma membrane of *Arabidopsis*. *Plant Physiology* 132(2):1041–52.

Qiu, Q.-S., Guo, Y., Dietrich, M.A., Schumaker, K.S., Zhu, J.-K. 2002. Regulation of SOS1, a plasma membrane Na^+/H^+ exchanger in *Arabidopsis thaliana*, by SOS2 and SOS3. *Proceedings of the National Academy of Science USA* 99(12):8436–41.

Quan, R., Lin, H., Mendoza, I., Zhang, Y., Cao, W., Yang, Y., Shang, M., Chen, S., Pardo, J.M., Guo, Y. 2007. SCABP8/CBL10, a putative calcium sensor, interacts with the protein kinase SOS2 to protect *Arabidopsis* shoots from salt stress. *Plant Cell* 19(4):1415–31.

Quintero, F.J., Ohta, M., Shi, H., Zhu, J.-K., Pardo, J.M. 2002. Reconstitution in yeast of the *Arabidopsis* SOS signaling pathway for Na^+ homeostasis. *Proceedings of the National Academy of Science USA* 99(13):9061–66.

Raczynska, K.D., Stepien, A., Kierzkowski, D., Kalak, M., Bajczyk, M., McNicol, J., Simpson, C.G., Szweykowska-Kulinska, Z., Brown, J.W., Jarmolowski, A. 2014. The SERRATE protein is involved in alternative splicing in *Arabidopsis thaliana*. *Nucleic Acids Research* 42(2):1224–44.

Rajagopalan, R., Vaucheret, H., Trejo, J., Bartel, D.P. 2006. A diverse and evolutionarily fluid set of microRNAs in *Arabidopsis thaliana*. *Genes Development* 20:3407–25.

Ramachandran, V., Chen, X. 2008. Degradation of microRNAs by a family of exoribonucleases in Arabidopsis. *Science* 321:1490–2.

Ren, X., Chen, Z., Liu, Y., Zhang, H., Zhang, M., Liu, Q., Hong, X., Zhu, J.K., Gong, Z. 2010. ABO3, a WRKY transcription factor, mediates plant responses to abscisic acid and drought tolerance in *Arabidopsis*. *Plant Journal* 63(3):417–29.

Ren, G., Chen, X., Yu, B. 2012a. Uridylation of miRNAs by hen1 suppressor1 in Arabidopsis. *Current Biology* 22:695–700.

Ren, G., Xie, M., Dou, Y., Zhang, S., Zhang, C., Yu, B. 2012b. Regulation of miRNA abundance by RNA binding protein TOUGH in Arabidopsis. *Proceedings of the National Academy of Science USA* 109:12817–21.

Reyes, J.L., Chua, N.H. 2007. ABA induction of miR159 controls transcript levels of two MYB factors during Arabidopsis seed germination. *Plant Journal* 9:592–606.

Rhoades, M.W., Reinhart, B.J., Lim, L.P., Burge, C.B., Bartel, B., Bartel, D.P. 2002. Prediction of plant microRNA targets. *Cell* 110:513–20.

Rubio-Somoza, I., Weigel, D. 2011. MicroRNA networks and developmental plasticity in plants. *Trends in Plant Science* 16:258–64.

Ruiz-Ferrer, V., Voinnet, O. 2009. Roles of plant small RNAs in biotic stress responses. *Annual Review in Plant Biology* 60:485–510.

Rutschmann, F., Stalder, U., Piotrowski, M., Oecking, C., Schaller, A. 2002. LeCPK1, a calcium-dependent protein kinase from tomato. Plasma membrane targeting and biochemical characterization. *Plant Physiology* 129(1):156–68.

Saijo, Y., Hata, S., Kyozuka, J., Shimamoto, K., Izui, K. 2000. Over-expression of a single Ca^{2+}-dependent protein kinase confers both cold and salt/drought tolerance on rice plants. *Plant Journal* 23(3):319–27.

Sailaja, B., Anjum, N., Prasanth, V.V., Sarla, N., Subrahmanyam, D., Voleti, S., Viraktamath, B., Mangrauthia, S.K. 2014. Comparative study of susceptible and tolerant genotype reveals efficient recovery and root system contributes to heat stress tolerance in rice. *Plant Molecular Biology Report* 32(6):1228–40.

Sakaguchi, J., Watanabe, Y. 2012. miR165/166 and the development of land plants. *Development, Growth and Differentiation* 54(1):93–9.

Sanders, D., Pelloux, J., Brownlee, C., Harper, J.F. 2002. Calcium at the crossroads of signaling. *Plant Cell* 14:S401–17.

Schommer, C., Bresso, E.G., Spinelli, S.V., Palatnik, J.F. 2012. Role of microRNA miR319 in plant development. In: *MicroRNAs in Plant Development and Stress Responses*, ed. R. Sunkar, 29–47. Springer, Berlin.

Seki, M., Narusaka, M., Ishida, J., Nanjo, T., Fujita, M., Oono, Y., Kamiya, A., Nakajima, M., Enju, A., Sakurai, T. 2002. Monitoring the expression profiles of 7000 *Arabidopsis* genes under drought, cold and high-salinity stresses using a full-length cDNA microarray. *Plant Journal* 31(3):279–92.

Serrano, R., Mulet, J.M., Ríos, G., Márquez, J.A., Larriona, D., Leube, I.F., Mendizabal, M.P., Pascual-Ahuir, I.A., Proft, M.R.R., Montesinos, C. 1999. A glimpse of the mechanism of ion homeostasis during salt stress. *Journal of Experimental Botany* 50:1023–36.

Sharma, N., Mittal, D., Sanan-Mishra, N. 2017. Micro-Regulators of hormones and stress. In: *Mechanism of Plant Hormone Signaling Under Stress*, ed. G.K. Pandey, 319–51. Wiley

Shi, H., Ishitani, M., Kim, C., Zhu, J.-K. 2000. The *Arabidopsis thaliana* salt tolerance gene SOS1 encodes a putative Na^+/H^+ antiporter. *Proceedings of the National Academy of Science USA* 97(12):6896–6901.

Shukla, L.I., Chinnusamy, V., Sunkar, R. 2008. The role of microRNAs and other endogenous small RNAs in plant stress responses. *Biochimica et Biophysica Acta* 1779(11):743–48.

Sivakumar, P., Sharmila, P., Saradhi, P.P. 1998. Proline suppresses rubisco activity in higher plants. *Biochemistry Biophysics Research Communication* 252(2):428–432.

Smirnoff, N. 1998. Plant resistance to environmental stress. *Current Opinion in Biotechnology* 9(2):214–19.

Song, J.B., Gao, S., Sun, D., Li, H., Shu, X.X., Yang, Z.M. 2013. miR394 and LCR are involved in Arabidopsis salt and drought stress responses in an abscisic acid-dependent manner. *BMC Plant Biology* 13:210.

Sorin, C., Declerck, M., Christ, A., Blein, T., Ma, L., Lelandais-Brière, C., Njo, M.F., Beeckman, T., Crespi, M., Hartmann, C. 2014. A miR169 isoform regulates specific NF-YA targets and root architecture in Arabidopsis. *New Phytologist* 202(4):1197–1211.

Sosa-Valencia, G., Palomar, M., Covarrubias, A.A., Reyes, J.L. 2017. The legume miR1514a modulates a NAC transcription factor transcript to trigger phasiRNA formation in response to drought. *Journal of Experimental Botany* 68(8):2013–26.

Stief, A., Altmann, S., Hoffmann, K., Pant, B.D., Scheible, W.-R., Bäurle, I. 2014. *Arabidopsis* miR156 regulates tolerance to recurring environmental stress through SPL transcription factors. *Plant Cell* 26(4):1792–1807.

Su, C., Li, Z., Cheng, J., Li, L., Zhong, S., Liu, L., Zheng, Y., Zheng, B. 2017. The protein phosphatase 4 and SMEK1 complex dephosphorylates HYL1 to promote miRNA biogenesis by antagonizing the MAPK cascade in Arabidopsis. *Developmental Cell* 41(5):527–39.

Sunkar, R., Kapoor, A., Zhu, J.K. 2006. Posttranscriptional induction of two Cu/Zn superoxide dismutase genes in Arabidopsis is mediated by downregulation of miR398 and important for oxidative stress tolerance. *Plant Cell* 18:2051–65.

Sunkar, R., Zhu, J.K. 2004. Novel and stress-regulated micro-RNAs and other small RNAs from arabidopsis. *Plant Cell* 16:2001–19.

Sun, X., Xu, L., Wang, Y., Yu, R., Zhu, X., Luo, X., Gong, Y., Wang, R., Limera, C., Zhang, K. 2015. Identification of novel and salt-responsive miRNAs to explore miRNA-mediated regulatory network of salt stress response in radish (*Raphanus sativus* L.). *BMC Genomics* 16(1):197.

Syvertsen, J.P., Garcia-Sanchez, F. 2014. Multiple abiotic stresses occurring with salinity stress in citrus. *Environmental and Experimental Botany* 103:128–37.

Tähtiharju, S., Sangwan, V., Monroy, A.F., Dhindsa, R.S., Borg, M. 1997. The induction of kin genes in cold-acclimating *Arabidopsis thaliana*. Evidence of a role for calcium. *Planta* 203(4):442–47.

Tamura, T., Hara, K., Yamaguchi, Y., Koizumi, N., Sano, H. 2003. Osmotic stress tolerance of transgenic tobacco expressing a gene encoding a membrane-located receptor-like protein from tobacco plants. *Plant Physiology* 131(2):454–62.

Tripathi, A., Chacon, O., Singla-Pareek, S.L., Sopory, S.K., Sanan-Mishra, N. 2017. Mapping the microRNA expression profiles in glyoxalase over-expressing salinity tolerant rice. *Current Genomics* 19(1):21–35.

Tripathi, D.K., Singh, V.P., Gangwar, S., Prasad, S.M., Maurya, J.N., Chauhan, D.K. 2014. Role of silicon in enrichment of plant nutrients and protection from biotic and abiotic stresses. In *Improvement of Crops in the Era of Climatic Changes*, ed. P. Ahmad, M. Wani, M. Azooz, L.S. Tran, 39–56. Springer.

Tu, B., Liu, L., Xu, C., Zhai, J., Li, S., Lopez, M.A., Zhao, Y., Yu, Y., Ramachandran, V., Ren, G., Yu, B., Li, S., Meyers, B.C., Mo, B., Chen, X. 2015. Distinct and cooperative activities of HESO1 and URT1 nucleotidyl transferases in microRNA turnover in Arabidopsis. *PLoS Genetics* 11(4):e1005119.

Urao, T., Katagiri, T., Mizoguchi, T., Yamaguchi-Shinozaki, K., Hayashida, N., Shinozaki, K. 1994. Two genes that encode Ca2+-dependent protein kinases are induced by drought and high-salt stresses in *Arabidopsis thaliana*. *Molecular Genetics and Genomics* 244(4):331–40.

Valliyodan, B., Nguyen, H.T. 2006. Understanding regulatory networks and engineering for enhanced drought tolerance in plants. *Current Opinion in Plant Biology* 9(2):189–95.

Vazquez, F., Gasciolli, V., Crete, P., Vaucheret, H. 2004. The nuclear dsRNA binding protein HYL1 is required for MicroRNA accumulation and plant development, but not posttranscriptional transgene silencing. *Current Biology* 14:346–51.

Vidal, E.A., Araus, V., Lu, C., Parry, G., Green, P.J., Coruzzi, G.M., Gutiérrez, R.A. 2010. Nitrate-responsive miR393/AFB3 regulatory module controls root system architecture in *Arabidopsis thaliana*. *Proceedings of the National Academy of Science USA* 107(9):4477–82.

Wang, T., Chen, L., Zhao, M., Tian, Q., Zhang, W.-H. 2011. Identification of drought-responsive microRNAs in *Medicago truncatula* by genome-wide high-throughput sequencing. *BMC Genomics* 12(1):367.

Wang, Q., Liu, N., Yang, X., Tu, L., Zhang, X. 2016. Small RNA-mediated responses to low-and high-temperature stresses in cotton. *Scientific Reports* 6:35558.

Wang, T., Pan, H., Wang, J., Yang, W., Cheng, T., Zhang, Q. 2013. Identification and profiling of novel and conserved microRNAs during the flower opening process in *Prunus mume* via deep sequencing. *Molecular Genetics and Genomics* 289(2):169–83.

Wang, Y., Sun, F., Cao, H., Peng, H., Ni, Z., Sun, Q., Yao, Y. 2012. TamiR159 directed wheat TaGAMYB cleavage and its involvement in anther development and heat response. *PLoS One* 7(11):e48445.

Wang, B., Sun, Y.F., Song, N., Wei, J.P., Wang, X.J., Feng, H., Yin, Z.Y., Kang, Z.S. 2014. MicroRNAs involving in cold, wounding and salt stresses in *Triticum aestivum* L. *Plant Physiology and Biochemistry* 80:90–6.

Wang, R., Xu, L., Zhu, X., Zhai, L., Wang, Y., Yu, R., Gong, Y., Limera, C. and Liu, L., 2015. Transcriptome-wide characterization of novel and heat-stress-responsive microRNAs in radish (Raphanus Sativus L.) using next-generation sequencing. *Plant molecular biology reporter,* 33, pp.867–880.

Warpeha, K.M., Upadhyay, S., Yeh, J., Adamiak, J., Hawkins, S.I., Lapik, Y.R., Anderson, M.B., Kaufman, L.S. 2003. The GCR1, GPA1, PRN1, NF-Y signal chain mediates both blue light and abscisic acid responses in Arabidopsis. *Plant Physiology* 143(4):1590–1600.

White, P.J., Broadley, M.R. 2003. Calcium in plants. *Annals of Botany* 92(4):487–511.

Willmann, M.R., Poethig, R.S. 2007. Conservation and evolution of miRNA regulatory programs in plant development. *Current Opinion in Plant Biology* 10(5):503–11.

Wohlbach, D.J., Quirino, B.F., Sussman, M.R. 2008. Analysis of the Arabidopsis histidine kinase ATHK1 reveals a connection between vegetative osmotic stress sensing and seed maturation. *Plant Cell* 20(4):1101–17.

Wu, A., Allu, A.D., Garapati, P., Siddiqui, H., Dortay, H., Zanor, M.I., Asensi-Fabado, M.A., Munné-Bosch, S., Antonio, C., Tohge, T., Fernie, A.R., Kaufmann, K., Xue, G.P., Mueller-Roeber, B., Balazadeh, S. 2012. JUNGBRUNNEN1, a reactive oxygen species–responsive NAC transcription factor, regulates longevity in Arabidopsis. *Plant Cell* 24(2):482–506.

Wu, G., Park, M.Y., Conway, S.R., Wang, J.-W., Weigel, D., Poethig, R.S. 2009. The sequential action of miR156 and miR172 regulates developmental timing in *Arabidopsis. Cell* 138(4):750–59.

Wu, G., Poethig, R.S. 2006. Temporal regulation of shoot development in *Arabidopsis thaliana* by miR156 and its target SPL3. *Development* 133:3539–47.

Xia, K., Wang, R., Ou, X., Fang, Z., Tian, C., Duan, J., et al. 2012. OsTIR1 and OsAFB2 downregulation via OsmiR393 overexpression leads to more tillers, early flowering and less tolerance to salt and drought in rice. *PLoS One* 7:e30039.

Xie, F., Jones, D.C., Wang, Q., Sun, R., Zhang, B. 2015. Small RNA sequencing identifies miRNA roles in ovule and fibre development. *Plant Biotechnology Journal* 13(3):355–69.

Xie, F., Stewart, C.N. Jr, Taki, F.A., He, Q., Liu, H., Zhang, B. 2014. High-throughput deep sequencing shows that micro RNA s play important roles in switchgrass responses to drought and salinity stress. *Plant Biotechnology Journal* 12(3):354–66.

Xing, D.-H., Lai, Z.-B., Zheng, Z.-Y., Vinod, K., Fan, B.-F., Chen, Z.-X. 2008. Stress-and pathogen-induced *Arabidopsis* WRKY48 is a transcriptional activator that represses plant basal defense. *Molecular Plant* 1(3):459–70.

Xin, M., Wang, Y., Yao, Y., Xie, C., Peng, H., Ni, Z., Sun, Q. 2010. Diverse set of microRNAs are responsive to powdery mildew infection and heat stress in wheat (*Triticum aestivum* l.). *BMC Plant Biology* 10(1):123.

Xiong, L., Wang, R.-G., Mao, G., Koczan, J.M. 2006. Identification of drought tolerance determinants by genetic analysis of root response to drought stress and abscisic acid. *Plant Physiology* 142(3):1065–74.

Xu, M.Y., Zhang, L., Li, W.W., Hu, X.L., Wang, M.B., Fan, Y.L., Zhang, C.Y., Wang, L. 2014. Stress-induced early flowering is mediated by miR169 in *Arabidopsis thaliana. Journal of Experimental Botany* 65(1):89–101.

Yamaguchi-Shinozaki, K., Shinozaki, K. 2006. Transcriptional regulatory networks in cellular responses and tolerance to dehydration and cold stresses. *Annual Review in Plant Biology* 57:781–803.

Yang, Y., He, M., Zhu, Z., Li, S., Xu, Y., Zhang, C., Singer, S.D., Wang, Y. 2012. Identification of the dehydrin gene family from grapevine species and analysis of their responsiveness to various forms of abiotic and biotic stress. *BMC Plant Biology* 12(1):140.

Yang, L., Liu, Z., Lu, F., Dong, A., Huang, H. 2006. SERRATE is a novel nuclear regulator in primary microRNA processing in Arabidopsis. *Plant Journal* 47:841–50.

Yang, S.D., Seo, P.J., Yoon, H.K., Park, C.M. 2011. The Arabidopsis NAC transcription factor VNI2 integrates abscisic acid signals into leaf senescence via the COR/RD genes. *Plant Cell* 23(6):2155–68.

Yang, H., Song, J., Yue, L., Mo, X., Song, J., Mo, B. 2017. Identification and expression profiling of *Oryza sativa* nucleotidyl transferase protein (NTP) genes under various stress conditions. *Gene* 628:93–102.

Yang, X.C., Sullivan, K.D., Marzluff, W.F., Dominski, Z. 2009. Studies of the 5' exonuclease and endonuclease activities of CPSF-73 in histone pre-mRNA processing. *Molecular Cell Biology* 29:31–42.

Yang, F., Yu, D. 2009. Overexpression of *Arabidopsis* MiR396 enhances drought tolerance in transgenic tobacco plants. *Acta Botanica Yunnanica* 31(5):421–26.

Yan, S., Tang, Z., Su, W., Sun, W. 2005. Proteomic analysis of salt stress-responsive proteins in rice root. *Proteomics* 5:235–44.

Yan, J., Wang, P., Wang, B., Hsu, C.C., Tang, K., Zhang, H., Hou, Y.J., Zhao, Y., Wang, Q., Zhao, C., Zhu, X., Tao, W.A., Li, J., Zhu, J.K. 2017. The SnRK2 kinases modulate miRNA accumulation in Arabidopsis. *PLoS Genetics* 139(4):e1006753.

Yan, J., Zhao, C., Zhou, J., Yang, Y., Wang, P., Zhu, X., Tang, G., Bressan, R.A., Zhu, J.K. 2016. The miR165/166 mediated regulatory module plays critical roles in ABA homeostasis and response in *Arabidopsis thaliana. PLoS Genetics* 12(11):e1006416.

Yeo, A. 1998. Molecular biology of salt tolerance in the context of whole-plant physiology. *Journal of Experimental Botany* 49(323):915–29.

Yuan, S., Li, Z., Li, D., Yuan, N., Hu, Q., Luo, H. 2015. Constitutive expression of rice microRNA528 alters plant development and enhances tolerance to salinity stress and nitrogen starvation in creeping bentgrass. *Plant Physiology* 169(1):576–93.

Yu, B., Bi, L., Zheng, B., Ji, L., Chevalier, D., Agarwal, M., Ramachandran, V., Li, W., Lagrange, T., Walker, J.C., Chen, X. 2008. The FHA domain proteins DAWDLE in Arabidopsis and SNIP1 in humans act in small RNA biogenesis. *Proceedings of the National Academy of Science USA* 105:10073–78.

Yu, X., Wang, H., Lu, Y.Z., de Ruiter, M., Cariaso, M., Prins, M., Van, T.A., He, Y.K. 2012. Identification of conserved and novel microRNAs that are responsive to heat stress in *Brassica rapa*. *Journal of Experimental Botany* 63(2):1025–38.

Yu, B., Yang, Z., Li, J., Minakhina, S., Yang, M., Padgett, R.W., Steward, R., Chen, X. 2005. Methylation as a crucial step in plant microRNA biogenesis. *Science* 307:932–5.

Zhang, B. 2015. MicroRNA: A new target for improving plant tolerance to abiotic stress. *Journal of Experimental Botany* 66(7):1749–61.

Zhang, J., Jia, W., Yang, J., Ismail, A.M. 2006. Role of ABA in integrating plant responses to drought and salt stresses. *Field Crops Research* 97(1):111–19.

Zhang, M., Smith, J.A.C., Harberd, N.P., Jiang, C. 2016. The regulatory roles of ethylene and reactive oxygen species (ROS) in plant salt stress responses. *Plant Molecular Biology* 91(6):651–59.

Zhang, X., Xia, J., Lii, Y., Barrera-Figueroa, B., Zhou, X., Gao, S., Lu, L., Niu, D., Liang, W., Chen, Z., Leung, C., Wong, T., Zhang, H., Guo, J., Li, Y., Liu, R., Zhu, J.K., Zhang, W., Jin, H. 2012. Genome-wide analysis of plant nat-siRNAs reveals insights into their distribution, biogenesis and function. *Genome Biology* 13:R20.

Zhao, B., Ge, L., Liang, R., Li, W., Ruan, K., Lin, H., Jin, Y. 2009. Members of miR-169 family are induced by high salinity and transiently inhibit the NF-YA transcription factor. *BMC Molecular Biology* 10:29.

Zhao, J., He, Q., Chen, G., Wang, L., Jin, B. 2016. Regulation of non-coding RNAs in heat stress responses of plants. *Frontiers in Plant Science* 7:1213.

Zheng, L., Liu, G., Meng, X., Liu, Y., Ji, X., Li, Y., Nie, X., Wang, Y. 2013. A WRKY gene from *Tamarix hispida*, ThWRKY4, mediates abiotic stress responses by modulating reactive oxygen species and expression of stress-responsive genes. *Plant Molecular Biology* 82(4–5):303–20.

Zheng, X.-Y., Spivey, N.W., Zeng, W., Liu, P.-P., Fu, Z.Q., Klessig, D.F., He, S.Y., Dong, X. 2012. Coronatine promotes *Pseudomonas syringae* virulence in plants by activating a signaling cascade that inhibits salicylic acid accumulation. *Cell Host and Microbe* 11(6):587–96.

Zhou, Y., Honda, M., Zhu, H., Zhang, Z., Guo, X., Li, T., Li, Z., Peng, X., Nakajima, K., Duan, L., Zhang, X. 2015. Spatiotemporal sequestration of miR165/166 by Arabidopsis Argonaute10 promotes shoot apical meristem maintenance. *Cell Reports* 10(11):1819–27.

Zhou, M., Li, D., Li, Z., Hu, Q., Yang, C., Zhu, L., Luo, H. 2013. Constitutive expression of a miR319 gene alters plant development and enhances salt and drought tolerance in transgenic creeping bentgrass. *Plant Physiology* 161(3):1375–91.

Zhou, J., Liu, M., Jiang, J., Qiao, G., Lin, S., Li, H., Xie, L., Zhuo, R. 2012. Expression profile of miRNAs in *Populus cathayana* L. and *Salix matsudana* Koidz under salt stress. *Molecular Biology Reports* 39(9):8645–54.

Zhou, L., Liu, Y., Liu, Z., Kong, D., Duan, M., Luo, L. 2010. Genome-wide identification and analysis of drought-responsive microRNAs in Oryza sativa. *Journal of Experimental Botany* 61(15):4157–68.

Zhou, M., Luo, H. 2014. Role of microRNA319 in creeping bentgrass salinity and drought stress response. *Plant Signaling and Behaviour* 9(4):1375–91.

Zhou, X., Ruan, J., Wang, G., Zhang, W. 2007. Characterization and identification of microRNA core promoters in four model species. *PLoS Computational Biology* 3:e37.

Zhu, J.-K. 2002. Salt and drought stress signal transduction in plants. *Annual Review in Plant Biology* 53(1):247–73.

Zhu, X., Leng, X., Sun, X., Mu, Q., Wang, B., Li, X., Wang, C., Fang, J. 2015. Discovery of conservation and diversification of miR171 genes by phylogenetic analysis based on global genomes. *Plant Genome* 8(2):1–11.

Zhu, Y., Song, D., Sun, J., Wang, X., Li, L. 2013. PtrHB7, a class III HD-Zip gene, plays a critical role in regulation of vascular cambium differentiation in Populus. *Molecular Plant* 6(4):1331–43.

5 MicroRNA-Mediated Strategies for Conferring Biotic Stress Tolerance in Plants

Priyanka Sharma[1]*, Shivaraj S.M.*[2]*, Varsha Arun Mahadik*[2,3]*, and Narendra Kadoo*[2,3]
[1]School of Bioengineering Sciences and Research, MIT-ADT University, Pune, India
[2]Biochemical Sciences Division, CSIR-National Chemical Laboratory, Pune, India
[3]Academy of Scientific and Innovative Research (AcSIR), Ghaziabad, India

5.1 INTRODUCTION

A class of small non-coding RNAs (~21 nt), called microRNAs (miRNAs), regulate target gene expression at the transcriptional and post-transcriptional levels (Bartel 2004; Kim et al. 2008; Khraiwesh et al. 2010). At the transcriptional level, they can regulate genes through methylation and histone modification (Chuang and Jones 2007; Chellappan et al. 2010; Wu et al. 2010; Singh and Campbell 2013), while at post-transcriptional level, they can regulate gene expression by directing transcript cleavage, translation repression, or both (Jones-Rhoades et al. 2006; Rogers and Chen 2013). miRNA-mediated transcript silencing depends on the extent of the base-pairing of the miRNA with the target mRNA and involves one or more of the following processes: (i) mRNA cleavage, (ii) mRNA destabilization by shortening its poly(A) tail, and (iii) suppression of translation of the mRNA (Bartel 2009). miRNA-mediated transcript silencing is more common in plants than in animals (Carrington and Ambros 2003). A single miRNA might regulate the expression of several target mRNAs, while multiple miRNAs could also regulate the expression of a single mRNA transcript (Brennecke et al. 2005). Lee et al. (1993) first discovered the miRNA *lin-4* in *Caenorhabditis elegans* via a forward genetics approach by screening mutants displaying temporal developmental defects. Later, it was confirmed that the 22 nt *lin-4* miRNA had partial complementarity to the 3′ untranslated region (UTR) of the *lin-14* mRNA, encoding a protein involved in larval development (Wightman et al. 1993). Translation inhibition of *lin-14* mRNA was confirmed when the *lin-4* miRNA caused a reduction in the levels of the LIN-14 protein without changing *lin-14* mRNA levels.

After their initial discovery, miRNAs have been identified in several plants, animals, and viruses. Based on their different characteristics, miRNAs are reported to have evolved independently in plants and animals (Millar and Waterhouse 2005). However, inter-kingdom conservation of some miRNAs has been shown. For example, Arteaga-Vazquez et al. (2006) showed that plants and animals share miRNAs of the miR854 family, which might suggest their common origin as regulators of basal transcriptional mechanisms. miRNAs regulate diverse biological processes such as growth and development, auxin signaling, reproductive development, biotic and abiotic stress tolerance, etc. (Sun 2012; Shriram et al. 2016; Brant and Budak, 2018). These mechanisms enable plants to regulate their growth and development under varied climatic conditions. miRNAs have been extensively studied in *Arabidopsis thaliana* and more than 160 miRNA genes have been identified. Plant miRNAs are synthesized from precursor miRNAs (pre-miRNAs), which have specific stem

DOI: 10.1201/9781003369288-5

regions, one or more single-stranded hairpins, and possess unique structures that distinguish miRNAs from other types of small RNAs. Several experimental and computational approaches have been employed for discovering miRNAs (Mathur et al. 2020). In this chapter, we discuss miRNA biogenesis in plants, methods of their identification and validation, as well as miRNA-based strategies for crop improvement.

5.1.1 miRNA Biogenesis in Plants

miRNAs are classified as either "intergenic" or "intronic" according to their locations in the genome. Intergenic miRNAs are located between two protein-coding genes, have their own promoters, and are transcribed as independent units by DNA-dependent RNA Polymerase II (Pol II). Intronic miRNAs are processed from the introns of their host transcripts (Millar and Waterhouse 2005; Budak and Akpinar 2015). RNA Pol II transcribes the MIR genes into primary transcripts (pri-miRNAs), which are 5′ capped, 3′ polyadenylated, and/or spliced (Xie et al. 2005; Rogers and Chen 2013). The stem-loops of pri-miRNAs are variable in length (from 60 nt to over 500 nt) and bear complex structures (Xie et al. 2005; Bologna and Voinnet 2014). As such, pri-miRNAs in plants can be processed from either the loop-distal site to the loop-proximal site or *vice versa*, primarily by Dicer-like RNase III endonuclease 1 (DCL1) to generate pre-miRNAs (Addo-Quaye et al. 2009; Bologna et al. 2009, 2013; Mateos et al. 2010; Song et al. 2010; Werner et al. 2010). The pre-miRNAs consist of a terminal loop, an upper stem, the miRNA/miRNA* region, a lower stem, and two arms, which can be recognized and processed by DCLs. The nascent miRNA/miRNA* duplex generated from pre-miRNA by DCL-mediated processing exhibits 2-nt 3′ overhangs at both strands, and each strand possesses a 5′ end phosphate and two 3′ end hydroxyl groups (2′ OH and 3′ OH). While both hydroxyl groups are essential, only the 2′-OH position is methylated by the small RNA methyltransferase HUA Enhancer 1 (HEN1) (Yu et al. 2005; Yang et al. 2006). miRNA/miRNA* duplex is transported to the cell cytoplasm via the nuclear pore complex with the help of HST1/HASTY protein.

In the cytoplasm, miRNA is loaded into the RNA-induced silencing complex (RISC), and miRNA* is degraded (Khvorova et al. 2003; Schwarz et al. 2003). The RISC complex contains the ARGONAUTE 1 (AGO1) protein, which has two domains: the N-terminal PAZ domain, which can bind to the 3′ end of the single-stranded RNA molecule, and the C-terminal PIWI domain that is structurally similar to RNAse H, which is responsible for the endonuclease activity of the RISC complex (Ma et al. 2004; Parker et al. 2004; Song et al. 2004). The RISC complex binds to the complementary sequence of target mRNA and causes its cleavage when perfect complementarity is present. However, if only a partial complementary sequence is present, it leads to the inhibition of translation. miRNA biogenesis and miRNA-mediated regulation of genes have been illustrated in Figure 5.1.

5.1.2 Role of miRNAs in Plants

Several studies have shown that miRNAs play crucial regulatory roles in plants, not only in growth and development, but also in other biological processes such as the maintenance of genome integrity, signal transduction, hormone signaling, hormone homeostasis, innate immunity, and responses to various biotic and abiotic stresses (Navarro et al. 2006; Sun 2012; Xie et al. 2015; Zhang and Wang, 2015). Next, we discuss the reported roles of identified miRNAs in various plants.

5.1.2.1 Plant Development

miRNAs play a major role in plant development. They regulate the expression of the highly conserved transcription factors as well as the regulatory genes involved in plant development. miRNAs also regulate plant metabolism, photosynthesis, root development, shoot development, fruit

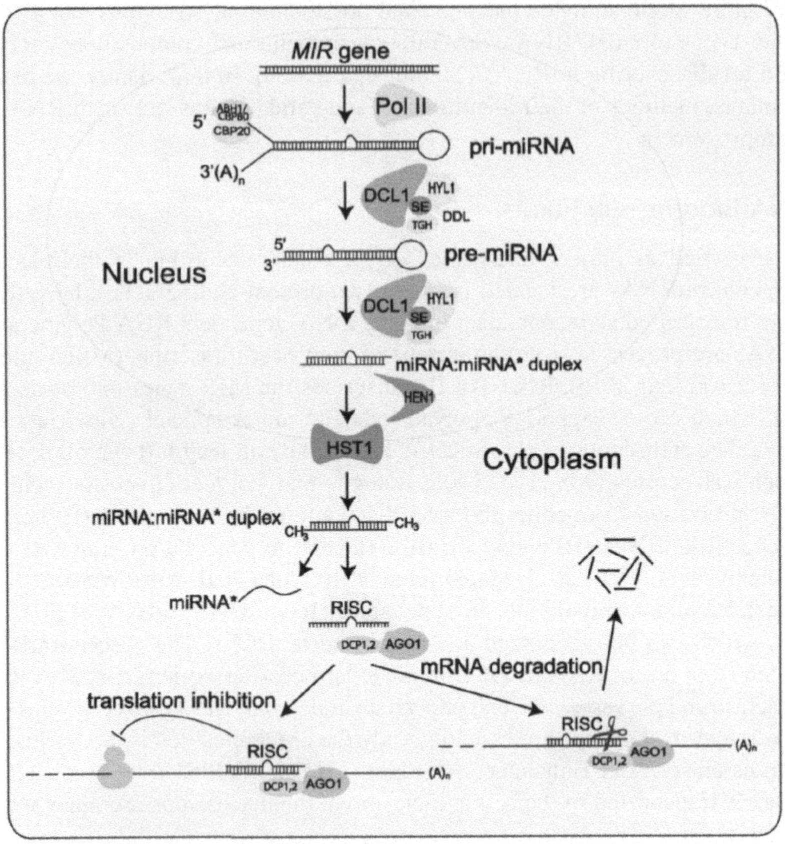

FIGURE 5.1 miRNA biogenesis and regulation of gene expression in plants.

Source: Yang and Li (2012); Creative Commons Attribution license (http://creativecommons.org/licenses/by/3.0/).

morphology, and fruit development. MiR171 is an important regulatory miRNA thought to be present in all lineages of land plants. In tomato, it regulates shoot branching, leaf morphogenesis, and development of male floral organs (Kravchik et al. 2019). Similarly, miR160 in orange targets the auxin response factor 10 (*ARF10*) gene, which is ultimately responsible for root development and is also involved in stress response (Song et al. 2012). Likewise, miR156 of the grapevine is important for fruit development as it targets the *SQUAMOSA-PROMOTER BINDING-LIKE* protein (Pantaleo et al. 2010). In apples, plant growth and development are enhanced by miR156, miR159, miR166, and miR167 by binding to various transcription factors.

5.1.2.2 Nutrient Availability

miRNAs also respond to changes in nutrient availability. miR397, miR398, miR408, and miR857 are the highly conserved plant miRNAs and are known to respond to changes in copper (Cu) levels. Shahbaz and Pilon (2019) created a short tandem target mimic (STTM) that inhibited the expression of miR397, miR398, and miR408 simultaneously, causing higher levels of their target mRNAs under Cu-limiting conditions. These targets were Cu-containing proteins, and failure to repress their expression under Cu-limiting conditions led to reduced levels of an unrelated Cu-containing protein, plastocyanin, in the STTM transgenic plant lines. As plastocyanin is an important component of photosynthetic machinery, reduction in its levels decreased the photosynthetic electron transport activity of the STTM lines under Cu-limiting conditions, finally resulting in a decrease in plant biomass.

Phosphorous is one of the plant-growth limiting nutrients worldwide. Regulation of the PHOSPHATE2 (PHO2) gene was investigated by Pegler et al. (2019). They found that miR399 regulated PHO2 and double-stranded RNA binding (DRB), by both mRNA cleavage and translation repression.

5.1.2.3 Abiotic and Biotic Stress Resistance

Exposure to abiotic stresses such as drought, salinity, cold, heavy metal toxicity, oxidative stress, UV radiations, etc., causes plants to synthesize or express various molecules, including miRNAs. Similarly, specific miRNAs are also expressed during biotic stresses such as infection of viruses, bacteria, fungi, etc. Plants show defense responses by synthesizing various miRNAs, which either target pathogen virulence genes, or activate plant defense genes. Several studies regarding the role of miRNAs in biotic and abiotic stress tolerance are discussed later in the chapter.

5.2 miRNA IDENTIFICATION AND CHARACTERIZATION

Identification of miRNAs is the primary step toward understanding their functions. Several experimental and computational methods have been employed for the discovery or identification of miRNAs. Initially, screening of mutant phenotypes and cloning were the most prevalent methods to discover miRNAs. Subsequently, based on the characteristic features of the identified miRNAs, bioinformatic tools were developed. With the advent of high-throughput sequencing technologies, large amounts of sequence data are being deposited in public databases, which can be analyzed using these bioinformatics tools for rapid identification of miRNAs from different plant species. The methods used to identify and validate miRNAs are illustrated in Figure 5.2.

5.2.1 Methods for miRNA Identification in Plants

5.2.1.1 Genetic Screening of Mutants

Screening of mutants is a forward genetics approach that involves identifying the gene(s) governing a particular phenotype. The first miRNA, *lin-4*, was identified through the classic forward genetics approach in *C. elegans* (Lee et al. 1993). However, in plants, very few miRNAs have been identified using this approach, owing to the existence of miRNAs in families and the functional redundancy

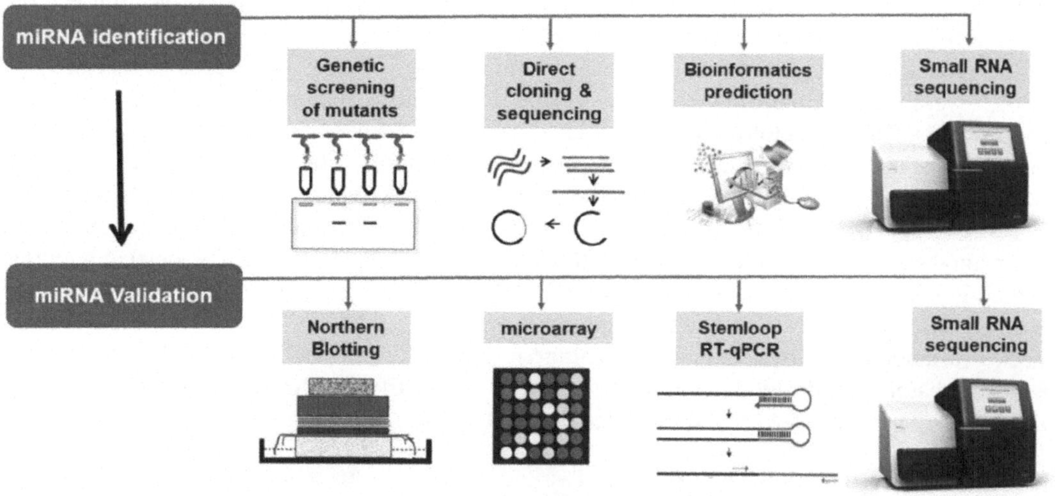

FIGURE 5.2 The methods used for identification and validation of miRNAs.

among the family members. Aukerman and Sakai (2003) screened mutants developed through the activation tagging approach and identified miRNA172-a, which governs flower development in *Arabidopsis*. Similarly, in another study, the screening of *Arabidopsis* mutants led to the identification of the *JAGGED and WAVY* (*JAW*) locus encoding miR-JAW, which mediates cleavage of the transcription factor gene, *TCP* (*TEOSINTE BRANCHED1*, *CYCLOIDEA*, and *PCF*), involved in the regulation of leaf development (Palatnik et al. 2003). Although successful, this approach is expensive and time-consuming compared to other approaches.

5.2.1.2 Cloning and Sequencing

This method involves cloning and sequencing small RNAs to identify miRNAs. This has been used as a method of choice to identify miRNAs even in plant species with no or little genomic information. This method facilitated the identification of tissue- and condition-specific miRNAs from several plant species. Many studies employed this approach to identify miRNAs in several plants including *Arabidopsis*, rice, and wheat (Sunkar and Zhu 2004; Sunkar et al. 2005; Yao et al. 2007). This method involves:

a. Isolation of total RNA from the target tissue
b. Separation of small RNA from total RNA using gel electrophoresis, column-based filtration, or other methods
c. Ligation of adapters to small RNA
d. Reverse transcription
e. Cloning the cDNA library
f. Sequencing the library (typically by Sanger sequencing) and data analysis to identify miRNAs.

While this method can identify novel miRNAs, only a limited set of the miRNAs present in an organism can be identified. This is because not all miRNAs will express in all tissues, growth stages, time-points, or experimental conditions, used for RNA isolation. Secondly, this method is likely to miss the miRNAs that are less abundant or that are very short-lived.

5.2.1.3 NGS-based miRNA Identification

The next-generation sequencing (NGS)-based high-throughput platforms such as Ion Torrent and Illumina can be used to efficiently sequence and quantify small RNAs. The NGS data are analyzed using bioinformatics tools to identify miRNAs. Thus, this method combines both the experimental and bioinformatic approaches of miRNA identification (Zhou et al. 2020). As cloning and Sanger sequencing are not involved, this method is more rapid than cloning and sequencing, and is more accurate than the pure bioinformatics approach. This method can capture even the less-abundant miRNAs, which is not possible with classical cloning and sequencing approaches (Banerjee et al. 2016). Several bioinformatics pipelines have been developed to analyze the NGS data and identify miRNAs. These pipelines require reference genome sequences to map the reads to identify potential precursors encoding miRNAs. The bioinformatics tools developed to analyze small RNA sequence data include miRDeep (Friedlander et al. 2008), miRExpress (Wang et al. 2009), miRA (Evers et al. 2015), and miPIE (Peace et al. 2019), among others. In addition to detecting miRNAs, this method can also be used to quantify miRNA levels. However, a limitation associated with this method is that it is less efficient in identifying miRNAs from non-model plants. For example, the small RNA sequencing of closely related species *A. thaliana*, and *Arabidopsis lyrata* showed inter-species variation in miRNA composition (Fahlgren et al. 2010; Chavez Montes et al. 2014). Hence, homology-based identification might not be sufficient to identify miRNAs in one species based on the homology with miRNAs found in a related species.

5.2.1.4 Bioinformatics-based Prediction

This is a highly efficient and cost-effective method, and has been widely used to identify miRNAs in diverse organisms, including plants. With the increasing amount of sequencing data being deposited in publicly accessible databases, this approach has become popular to analyze the data to identify miRNAs, as well as determine their expression levels in various tissues, in several organisms including plants. The principle behind this method is that most of the miRNAs are evolutionarily conserved, and they display sequence and structural similarity among related species (Lagos-Quintana et al. 2001; Lee and Ambros 2001; Adai et al. 2005). The conserved miRNAs are identified in different plants using homology analysis based on the Basic Local Alignment Search Tool (BLAST) using known miRNAs as a query from the genome sequence surveys (GSS), expressed sequence tags (ESTs), and whole-genome sequences (Zhang et al. 2009; Katiyar et al. 2012; Usha et al. 2017). Other than the homology search, several programs such as MIRFINDER (Bonnet et al. 2004), mirCHECK (Jones-Rhoades and Bartel 2004), and micro-HARVESTER (Dezulian et al. 2006) have been developed to identify the miRNAs from plants. These methods predict sequences with similar RNA secondary structures to known miRNAs based on the features such as folding energy of the hairpin structure, number of paired bases in the miR:miR* duplex, sequence composition, and entropy. Computational prediction can reveal the entire repertoire of miRNAs in an organism's genome (Mathur et al. 2020), as the identification of putative miRNAs is not dependent on their expression. The miRNAs identified through bioinformatics tools can be validated by experimental techniques. This method was used to identify miRNAs from wheat (Han et al. 2013), cotton (Zhang et al. 2007), flax (Barvkar et al. 2013), potato (Xie et al. 2011), and brassicas (Shivaraj et al. 2014), among others. However, miRNA identification using the bioinformatics approach has its limitations. This approach cannot identify novel miRNAs, and most of the programs available for miRNA prediction may provide many false-positive results, which need further experimental validation (Bentwich 2005; Tan Gana et al. 2012).

5.2.2 Methods of miRNA Validation

After miRNAs are identified, they need to be validated for their expression, which will aid in predicting their function. Several methods have been developed for miRNA validation. These methods can be grouped into hybridization-based or PCR-based methods. The hybridization-based methods include Northern blotting analysis and microarrays, while PCR-based methods include qRT-PCR and small RNA sequencing (Ouyang et al. 2019). Small RNA sequencing is used for both identification as well as quantification of miRNA levels in the species of interest.

5.2.2.1 Northern Blotting

Northern blotting is a classical technique used for expression analysis of genes by measuring the abundance of RNA (Lagos-Quintana et al. 2001; Bagga et al. 2005). Later, this technique was adopted for the detection of miRNAs. miRNAs are separated using denaturing polyacrylamide gel electrophoresis. Subsequently, these molecules are transferred to a nylon membrane, immobilized, and hybridized with the radiolabeling oligonucleotide probe. The sequences are not subjected to any kind of amplification during the whole process, which may introduce modifications. Hence, it remains a more accurate technique (Ouyang et al. 2019). However, although Northern blotting can detect target sequences accurately, it has limitations such as cumbersome procedure, need for optimization, need for a high amount of sample for analysis, and low sensitivity (Cheng et al. 2018; Ouyang et al. 2019). Many modifications have been proposed in Northern blotting to circumvent the limitations associated with miRNA detection. The modification in probes and the crosslinking of the immobilized RNA to the membrane played an important role in improving the specificity and sensitivity of Northern blotting. The modified probes based on locked nucleic acid (LNA) improved the sensitivity by 10-fold compared to traditional DNA probes. The traditional UV-based crosslinking of RNA to the nylon membrane is effective for the long RNA molecules (~70 nt); however, it is unsuitable for small RNAs. Hence 1-ethyl-3-(3-dimethylaminopropyl) carbodiimide (EDC) was proposed for crosslinking of miRNA, which increased the sensitivity by 25–50-fold (Valoczi et al. 2004). Furthermore, the combined use of LNA probes labeled with digoxigenin (DIG) and EDC-mediated crosslinking of miRNA to the nylon membrane achieved high sensitivity of miRNA detection even with the RNA amounts as low as 0.05 fmol (Kim et al. 2010). Recently, the Northern blotting technique based on the mechanism of hybridization chain reaction (HCR) has been developed. It combines the isothermal nucleic acid amplification method with multiplex detection of all targets (Schwarzkopf and Pierce 2016). The authors demonstrated simultaneous detection of three endogenous miRNAs in the same RNA sample, enabling relative and absolute miRNA quantification.

5.2.2.2 Microarrays

The microarray is a high-throughput technique that can analyze the expression of thousands of miRNAs simultaneously. miRNA-based microarray analysis can quantify miRNA levels between species, or between different tissues of the same species (Yin et al. 2008; Pritchard et al. 2012). It involves anchoring known DNA sequences on a solid surface like a glass or nylon membrane and interacting with the miRNAs isolated from the target tissue/species following the detection of the signal of hybridization between target miRNA and complementary probe. However, several limitations associated with this technique reduced its popularity. The limitations are it is expensive, less sensitive, the short sequence of the small RNAs affects labeling, it cannot identify novel miRNAs, and often produces false-positive signals (Dong et al. 2013). Several improved detection methods have been adopted to enhance the sensitivity of miRNA microarray, which includes chemiluminescent, electrochemical, and nanoparticle-based detection (Nelson et al. 2004; Fang et al. 2006; Fan et al. 2007; Cissell et al. 2008). The spherical nucleic acid (SNA) gold nanoparticle-coupled method called Scanometric MicroRNA (Scano-miR) array profiling was developed, which increased the sensitivity of miRNA detection. It can detect miRNAs at levels as low as 30 fmol (Alhasan et al. 2012). Another modified version of an miRNA microarray called ligase-assisted sandwich hybridization (LASH) was developed, which involves a label-free and amplification-free miRNA assay to increase the sensitivity of miRNA detection (Ueno and Funatsu 2014).

5.2.2.3 qRT-PCR

The quantitative reverse transcription-polymerase chain reaction (qRT-PCR, or RT-qPCR) is a highly sensitive technique used for quantification of miRNA expression, involving PCR-based amplification of the target sequence. qRT-PCR can successfully detect even the rare miRNAs that could not be detected by techniques such as Northern hybridization (Lim et al. 2003) and microarray analysis (Liu et al. 2004). The small size of the miRNA posed a significant challenge while designing the probe or primer for detection using the PCR. Chen et al. (2005) used a stem-loop probe for PCR miRNA detection, which is suitable for analyzing small RNA molecules. In this procedure, the stem-loop probe initially binds to the target miRNA and converts it into cDNA by reverse transcriptase. Subsequently, PCR is performed using the stem-loop probe attached to cDNA as a template and the TaqMan probe for real-time fluorescence detection. It is a more sensitive, accurate, and automated technique and is feasible to carry out using the existing real-time PCR instruments (Pritchard et al. 2012). Other than stem-loop PCR, several PCR-based techniques have been developed to detect miRNAs, including the ligation-based PCR assay (Zhang et al. 2011a), multiplex ligation-dependent probe amplification (MLPA) technique (Zhang et al. 2013), and the droplet digital PCR (ddPCR)-based miRNA detection (Hindson et al. 2013).

5.3 STRESSES AFFECTING miRNA EXPRESSION IN PLANTS

Plants are exposed to various biotic and abiotic factors in the environment in which they grow. Abiotic stresses such as high salinity, drought, heat, and cold, and biotic factors such as insects, pests, and microbial pathogens significantly alter the growth and development of crop plants resulting in reduced or no yield (Mittler 2006). These factors produce such results by modifying the cellular, biochemical, physiological, and molecular processes in plants (Fernandez 2014; Mathur et al. 2014; Zhang 2015). However, plants have evolved sophisticated mechanisms to deal with various stresses. Many studies have shown differential expression of specific genes in plants

TABLE 5.1
Plant miRNA Families Affected by Different Stresses

Environmental Stress	Plant	miRNA Families	Target Genes	Reference
Drought	*Arabidopsis*	miR393	Auxin receptor encoding genes, *TIR1,* and *AFB2*	Chen et al. 2012
	Rice	miR393	Auxin receptor encoding genes, *TIR1,* and *AFB2*	Xia et al. 2012
	Chickpea	miR408	Plantacyanin repression, which regulates the Drought-Responsive Element Binding Protein (DREB)	Hajyzadeh et al. 2015
	Tomato	miR169	Nuclear factor Y subunit genes (SlNF-YA1/2/3) and multidrug resistance-associated protein gene (SlMRP1)	Zhang et al. 2011b
	Tomato and tobacco	miR1916	Histone deacetylases (HDAC) and strictosidine synthase (STR)	Chen et al. 2019
Salinity	*Arabidopsis*	miR399f	*ABF3* and *CSP41b*	Baek et al. 2016
	Cotton	miR414c	Iron superoxide dismutase gene, *GhFSD1*	Wang et al. 2019
	Rice	miR164	*NAC2*	Jiang et al. 2019
	Tomato	miR398	Superoxide dismutase encoding gene, *CSD*	He et al. 2021
Heat stress	*Arabidopsis*	miR156	*SQUAMOSA-PROMOTER BINDING-LIKE (SPL)* transcription factors	Stief et al. 2014
	Wheat	miR159	*GAMYB*-like family transcription factors	Wang et al. 2012
	Alfalfa	miR156	*SQUAMOSA-PROMOTER BINDING-LIKE (SPL)* transcription factors	Matthews et al. 2019
Cold stress	*Arabidopsis*	miR397	*CBFs* (*CRT/DRE binding factor*) transcription factors	Dong and Pei 2014
	Arabidopsis	miR394	*LEAF CURLING RESPONSIVENESS (LCR)*	Song et al. 2016
	Rice	miR319	*TEOSINTE BRANCHED/CYCLOIDEA/PCF (TCP)* genes, *PCF5,* and *PCF8*	Yang et al. 2013
	Rice	miR319	*TEOSINTE BRANCHED/CYCLOIDEA/PCF (TCP)* genes, *PCF6,* and *TCP21*	Wang et al. 2014

exposed to various biotic and abiotic stresses. Many important genes, including transcription factors, have been functionally characterized in different crop plants. Overexpression of these genes has shown improved tolerance to stresses. After miRNA discovery, further identification and characterization of miRNAs in plants provided evidence for their role in regulating stress-responsive genes (Fujii et al. 2005; Zhao et al. 2007; Zhou et al. 2008). The first such evidence was obtained from a study in *Arabidopsis*, where the miRNAs cloned from the stressed plants were found to be different from those reported earlier from unstressed plants (Bao et al. 2004; Wu et al. 2010). miRNAs control target genes by mediating mRNA cleavage, translational repression, and DNA methylation. In general, miRNAs induced by stress will downregulate their target mRNAs, while their downregulation causes increased levels of the positive regulators (Chinnusamy et al. 2007). The details of studies showing the role of miRNAs in various abiotic stresses are provided in Table 5.1. A detailed description of the response of miRNAs to biotic stresses is discussed in the following section.

5.4 POTENTIAL OF miRNAs IN RESPONSE TO BIOTIC STRESSES

Over the past decade, small non-coding RNAs, particularly miRNAs, emerged as vital and highly conserved gene regulators for biotic stress tolerance. Recent literature reveals that miRNAs can enhance key agronomic properties of crops by altering the levels of target mRNA transcripts and differential gene regulation. These miRNAs regulate gene expression via cleavage/repression/silencing of the target mRNAs and function at the forefront in conserving genome integrity and metabolism in response to various stresses. Abiotic stresses are also known to affect the spread, occurrence of pathogens, and their interactions with plants.

Biotic stresses caused by various plant pathogens such as bacteria, fungi, viruses, insects, parasites, nematodes, etc., are the major constraints of crop production worldwide and account for nearly 30% yield losses (Bebber and Gurr 2015; Das and Rakshit 2016; Madani et al. 2019). Plants, being sessile, have specific adaptive defense mechanisms to combat such stresses through processes such as induction, differential regulation, and inhibition/repression of gene expression (Foyer et al. 2016). Current progress in plant miRNA research has also confirmed their functions in crop responses to biotic stresses. Many researchers found that plants, at the time of microbial infection, respond via primary and secondary defense mechanisms. The primary mechanism includes detection of microbe-associated molecular patterns (MAMPs), while the secondary mechanisms include hypersensitivity. These responses are recognized at the plasma membrane, where the pattern recognition receptors (PRR) activate pattern-triggered immunity (PTI) in plants (Macho and Zipfel 2015; Trda et al. 2015), and enhance the response by regulating small non-coding RNAs in provision of resistance (Djami-Tchatchou and Dubery 2015). miRNAs play a significant role in regulation of defense signals through various pathways such as, hormone signals, NBS-LRR gene expression, reactive oxygen species (ROS) production, protein phosphorylation, and gene silencing (Baldrich et al. 2015; Sattar et al. 2016). Insights into the potential and mode of action of miRNAs in plant responses to pathogen attack and plant-pathogen interactions will enable the design of suitable strategies to develop resistant crop varieties.

5.4.1 Bacterial Pathogens

Navarro et al. (2006) found that the expression of miR393 induces antibacterial resistance by repressing the auxin signaling receptor in response to detection of the bacterial pattern associated molecular pattern (PAMP). Several upregulated and downregulated miRNAs were identified (upregulated: miR158, miR159, miR160, miR167, and miR393; downregulated: miR482, miR825, miR390, miR398, miR408, and miR2118b) in *Arabidopsis* and tomato in response to the infection by *Pseudomonas syringae*, which enhanced the resistance against the bacterium (Canto-Pastor et al. 2019). The expression/repression of auxin perception and signaling at the site of bacterial pathogenesis promotes host-plant resistance. The miRNA-deficient *Arabidopsis* mutants, i.e., *dcl1* and *hen1*, were more susceptible to the pathogenic bacteria *Pseudomonas syringae* pv. *tomato DC3000 hrcC* and even the non-infectious strains of *Escherichia coli*, *Pseudomonas fluorescens*, and *P. syringae* pv. *phaseolicola*.

Xanthomonas axonopodis pv. *manihotis* (Xam), a gram-negative bacterium, causes bacterial blight in cassava (*Manihot esculenta*). Xam infection in resistant lines caused the upregulation of many miRNA families (miR160, miR165, miR167, miR171, miR390, miR393, and miR394), whereas the expression of other miRNA families was downregulated (miR397, miR399, miR535, miR395, miR408, miR482, miR408, and miR1507). The miRNAs families (miR395, miR397, miR398, miR408, miR482, and miR535) were downregulated and the corresponding target genes were upregulated as an Effector Triggered Immunity (ETI) response upon bacterial infection (Abdel-Ghany and Pilon 2008; Liu et al. 2008; Sunkar and Zhu 2004). In *A. thaliana*, it was found that miR165, miR394, and miR171 inhibited auxin receptors by negatively targeting the expression of F-box family auxin receptors, homeobox (ATHB), and a scarecrow-like transcription factors, which resulted in enhanced resistance against the pathogens. Likewise, in *M. esculenta*, miR395, miR482, and miR535 targeted regulation of the nucleotide-binding site, leucine-rich (NBS-LRR), or receptor-like pseudokinase (ARLPK) genes, which are vital for plant defense (Perez-Quintero et al. 2012; Jodder et al. 2017; Snyman et al. 2017; Rajwanshi et al. 2019).

It was found that in tomato and other *Solanaceae* family members, the expression of several NBS-LRR genes is normally silenced by miR482/2118 family miRNAs. However, this miRNA-mediated silencing is lost in virus- and bacteria-infected plants, resulting in pathogen-induced expression of the NBS-LRR genes (Shivaprasad et al. 2012), resulting in a resistance response. Niu et al. (2016) also demonstrated the defense mechanism of miRNAs in plants in response to *P. syringae* infection. Their results indicated downregulation of miR825 and miR825* and upregulation of target genes, such as NBS-LRR, toll-interleukin-like receptor (TIR), and ubiquitin-protein ligase in *Bacillus cereus* pre-treated plants, suggesting bacterial-induced immunity against other bacteria. In addition to auxin hormone regulation, the miRNAs miR159 and miR319 also had a major impact on the expression of other signaling hormones such as jasmonic acid (JA), gibberellic acid (GA), salicylic acid (SA), and abscisic acid (ABA) during bacterial infection (Li et al. 2010; Mishra et al. 2018). Naseem et al. (2015) found that auxin suppresses SA but not the JA cascade. Hence, overexpression of miR393-induced repression of auxin signaling receptors, which in turn resulted in high levels of SA, ultimately caused resistance to biotrophic microbes. Gupta et al. (2020) found that the loss-of-function mutation in ath-miR164c in *Arabidopsis* resulted in resistance to bacterial pathogens. During the nitrogen fixing process, the miRNAs miR168, miR172, miR482, and miR1512 regulated the nodule growth in *Glycine max* roots in response to *Bradyrhizobium japonicum* symbiotic association (Yan et al. 2016).

5.4.2 Fungal Pathogens

Plants, upon fungal infection, respond through several cellular, physiological, and molecular changes. The resistance response to fungal attack comprises the expression of hydrolytic enzymes, hydrolysis of chitin, b-1,3-glucan of the cell wall, and overexpression of receptor-like kinase genes and small non-coding RNAs (Li et al. 2014, 2017; Baldrich et al. 2015; Soto-Suarez et al. 2017). In plants, miRNAs, as post-transcriptional gene regulatory molecules, have a central position in PTI and ETI immune response to microbial infection (Katiyar-Agarwal and Jin 2010). High-throughput sequencing is a common method to profile miRNA expression in response to fungal infection. During *Melampsora larici-populina* infection in poplar, of the total 90 known miRNA families, 42 families, and 378 novel miRNAs were identified (Chen and Cao 2015). Similarly, Campo et al. (2013) manifested the influence of upregulated Osa-miR7696 to confer tolerance to blast infection in rice.

Overexpression of miR393 in tobacco, *Arabidopsis*, soybean, and maize ameliorated the intensity of infection by *Phytophthora dahliae* (Wong et al. 2014; Balmer et al. 2017; Hou et al. 2019). Zhang et al. (2016) highlighted the involvement of two miRNAs, miR159 and miR166, in cotton against *Verticillium dahliae* infection by targeting a definite virulence-transcript encoding protein. This protective mechanism was also conserved in *Arabidopsis* due to the upregulation of miR159 and miR166 induced by *V. dahliae* infection (Zhang et al. 2016). Likewise, miR164 and miR477 in cotton significantly cleaved their targets *viz.* GhCBP60A and GhNAC10, and helped regulate the resistance to *V. dahliae* (Hu et al. 2020). *Botryosphaeria dothidea* infection was found to induce the expression of miR168a in *Malus hupehensis*, which helped in gaining resistance (Yu et al. 2017).

The mutants of miR166 confirmed the role of this miRNA in imparting resistance to fungal pathogens, possibly through trans-species gene silencing. In eggplant, the loss-of-function mutant of miR393 was found when challenged with *V. dahliae* (Yang et al. 2013). It was proposed that MAMPs overexpress miR393 against oomycetes and fungi with biotrophic and hemibiotrophic infection, and downregulate it during infection by fungi having a necrotrophic lifestyle.

In summary, miRNAs impart resistance through modulation of auxin signaling pathways and PAMP-triggered immunity, which are negatively regulated by miR396, miR773, and miR858 in response to infection by necrotrophic or hemibiotrophic fungi (Salvador-Guirao et al. 2018; Camargo-Ramirez et al. 2018; Hu et al. 2020). Likewise, upregulation of miR396 and miR773 represses their targets, growth regulating factor (GRF) and methyltransferase 2, to boost resistance against *Magnaporthe oryzae* (Liebsch and Palatnik 2020) in rice and *Arabidopsis* (Salvador-Guirao et al. 2018). Overexpression of miR160, miR162a, miR398, miR7695, and suppression of miR169 indicated greater production of ROS and expression of other defense genes in rice during *M. oryzae* infection (Li et al. 2019a, 2019b, 2020). The knock-out mutants deficient in the SlymiR482e-3p gene in tomato showed resistance to *Fusarium oxysporum* by targeting NBS-LRR proteins (Gao et al. 2021). Studies show that various fungal pathogens alter the expression of DCL1 and AGO1 genes by over-expressing miR162 and miR168 to maintain the homeotic balance of DCL1 and AGO1 as host-derived PTI immunity in response to fungal elicitors (Baldrich et al. 2015).

Fungi have also been shown to synthesize miRNA-like RNAs (milRNAs), which also regulate target gene expression in their respective genomes. The milRNAs have been reported in *Neurospora crassa* (Lee et al. 2010), *Metarhizium anisopliae* (Zhou et al. 2012), *Trichoderma reesei* (Kang et al. 2013),

F. oxysporum (Chen et al. 2014), etc. Using bioinformatics tools, Mathur et al. (2020) predicted milRNAs in 13 fungi and identified their targets in transcriptomes of the respective fungi. Interestingly, they showed miRNA-mediated trans-kingdom gene regulation and the roles of miRNAs in plant-pathogen interactions. They found that plant miRNAs regulated virulence genes in fungal pathogens, while fungal milRNAs targeted plant resistance genes.

5.4.3 Viruses

Viruses are obligate pathogens that affect nearly all living organisms by transferring their genetic material into the host cytoplasm, where they utilize the host machinery to replicate. They modify the host genomic and transcriptome levels and alter small RNA levels (Du et al. 2011; Sunkar et al. 2012; Pradhan et al. 2015; Reyes et al. 2016). Plants have also developed an adaptive resistance mechanism to counter virus attacks such as hormone-mediated defense, regulation of metabolism, RNA silencing, immune receptor signaling, etc. (Ramesh et al. 2014; Calil and Fontes 2016; Liu et al. 2017). Previous studies showed that post-transcriptional regulation contributed to miRNA alteration upon viral infection or expression of viral proteins in the host. Until now, many miRNAs have been shown to play a vital role in antiviral RNA silencing pathways to protect crops (Ghoshal and Sanfacon 2015; Khalid et al. 2017). In plants, mainly two defense modes of miRNAs contribute to antiviral defense: (i) directly targeting viral RNAs, and (ii) triggering the biogenesis of miRNAs responsible for the antiviral response (Liu et al. 2017). Most studies provide indirect evidence for the first mode of plant miRNA function, and more studies are needed to clarify this mode of action.

A plant miRNA target for plum pox virus (PPV) was reported in *Arabidopsis* (Simon-Mateo and Garcia 2006). Several studies were conducted to understand the miRNA-mediated plant defense responses against viral diseases (Li et al. 2012; Pacheco et al. 2012). It was found that alteration in the levels of miR156, miR160, miR164, miR166, miR169, and miR171 expression in tobacco (*Nicotiana tabacum*) correlated with severity in symptoms (Bazzini et al. 2007; Guo et al. 2017b, 2017a). In *Brassica*, miR1885 was induced by turnip mosaic virus (TuMV) infection, which encoded RNA silencing/downregulation of its targets, i.e., TIR-NBS-LRR class of proteins (He et al. 2008). Similarly, in rice, miR444 exhibits a crucial role by targeting the MADS-box proteins (MADS23, MADS27, and MADS5) that are transcriptional repressors of RNA-dependent RNA polymerase 1 of antiviral pathways and promote rice stripe virus (RSV) infection (Wang et al. 2016). Likewise, Wu et al. (2017) observed that miR528 suppresses the expression of the target L-ascorbate oxidase (AO) and decreases the amount of ROS during infection by RSV and rice black-streaked dwarf virus. Similarly, Xia et al. (2018) reported the same function of miR444 and miR528 in maize in response to sugarcane mosaic virus infection. MiR159/miR319 expressed in dicot plants exhibited a positive action over the GA-signaling pathway to viral response (Du et al. 2014). Recent results indicate that overexpression or downregulation of miR156 compared to its target SPL9 contributed to innate immunity via ROS generation and SA signaling pathways in *Arabidopsis* (Yin et al. 2019). In case of African mosaic virus-infected cassava plants, the viral miRNAs were shown to target host resistance-related genes. Similarly, the cassava miRNAs were also shown to have potential targets in the viral genome (Maghuly et al. 2014). In *N. crassa* too, milRNAs were shown to be involved in defense against viruses and transposons (Lee et al. 2010).

Many other miRNAs have also been shown to act as essential epigenetic regulators in plants against virus infection. For example, miR156, miR171, miR398, and miR168 were involved in infection of Potato virus X and plum pox virus, miR168 to *B. dothidea* in rice (Wu et al. 2015), and *M. hupehensis* (Yu et al. 2017). In the case of susceptible plants, the miRNAs miR156, miR160, miR166, miR398, miR1514, and miR2118 were significantly upregulated together with oxidative stress-responsive genes. In response to viral infection, they altered the target genes such as auxin metabolic and auxin response factor (ARF), NBS-LRR, copper superoxide dismutase, etc., and various transcription factors (Pradhan et al. 2015; Kundu et al. 2017; Ramesh et al. 2017a).

Table 5.2 lists various studies on the role of miRNAs in biotic stresses caused by pathogenic fungi, bacteria and viruses in several plant species.

TABLE 5.2
Studies Showing the Role of miRNAs under Biotic Stresses

Crop	Pathogen	miRNA	Target Gene	Reference
		Fungal infection		
Tomato	*Phytophthora infestans*	miR482 and miR2118b MiR482/2118	TCP4, NB domain NBS–LRR	Shivaprasad et al. 2012; Jiang et al. 2018; Canto-Pastor et al. 2019
	Botrytis cinerea	miR319c	miR319c	Wu et al. 2020
	Fusarium oxysporum f.sp. *lycopersici*	slmiR482f and slmiR5300	*LRR1* mRNA as a siRNA-mediated secondary target	Ouyang et al. 2014
Rice	*Rhizoctonia solani*	miR444b.2, miR531a, mir1861i, novel_miR1956 miR135	Nucleotide-binding/leucine-rich repeat proteins, hormonal signal transduction factors	Cao et al. 2020
		miR160	ARF10, ARF16, ARF17 overexpression	Li et al. 2014
	Fungal, insect resistance	miR396	GRF6	Soto-Suarez et al. 2017; Dai et al. 2019
	Albino	osa-MIR528	Pds	Warthmann et al. 2008
Arabidopsis	*Puccinia cucumerina*	miR396	*GROWTH-REGULATING FACTOR*	Soto-Suarez et al. 2017; Dai et al. 2019
Garlic	*Fusarium oxysporum* f.sp. *cepae*	miR164a, miR168a, and miR393	F-box domain and cytochrome P450 (CYP450) family protein	Chand et al. 2016; 2017
Potato; cotton	*Verticillium dahliae*	miR482 miR482b	Ca^{2+}-dependent, isotrichodermin C CIp-1 and HiC-15 NBS-LRR genes	Zhu et al. 2013; Yang et al. 2015; Zhang et al. 2016; 2018
Oncidium	*Piriformospora indica*	miR393	Transport inhibitor response 1 (TIR)	Ye et al. 2014
Apple	*Alternaria* f.sp. *mali*	Md-miR156ab	WRKY transcription factor, MdWRKYN1	Zhang et al. 2017
		Bacterial infection		
Arabidopsis	*Pseudomonas syringae*	miR393 At-miR393a	Auxin receptors (AFB2, AFB3, and TIR1) Auxin signaling F-Box proteins	Navarro et al. 2006
Rice	*Xanthomonas oryzae*	Osa-miR1876 sa-miR167h-3p	NBS-ARC gene Flavonoid, ROS, and hormone signal transduction pathways	Jiang et al. 2020 Lu et al. 2021
Cassava	*Xanthomonas axonopodis*	miR393	TIR1, AFB	Perez-Quintero et al. 2012
Tomato	*P. syringae* pv. *tomato*	miR160 and miR167 miR398	Auxin response factors Copper/Zinc superoxide dismutase genes	Zang et al. 2011
	P. syringae	miR482 and miR2118b	NLR proteins	Canto-Pastor et al. 2019
		Viral infection		
Cassava	South African cassava mosaic virus	miR162, miR168, miR167	PTGS target Auxin responsive factor	Bizabani et al. 2021

(Continued)

TABLE 5.2 (*Continued*)
Studies Showing the Role of miRNAs under Biotic Stresses

Crop	Pathogen	miRNA	Target Gene	Reference
Cucumber	Cucumber green mottle mosaic virus (CGMMV)	Some key miR families miR159, miR854, miR5658 and csa-miRn1-3p	Transcription factor MYB29 WRKY transcription factor 21 LRR receptor	Liang et al. 2019
Beans	Mungbean yellow mosaic India virus	miR393	TIR1, AFB2, and AFB3	Kundu et al. 2017
Phaseolus vulgaris	Mungbean yellow mosaic India virus	pvu-miR160j pvu-miR390d pvu-miR395f	Auxin response factor-16, LRR kinase, NB-ARC domain	Patwa et al. 2019

5.4.4 Nematodes

Nematodes are the root parasites that cause acute damage to crops and massive agricultural losses worldwide (Lu et al. 2014; Jaubert-Possamai et al. 2019). After invasion into the host plant roots, nematodes induce genetic reprogramming of vascular cells, turning them into giant feeding cells and forming root galls. Nematode-responsive conserved and novel miRNAs were identified in many plant species using computational tools (Medina et al. 2017; Tian et al. 2017). Incidentally, miRNAs were first discovered in the nematode *C. elegans* (Lee et al. 1993). In the initial years, miRNAs were considered non-protein-coding RNAs (lin-4), which regulated the target mRNA (lin-14) via base-pairing complementarity (Zhang et al. 2007). miRNAs have been shown to regulate gene expression throughout development and plant-nematode interaction. High throughput sequencing tools identified 30 miRNAs in *A. thaliana* and 40 miRNAs in soybean that were differentially expressed in response to cyst nematode infection (Li et al. 2012). Functional analysis showed the importance of two miRNAs (miR396 and miR827) during reprogramming of root cells to formulate syncytial cells and, hence, in the plant defenses in *A. thaliana* (Hewezi et al. 2016).

Disruption of post-transcriptional gene silencing in *Arabidopsis* AGO mutants (*ago1* or *ago2*) was found to result in a reduced infection rate of the root knot nematode (RKN) (*Meloidogyne incognita*), signifying the role of miRNAs in the plant-nematode interaction. This defense mechanism at the plant-nematode interface was cross-checked by sequencing small RNAs from uninfected *Arabidopsis* root cells and root galls infected with *M. incognita*. The studies indicated differential expression of 24 miRNAs in gall as putative regulators of gall development (Medina et al. 2017). Similarly, it was confirmed that the *Arabidopsis* miR159abc mutant had lower susceptibility to RKN, explaining the role of the miR159 gene family in plants against *M. incognita*. Later, the experimental analysis showed overexpression of miR159 in galls at 14 days post-inoculation with post-transcriptional repression of the miR159 target MYB33, suggesting a key role of miR159 in plant response to RKN (Piya et al. 2017).

Similar upregulated expression of miR396 and GRF9 RNAi was observed in transgenic soybean that displayed a reduction in the number of cysts produced by *Heterodera glycines*, showing resistance to the cyst nematode. These results specify that the miR396/GRF module is crucial for *H. glycines* infection in *Arabidopsis* and soybean (Hewezi et al. 2016). Overexpression of miR827 upon *Heterodera schachtii* infection in *Arabidopsis* causes the expression of the target gene Nitrogen Limitation Adaptation (NLA) that encodes ubiquitin E3 ligase enzyme to mute in all parasitic stages, resulting in resistance to the cyst nematode (Hewezi et al. 2016).

A recent study on tomato plants infected with *Globodera rostochiensis* revealed 200–300 differentially expressed miRNAs at each stage of infection (Koter et al. 2018). Pan et al. (2019) suggested that three miRNA-target pairs *viz.* miR159-MYB, miR319-TCP4, and miR167-ARF8 indicated an inversely expressed pattern of the expression of miRNAs and their targets, which confirms the miRNA-target relationships. They found that these inverse expression patterns accounted for *M. incognita* interaction, both susceptible and resistant. Likewise, the upregulated pattern of gma-miR5032 and gma-miR5770b enhanced resistance against soybean cyst nematodes in soybean via altering methylation patterns (Rambani et al. 2020). These reports show that modulation in gene expression through gain- or loss-of-function methods altered the plants' response to nematode infection.

5.4.5 Insect Resistance

Interaction between plants and insects occurs in a diverse and multifaceted way (Lucas-Barbosa 2016). Plants, being sessile, need to adapt themselves against insect herbivory. In response to herbivory or sap-sucking insects, plants have developed an inducive defense mechanism to minimize tissue losses (Lucas-Barbosa 2016). Recent studies show the involvement of miRNAs in modulating effective plant defense strategy (Sattar et al. 2016) and may activate other processes such as JA, SA, ABA, ET, and GA hormonal signaling for resistance to insect herbivory (Morkunas et al. 2011; Louis and Shah 2013).

Jeyaraj et al. (2017) documented differential gene expression of about 150 miRNAs in *Camellia sinensis* during the attack of caterpillars of the geometrid moth *Ectropis*. It was also found that in tobacco leaf, nicotine and trypsin proteinase inhibitors play significant roles by inducing jasmonate, the main signaling molecule, and muting the vital proteins involved in miRNA biogenesis (RNA-directed RNA polymerase 1 [RdR1] and Dicer-like 3 & 4 [DCL3 and DCL4] in response to damage caused by caterpillar herbivory. Recent studies demonstrate that detoxification enzymes such as esterases, cytochrome P450s (CYP), or glutathione S-transferases play major modulation roles through the endogenous miRNA, miR656a-3p, which binds at the 3′-UTR of CYP transcripts to counter insect damage. Ma et al. (2017) characterized differentially expressed aphid miRNAs in response to plant allelochemicals by bioinformatically predicting silencing of the endonuclease Dicer-1 and core components of the RISC complex AGO1. They found that this increases the sensitivity of cotton aphids (*Aphis gossypii*) to plant allelochemicals.

Overexpression of miR166b in cotton resulted in resistance to the silverleaf whitefly (Wamiq and Khan 2018). However, overexpression of miR396 in rice silenced the target mimicry, i.e., MIM396, and exhibited susceptibility to the brown planthopper (Dai et al. 2019). Saini et al. (2018) designed an artificial miRNA (amiRNA) by replacing the miR164b/miR164b* sequences of *Arabidopsis* with those of HaAce1 of *Helicoverpa armigera* insects. This amiRNA was expressed in tobacco, which showed increased tolerance to the insect attack. Similarly, amiRHaEcR (*H. armigera*) and amiR-Ace1 (*Myzus persicae*) expression in tomato downregulated the target gene transcripts, affecting the overall development of the insects (Yogindran and Rajam, 2021; Faisal et al. 2021).

5.5 miRNA-BASED STRATEGIES FOR CROP IMPROVEMENT

5.5.1 Transgenic Strategy

As mature miRNAs are very small (21–24 nt), it is difficult to directly transform them to develop transgenics. However, transforming the long pri-miRNA, instead of the mature miRNA, is a better approach to express the desired miRNAs in transgenic plants. This strategy can result in direct expression of the mature miRNA, rather than manipulating the MIR gene. Thus, miRNA-based gene regulation can be manipulated by various transgenic methods (Franco-Zorrilla et al. 2007; Gupta 2015) to achieve the desired outcome. For instance, inserting mutations in the *SPL14* target site for improved miR156 binding increased growth and grain yield in rice (Miura et al. 2010). Similarly, overexpression of soybean *NF-YA* (*HAP2*) with the absence of 3′ UTR at the miR169 site conferred increased drought tolerance in *Arabidopsis* (Ni et al. 2013). Recent studies demonstrating overexpression of miRNA families miR160, miR398, miR393, and miR397 in rice (Li et al. 2014; 2019a) and miR482 in tomato (Jiang et al. 2018) imparted a higher tolerance to pathogen infection. Similarly, overexpression of miR393 in *Arabidopsis* enhanced resistance against biotrophic pathogens (Navarro et al. 2006; Robert-Seilaniantz et al. 2011; Hou et al. 2019).

5.5.2 Artificial miRNA (amiRNA) Strategy

AmiRNAs are the synthetic miRNAs that mimic the naturally occurring pri-miRNAs, for which the Dicer cleavage sites have been mapped and experimentally validated. The amiRNAs have opened a new prospect for advanced research. Using amiRNA, a particular gene can be forcefully silenced and inhibited. Niu et al. (2006) modified the miR59 precursor and expressed the amiRNA to gain virus resistance in *A. thaliana.* Since then, the strategy has been used against various plant diseases. In this strategy, first, the target gene involved in causing disease in plants needs to be identified. An amiRNA targeting this gene needs to be designed and cloned into a vector, and the vector is then transferred into the host plant. Upon attack by the disease-causing agent, the amiRNA would bind to the target gene, and inhibit its expression. This would silence the gene, which would mainly be responsible for the diseased condition in plants.

In the case of miRNA-mediated inter-kingdom gene regulation in plant-pathogen or plant-insect interactions, an amiRNA targeting virulence gene of the pathogen, or development, reproduction, or other essential process governing gene of the insect can be designed and transformed in the host plant. When the pathogen or the pest attacks the plant, the amiRNA would express, translocate to the pathogen or the pest, and silence the target pathogen or pest gene, thereby making the plant resistant to the pathogen or the pest. The amiRNA strategy has been successfully used to knock down the target genes from insect pests, nematodes, viruses, and other phytopathogens. According to Manavella et al. (2012), amiRNAs are stable and heritable. However, the systemic movement of these mature amiRNAs is restricted in some cases. The main hurdles of this strategy are (i) selecting the pre-amiRNA sequences for effective silencing in the target organism without any off-target effects (Carbonell et al. 2015), and (ii) ensuring uptake of the amiRNA by the target organism without degradation.

5.5.3 CRISPR/Cas9-Associated Approach

Among the genome editing techniques, the clustered regularly interspaced short palindromic repeat (CRISPR)/CRISPR-associated protein 9 (Cas9) proved to be an efficient method than zinc finger nucleases (ZFNs) and transcription activator-like effector nucleases (TALENs) (Mao et al. 2013). The Cas9 nuclease creates double-strand breaks at the target site in the genome, which triggers the activation of the non-homologous end-joining repair mechanism. During the strand-break repair, the insertion or deletion of nucleotides may occur, causing the loss of gene function. Alternatively, a homology-directed repair can be induced to achieve targeted base modification, targeted knock-in, or replacement of miRNA encoding genes (Zhao et al. 2016). CRISPR/Cas9-mediated introduction of the indels at the precursor miRNA region showed impaired miRNA biogenesis (Chang et al. 2016; Zhou et al. 2017). Similarly, the modifications in target genes hinder the interaction between miRNA-target pairing (Basso et al. 2019). Although genome editing is a promising strategy, certain features associated with miRNA hinder its applicability, such as the short length of miRNA genes decreases the number of possible gRNA/Cas9 targets (Jacobs et al. 2015), miRNAs are a part of a complex regulatory network and many of them are found within the introns, which may affect the editing results (Barrangou et al. 2015).

Nevertheless, few studies have demonstrated the applicability of the CRISPR/Cas9-mediated genome editing technique to develop crop varieties tolerant to biotic and abiotic stresses (Zhao et al. 2016; Hong et al. 2016). Recently, Hong et al. (2016) used this strategy to develop tomato plants resistant to *Phytophthora infestans*, which causes the late blight disease. CRISPR/Cas9 was used to knock down miR482b and miR482c simultaneously. Two plants showed silencing of both miRNAs, while one plant showed silencing of miR482b alone. All the transgenic plants were tolerant to *P. infestans* compared to the wild-type plants. The transgenic plants also showed enhanced levels of the common target NBS-LRR genes and reduced accumulation of ROS.

5.6 CONCLUSIONS AND FUTURE PROSPECTS

Several studies have confirmed the miRNA-mediated gene regulation in the plant response against various abiotic and biotic stresses. miRNAs modulate target gene expression either by cleavage or translation-inhibition of the mRNA transcript, resulting in improvement in key agronomic traits such as biotic and abiotic stress tolerance and productivity in various crops. These small regulatory RNAs have also been shown to play critical roles in the regulation of defense signals via modulation of hormone levels, NBS-LRR gene expression, ROS production, protein phosphorylation, and gene silencing. When a suitable miRNA cannot be found to regulate certain target genes, either in the plant or in the pathogen, artificial miRNAs can be designed and used to silence the target gene. Using this approach, it is possible to target multiple genes of either a single, or multiple pathogens and pests. Thus, the amiRNA strategy can enable development of crops resistant to multiple biotic and abiotic stresses. Recent advances in genome editing have opened a new way to suitably modify either the miRNAs, or their targets for improved gene regulation, thereby improving specific agronomic traits in crops. A comprehensive understanding of the advanced molecular networking strategies regulated by miRNA would empower researchers to decode the miRNA-mediated regulatory networks for developing improved crops.

ACKNOWLEDGEMENTS

PS acknowledges the facilities provided by the School of Bioengineering Sciences and Research, MIT-ADT University, Pune Maharashtra.

SSM acknowledges the Council of Scientific and Industrial Research (CSIR), India, for Senior Research Associateship. VAM acknowledges the University Grants Commission (UGC), India, for Junior Research Fellowship and the Academy of Scientific and Innovative Research (AcSIR), India, for the Ph.D. degree program. The authors acknowledge CSIR, India for funding (Project Code: MLP101226).

REFERENCES

Abdel-Ghany SE, Pilon M (2008) MicroRNA-mediated systemic down-regulation of copper protein expression in response to low copper availability in *Arabidopsis*. The Journal of Biological Chemistry 283 (23):15932–15945

Adai A, Johnson C, Mlotshwa S, Archer-Evans S, Manocha V, Vance V, Sundaresan V (2005) Computational prediction of miRNAs in *Arabidopsis thaliana*. Genome Research 15 (1):78–91

Addo-Quaye C, Miller W, Axtell MJ (2009) Cleave land: A pipeline for using degradome data to find cleaved small RNA targets. Bioinformatics 25 (11):130–131

Alhasan AH, Kim DY, Daniel WL, Watson E, Meeks JJ, Thaxton CS, Mirkin CA (2012) Scanometric microRNA array profiling of prostate cancer markers using spherical nucleic acid–gold nanoparticle conjugates. Analytical Chemistry 84 (9):4153–4160

Arteaga-Vazquez M, Caballero-Perez J, Vielle-Calzada JP (2006) A family of microRNAs present in plants and animals. Plant Cell 18 (12):3355–3369

Aukerman MJ, Sakai H (2003) Regulation of flowering time and floral organ identity by a microRNA and its APETALA2-like target genes. The Plant Cell 15 (11):2730–2741

Baek D, Chun HJ, Kang S, Shin G, Park SJ, Hong H, Kim C, Kim DH, Lee SY, Kim MC (2016) A role for *Arabidopsis* miR399f in salt, drought, and ABA signaling. Molecules and Cells 39 (2):111

Bagga S, Bracht J, Hunter S, Massirer K, Holtz J, Eachus R, Pasquinelli AE (2005) Regulation by let-7 and lin-4 miRNAs results in target mRNA degradation. Cell 122 (4):553–563

Baldrich P, Campo S, Wu MT, Liu TT, Hsing YI, San Segundo B (2015) MicroRNA-mediated regulation of gene expression in the response of rice plants to fungal elicitors. RNA Biol 12(8):847–863

Balmer A, De Paoli E, Si-Ammour A, Mauch-Mani B, Balmer D (2017) Signs of silence: Small RNAs and antifungal responses in *Arabidopsis thaliana* and *Zea mays*. Plant Engineering InTech. https://dx.doi.org/10.5772/intechopen.69795

Banerjee A, Roychoudhury A, Krishnamoorthi S (2016) Emerging techniques to decipher microRNAs (miRNAs) and their regulatory role in conferring abiotic stress tolerance of plants. Plant Biotechnology Reports 10 (4):185–205

Bao N, Lye KW, Barton MK (2004) MicroRNA binding sites in *Arabidopsis* class III HD-ZIP mRNAs are required for methylation of the template chromosome. Developmental Cell 7 (5):653–662

Barrangou R, Birmingham A, Wiemann S, Beijersbergen RL, Hornung V, Smith AvB (2015) Advances in CRISPR-Cas9 genome engineering: Lessons learned from RNA interference. Nucleic Acids Research 43 (7):3407–3419

Bartel DP (2004) MicroRNAs: Genomics biogenesis mechanism and function. Cell 2116(2):281–297

Bartel DP (2009) MicroRNAs: Target recognition and regulatory functions. Cell 136(2):215–233

Barvkar VT, Pardeshi VC, Kale SM, Qiu S, Rollins M, Datla R, Gupta VS, Kadoo NY (2013) Genome-wide identification and characterization of microRNA genes and their targets in flax (*Linum usitatissimum*). Planta 237(4):1149–1161

Basso MF, Ferreira PCG, Kobayashi AK, Harmon FG, Nepomuceno AL, Molinari HBC, Grossi-de-Sa MF (2019) MicroRNAs and new biotechnological tools for its modulation and improving stress tolerance in plants. Plant Biotechnology Journal 17(8):1482–1500

Bazzini AA, Hopp HE, Beachy RN, Asurmendi S (2007) Infection and coaccumulation of tobacco mosaic virus proteins alter microRNA levels, correlating with symptom and plant development. Proceedings of the National Academy of Sciences 104 (29):12157–12162

Bebber DP, Gurr SJ (2015) Crop-destroying fungal and oomycete pathogens challenge food security. Fungal Genetics and Biology 74:62–64

Bentwich I (2005) Prediction and validation of microRNAs and their targets. FEBS Letters 579 (26):5904–5910

Bizabani C, Rogans SJ, Rey MEC (2021) Differential miRNA profiles in South African cassava mosaic virus-infected cassava landraces reveal clues to susceptibility and tolerance to cassava mosaic disease. Virus Research 303:198400.

Bologna NG, Mateos JL, Bresso EG, Palatnik JF (2009) A loop-to-base processing mechanism underlies the biogenesis of plant microRNAs miR319 and miR159. EMBO 28(23):3646–3656

Bologna NG, Voinnet O (2014) The diversity, biogenesis and activities of endogenous silencing small RNAs in *Arabidopsis*. Annu Rev Plant Biol 65:473–503

Bonnet E, Wuyts J, Rouze P, Van de Peer Y (2004) Detection of 91 potential conserved plant microRNAs in *Arabidopsis thaliana* and *Oryza sativa* identifies important target genes. Proceedings of the National Academy of Sciences 101 (31):11511–11516

Brant EJ, Budak H (2018) Plant small non-coding RNAs and their roles in biotic stresses. Front Plant Sci 9:1038–103389

Brennecke J, Stark A, Russell RB, Cohen SM (2005) Principles of microRNA-target recognition. PLoS Biology 2005 (3):e85.

Budak H, Akpinar BA (2015) Plant miRNAs: Biogenesis, organization and origins. *Functional & Integrative Genomics* 15(5):523–31

Calil IP, Fontes EP (2016) Plant immunity against viruses: Antiviral immune receptors in focus. Annals of Botany 119:711–723

Camargo-Ramirez R, Val-Torregrosa B, San Segundo B (2018) MiR858-mediated regulation of flavonoid-specific MYB transcription factor genes controls resistance to pathogen infection in *Arabidopsis*. Plant Cell Physiology 59 (1):190–204

Campo S, Peris-Peris C, Sire C, Moreno AB, Donaire L, Zytnicki M et al (2013) Identification of a novel microRNA (miRNA) from rice that targets an alternatively spliced transcript of the Nramp6 (Natural resistance-associated macrophage protein 6) gene involved in pathogen resistance. New Phytologist 199:212–227

Canto-Pastor A, Santos B, Valli AA, Summers W, Schornack S, Baulcombe DC (2019) Enhanced resistance to bacterial and oomycete pathogens by short tandem target mimic RNAs in tomato. Proceedings of the National Academy of Sciences U S A 116 (7):2755–2760

Cao W, Cao X, Zhao J, Zhang Z, Feng Z, Ouyang S, Zuo S (2020) Comprehensive characteristics of microRNA expression profile conferring to *Rhizoctonia solani* in rice. Rice Science 27(2):101–112

Carbonell A, Fahlgren N, Mitchell S, Cox KL Jr, Reilly KC, Mockler TC, Carrington JC (2015) Highly specific gene silencing in a monocot species by artificial microRNAs derived from chimeric miRNA precursors. The Plant Journal 82(6):1061–1075

Carrington JC, Ambros V (2003) Role of microRNAs in plant and animal development. Science 18:336–338

Chand SK, Nanda S, Rout E, Mohanty, Mishra R, Joshi RK (2016) Identification and characterization of microRNAs in turmeric (*Curcuma longa* L.) responsive to infection with the pathogenic fungus *Pythium aphanidermatum*. Physiological and Molecular Plant Pathology 93:119–128

Chang H, Yi B, Ma R, Zhang X, Zhao H, Xi Y (2016) CRISPR/cas9, a novel genomic tool to knock down microRNA *in vitro* and *in vivo*. Scientific Reports 6 (1):1–12

Chavez Montes RA, Rosas-Cárdenas DF, De Paoli E, Accerbi M, Rymarquis LA, Mahalingam G, Marsch-Martínez N, Meyers BC, Green PJ, De Folter S (2014) Sample sequencing of vascular plants demonstrates widespread conservation and divergence of microRNAs. Nature Communications 5(1): 3722

Chellappan P, Xia J, Zhou X, Gao S, Zhang X, Coutino G, Vazquez F, Zhang W, Jin H (2010) siRNAs from miRNA sites mediate DNA methylation of target genes. Nucleic Acids Research 38: 6883–6894

Chen M, Cao Z (2015) Genome-wide expression profiling of microRNAs in poplar upon infection with the foliar rust fungus *Melampsora larici-populina*. BMC Genomics 16:696

Cheng Y, Dong L, Zhang J, Zhao Y, Li Z (2018) Recent advances in microRNA detection. Analyst 143 (8):1758–1774

Chen R, Jiang N, Jiang Q, Sun X, Wang Y, Zhang H, Hu Z (2014) Exploring microRNA-like small RNAs in the filamentous fungus *Fusarium oxysporum*. PLoS ONE2014 9:e104956.

Chen H, Li Z, Xiong L (2012) A plant microRNA regulates the adaptation of roots to drought stress. FEBS Letters 586 (12):1742–1747

Chen L, Meng J, Luan Y (2019) miR1916 plays a role as a negative regulator in drought stress resistance in tomato and tobacco. Biochemical and Biophysical Research Communications 508 (2):597–602

Chen C, Ridzon DA, Broomer AJ, Zhou Z, Lee DH, Nguyen JT, Barbisin M, Xu NL, Mahuvakar VR, Andersen MR (2005) Real-time quantification of microRNAs by stem–loop RT–PCR. Nucleic Acids Research 33 (20):e179

Chinnusamy V, Zhu J, Zhou T, Zhu JK (2007) Small RNAs: Big role in abiotic stress tolerance of plants. *In:* Advances in molecular breeding toward drought and salt tolerant crops. Springer, pp 223–260

Chuang JC, Jones PA (2007) Epigenetics and microRNAs. Pediatric Research 61:24R–29R

Cissell KA, Rahimi Y, Shrestha S, Hunt EA, Deo SK (2008) Bioluminescence-based detection of microRNA, miR21 in breast cancer cells. Analytical Chemistry 80 (7):2319–2325

Dai Z, Tan J, Zhou C, Yang X, Yang F, Zhang S, Sun S, Miao X, Shi Z (2019) The OsmiR396–Os GRF 8–OsF3H-flavonoid pathway mediates resistance to the brown planthopper in rice (*Oryza sativa*). Plant Biotechnology Journal 17(8):1657–1669

Das IK, Rakshit S (2016) Millets their importance and production constraints. *In*: Biotic Stress Resistance in Millets. Academic Press, pp 3–19 ISBN 9780128045497

Dezulian T, Remmert M, Palatnik JF, Weigel D, Huson DH (2006) Identification of plant microRNA homologs. Bioinformatics 22 (3):359–360

Djami-Tchatchou AT, Dubery IA (2015) Lipopolysaccharide perception leads to dynamic alterations in the microtranscriptome of *Arabidopsis thaliana* cells and leaf tissues. BMC Plant Biology 15:79

Dong H, Lei J, Ding L, Wen Y, Ju H, Zhang X (2013) MicroRNA: Function, detection, and bioanalysis. Chemical Reviews 113 (8):6207–6233

Dong CH, Pei H (2014) Over-expression of miR397 improves plant tolerance to cold stress in *Arabidopsis thaliana*. Journal of Plant Biology 57 (4):209–217

Du Z, Chen A, Chen W, Westwood JH, Baulcombe DC, Carr JP (2014) Using a viral vector to reveal the role of microRNA159 in disease symptom induction by a severe strain of cucumber mosaic virus. Plant Physiology 164 (3):1378–1388

Evers M, Huttner M, Dueck A, Meister G, Engelmann JC (2015) miRA: Adaptable novel miRNA identification in plants using small RNA sequencing data. BMC Bioinformatics 16 (1):1–10

Fahlgren N, Jogdeo S, Kasschau KD, Sullivan CM, Chapman EJ, Laubinger S, Smith L et al (2010) MicroRNA gene evolution in *Arabidopsis lyrata* and *Arabidopsis thaliana*. The Plant Cell 22(4): 1074–1089.

Faisal M, Abdel-Salam EM, Alatar AA (2021) Artificial microRNA-based RNA interference and specific gene silencing for developing insect resistance in *Solanum lycopersicum*. Agronomy 11 (1):136

Fan Y, Chen X, Trigg AD, Tung C-h, Kong J, Gao Z (2007) Detection of microRNAs using target-guided formation of conducting polymer nanowires in nanogaps. Journal of the American Chemical Society 129 (17):5437–5443

Fang S, Lee HJ, Wark AW, Corn RM (2006) Attomole microarray detection of microRNAs by nanoparticle-amplified SPR imaging measurements of surface polyadenylation reactions. Journal of the American Chemical Society 128 (43):14044–14046

Fernandez JE (2014) Understanding olive adaptation to abiotic stresses as a tool to increase crop performance. Environmental and Experimental Botany 103:158–179

Foyer CH, Rasool B, Davey JW, Hancock RD (2016) Cross-tolerance to biotic and abiotic stresses in plants: A focus on resistance to aphid infestation. Journal of Experimental Botany 67:2025–2037

Franco-Zorrilla JM et al (2007) Target mimicry provides a new mechanism for regulation of microRNA activity. Nature Genetics 39:1033–1037

Friedlander MR, Chen W, Adamidi C, Maaskola J, Einspanier R, Knespel S, Rajewsky N (2008) Discovering microRNAs from deep sequencing data using miRDeep. Nature Biotechnology 26 (4):407–415

Fujii H, Chiou TJ, Lin SI, Aung K, Zhu JK (2005) A miRNA involved in phosphate-starvation response in *Arabidopsis*. Current Biology 15 (22):2038–2043

Gao Y, Li SJ, Zhang SW, Feng T, Zhang ZY, Luo SJ, Mao HY, Borkovich KA, Ouyang SQ (2021) SlymiR482e-3p mediates tomato wilt disease by modulating ethylene response pathway. Plant Biotechnology Journal 19(1):17–19

Ghoshal B, Sanfacon H (2015) Symptom recovery in virus-infected plants: Revisiting the role of RNA silencing mechanisms. Virology 479:167–179

Guo JE, Hu Z, Li F, Zhang L, Yu X, Tang B, Chen G (2017a) Silencing of histone deacetylase SlHDT3 delays fruit ripening and suppresses carotenoid accumulation in tomato. Plant Science 265:29–38

Guo C, Xu Y, Shi M, Lai Y, Wu X, Wang H, Zhu Z, Poethig RS, Wu G (2017b) Repression of miR156 by miR159 regulates the timing of the juvenile-to-adult transition in *Arabidopsis*. Plant Cell 29 (6):1293–1304

Gupta PK (2015) MicroRNAs and target mimics for crop improvement. Current Science 108:1624–1633

Gupta A, Patil M, Qamar A, Senthil-Kumar M (2020) Ath-miR164c influences plant responses to the combined stress of drought and bacterial infection by regulating proline metabolism. Environmental and Experimental Botany 172:103998

Hajyzadeh M, Turktas M, Khawar KM, Unver T (2015) miR408 overexpression causes increased drought tolerance in chickpea. Gene 555 (2):186–193

Han J, Kong M, Xie H, Sun Q, Nan Z, Zhang Q, Pan J (2013) Identification of miRNAs and their targets in wheat (*Triticum aestivum* L.) By EST analysis. The Genetics and Molecular Research Journal 12 (3793):805

He XF, Fang YY, Feng L, Guo HS (2008) Characterization of conserved and novel microRNAs and their targets including a TuMV-induced TIR–NBS–LRR class R gene-derived novel miRNA in *Brassica*. FEBS Letters 582 (16):2445–2452

Hewezi T, Piya S, Qi M, Balasubramaniam M, Rice JH, Baum TJ (2016) *Arabidopsis* miR827 mediates post-transcriptional gene silencing of its ubiquitin E3 ligase target gene in the syncytium of the cyst nematode *Heterodera schachtii* to enhance susceptibility. The Plant Journal 88(2):179–192.

He Y, Zhou J, Hu Y, Fang C, Yu Y, Yang J, Zhu B, Ruan YL, Zhu Z (2021) Overexpression of sly-miR398b increased salt sensitivity likely via regulating antioxidant system and photosynthesis in tomato. Environmental and Experimental Botany 181:104273

Hindson CM, Chevillet JR, Briggs HA, Gallichotte EN, Ruf IK, Hindson BJ, Vessella RL, Tewari M (2013) Absolute quantification by droplet digital PCR versus analog real-time PCR. Nature Methods 10 (10):1003–1005

Hong J, Yang L, Zhang D, Shi J (2016) Plant metabolomics: An indispensable system biology tool for plant science. International Journal of Molecular Sciences 17 (6):767

Hou Y, Zhai Y, Feng L, Karimi HZ, Rutter BD, Zeng L, Choi DS, Zhang B, Gu W, Chen X, Ye W, Innes RW, Zhai J, Ma W (2019) A phytophthora effector suppresses trans-kingdom RNAi to promote disease susceptibility. Cell Host Microbe 25:153–165

Hu G, Lei Y, Liu J, Hao M, Zhang Z, Tang Y, Chen A, Wu J (2020) The GHR-miR164 and GhNAC100 modulate cotton plant resistance against *Verticillium dahliae*. Plant Sci 293:110438

Jacobs TB, La Fayette PR, Schmitz RJ, Parrott WA (2015) Targeted genome modifications in soybean with CRISPR/Cas9. BMC Biotechnology 15 (1):1–10

Jaubert-Possamai S, Noureddine Y, Favery B (2019) MicroRNAs new players in the plant–nematode interaction. Frontiers in Plant Science 10:1180

Jeyaraj A, Liu S, Zhang X et al (2017) Genome-wide identification of microRNAs responsive to *Ectropis oblique* feeding in tea plant (*Camellia sinensis* L). Scientific Reports 7:13634

Jiang G, Liu D, Yin D, Zhou Z, Shi Y, Li C, Zhu L, Zhai W (2020) A rice NBS-ARC Gene conferring quantitative resistance to bacterial blight is regulated by a pathogen effector-inducible miRNA. Molecular Plant 13(12):1752–1767

Jiang N, Meng J, Cui J, Sun G, Luan Y (2018) Function identification of miR482b a negative regulator during tomato resistance to *Phytophthora infestans*. Horticulture Research 5:9

Jiang D, Zhou L, Chen W, Ye N, Xia J, Zhuang C (2019) Overexpression of a microRNA-targeted NAC transcription factor improves drought and salt tolerance in rice via ABA-mediated pathways. Rice 12 (1):1–11

Jodder J, Basak S, Das R, Kundu P (2017) Coherent regulation of miR167a biogenesis and expression of auxin signaling pathway genes during bacterial stress in tomato. Physiological *and* Molecular Plant *Pathology* 100:97–105

Jones-Rhoades MW, Bartel DP (2004) Computational identification of plant microRNAs and their targets, including a stress-induced miRNA. Molecular Cell 14 (6):787–799

Jones-Rhoades MW, Bartel DP, Bartel B (2006) MicroRNAs and their regulatory roles in plants. Annual Review of Plant Biology 57:19–53.

Kang K, Zhong J, Jiang L, Liu G, Gou CY, Wu Q, Wang Y, Luo J, Gou D (2013) Identification of microRNA-like RNAs in the filamentous fungus *Trichoderma reesei* by Solexa sequencing. PLoS ONE 8:e76288.

Katiyar-Agarwal S, Jin H (2010) Role of small RNAs in host-microbe interactions. The Annual Review of Phytopathology 48:225–46

Katiyar A, Smita S, Chinnusamy V, Pandey DM, Bansal K (2012) Identification of miRNAs in sorghum by using bioinformatics approach. Plant Signaling and Behavior 7 (2):246–259

Khalid A, Zhang Q, Yasir M, Li F (2017) Small RNA based genetic engineering for plant viral resistance: Application in crop protection. Frontiers in Microbiology 8:43

Khraiwesh B, Arif MA, Seumel GI, Ossowski S, Weigel D, Reski R, Frank W (2010) Transcriptional control of gene expression by microRNAs. Cell 140:111–122

Khvorova A, Reynolds A, Jayasena SD (2003) Functional siRNAs and miRNAs exhibit strand bias. Cell 115:505–505

Kim SW, Li Z, Moore PS, Monaghan AP, Chang Y, Nichols M, John B (2010) A sensitive non-radioactive northern blot method to detect small RNAs. Nucleic Acids Research 38 (7):e98–e98

Kim DH, Saetrom P, Snove O Jr, Rossi JJ (2008) MicroRNA-directed transcriptional gene silencing in mammalian cells. Proceedings of the National Academy of Sciences of the United States of America 105:16230–16235

Koter MD, Swięcicka M, Matuszkiewicz M, Pacak A, Derebecka N, Filipecki M (2018) The miRNAome dynamics during developmental and metabolic reprogramming of tomato root infected with potato cyst nematode. Plant Science 268:18–29

Kravchik M, Stav R, Belausov E, Arazi T (2019) Functional characterization of microRNA171 family in tomato. Plants 8:10

Kundu A, Paul S, Dey A, Pal A (2017) High throughput sequencing reveals modulation of microRNAs in *Vigna mungo* upon mungbean yellow mosaic India virus inoculation highlighting stress regulation. Plant Science 257:96–105

Lagos-Quintana M, Rauhut R, Lendeckel W, Tuschl T (2001) Identification of novel genes coding for small expressed RNAs. Science 294 (5543):853–858

Lee RC, Ambros V (2001) An extensive class of small RNAs in *Caenorhabditis elegans*. Science 294 (5543):862–864

Lee RC, Feinbaum RL, Ambros V (1993) The *C. elegans* heterochronic gene lin-4 encodes small RNAs with antisense complementarity to lin-14. Cell 75(5):843–854

Lee HC, Li L, Gu W, Xue Z, Crosthwaite SK, Pertsemlidis A, Lewis ZA, Freitag M, Selker EU, Mello CC, Liu Y (2010) Diverse pathways generate MicroRNA-like RNAs and Dicer independent small interfering RNAs in fungi. Molecular Cell 38:803–814

Liang C, Liu H, Hao J, Li J, Luo L (2019) Expression profiling and regulatory network of cucumber miRNAs and their putative target genes in response to cucumber green motte mosaic infection. The Archives of Virology 164:121–1134

Li Y, Cao XL, Zhu Y, Yang XM, Zhang KN et al (2019a) Osa-miR398b boosts H_2O_2 production and rice blast disease-resistance via multiple superoxide dismutases. New Phytologist 222 (3):1507–1522

Li SJ, Castillo-Gonzalez C, Yu B, Zhang XR (2017) The functions of plant small RNAs in development and in stress responses. The Plant Journal 90:654–670

Liebsch D, Palatnik JF (2020) MicroRNA miR396 GRF transcription factors and GIF co-regulators: A conserved plant growth regulatory module with potential for breeding and biotechnology. Current Opinion in Plant Biology 53:31–42

Li K, Liu Z, Xing L, Wei Y, Mao J, Meng Y, Bao L, Han M, Zhao C, Zhang D (2019b) miRNAs associated with auxin signaling stress response and cellular activities mediate adventitious root formation in apple rootstocks. Plant Physiology and Biochemistry 139:66–81

Li Y, Lu YG, Shi Y, Wu L, Xu YJ, et al (2014) Multiple rice microRNAs are involved in immunity against the blast fungus *Magnaporthe oryzae*. Plant Physiology 164 (2):1077–1092

Lim LP, Glasner ME, Yekta S, Burge CB, Bartel DP (2003) Vertebrate microRNA genes. Science 299 (5612):1540–1540

Li F, Pignatta D, Bendix C, Brunkard JO, Cohn MM, Tung J et al (2012) MicroRNA regulation of plant innate immune receptors. Proceedings of the National Academy of Sciences of the U S A 109(5):1790–1795

Liu HH, Tian X, Li YJ, Wu CA et al (2008) Microarray based analysis of stress regulated microRNAs in *Arabidopsis thaliana*. RNA 14:836–843

Liu CG, Calin GA, Meloon B, Gamliel N, Sevignani C, Ferracin M, Dumitru CD, Shimizu M, Zupo S, Dono M (2004) An oligonucleotide microchip for genome-wide microRNA profiling in human and mouse tissues. Proceedings of the National Academy of Sciences 101 (26):9740–9744

Liu SR, Zhou JJ, Hu CG, Wei CL, Zhang JZ (2017) MicroRNA-mediated Gene silencing in plant defense and viral counter-defense. Frontiers Microbiology 8:1801

Li X, Wang X, Zhang S, Liu D, Duan Y, Dong W (2012) Identification of soybean microRNAs involved in soybean cyst nematode infection by deep sequencing. PLoS ONE 7(6):e39650

Li Y, Zhang Q, Zhang J, Wu L, Qi Y, Zhou JM (2010) Identification of microRNAs involved in pathogen-associated molecular pattern-triggered plant innate immunity. Plant Physiology 152 (4): 2222–2231

Li S, Zhang N, Zhu X, Ma R, Yang J, Tang X, Si H (2020) Enhanced drought tolerance with artificial microRNA-mediated StProDH1 gene silencing in potato. Crop Science 60(3):1462–1471

Louis J, Shah J (2013) *Arabidopsis thaliana–Myzus persicae* interaction: Shaping the understanding of plant defense against phloem-feeding aphids. Frontiers in Plant Science 4:213

Lucas-Barbosa D (2016) Integrating studies on plant–pollinator and plant–herbivore interactions. Trends in Plant Sciences 21:125–133

Lu P, Davis RF, Kemerait RC, van Iersel MW, Scherm H (2014) Physiological effects of *Meloidogyne incognita* infection on cotton genotypes with differing levels of resistance in the greenhouse. Journal of Nematology 46:352

Lu J, Wang C, Zhang F, Zeng D, Zhou Y (2021) Comparative microRNA profiling reveals microRNAs involved in rice resistant response to bacterial blight. The Crop Journal, 9(4):834–842

Madani B, Mirshekari A, Imahori Y (2019) Physiological Responses to Stress. *In:* Elhadi M. Yahia (eds) Postharvest Physiology and Biochemistry of Fruits and Vegetables. Woodhead Publishing, pp 405–423, ISBN 9780128132784

Manavella PA, Koenig D, Weigel D (2012). Plant secondary siRNA production determined by microRNA-duplex structure. Proceedings of the National Academy of Sciences 109(7):461–2466

Maghuly F, Ramkat RC, Laimer M (2014) Virus versus host plant microRNAs: Who determines the outcome of the interaction? PLoS One 9:e98263

Mao Y, Zhang H, Xu N, Zhang B, Gou F, Zhu JK (2013) Application of the CRISPR–Cas system for efficient genome engineering in plants. Molecular Plant 6:2008

Mateos JL, Bologna NG, Chorostecki U, Palatnik JF (2010) Identification of microRNA processing determinants by random mutagenesis of *Arabidopsis* MIR172a precursor Current Biology 20:49–54

Mathur S, Agrawal D, Jajoo A (2014) Photosynthesis: Response to high temperature stress. Journal of Photochemistry and Photobiology B: Biology 137:116–126

Mathur M, Nair A, Kadoo N (2020) Plant-pathogen interactions: MicroRNA-mediated trans-kingdom gene regulation in fungi and their host plants. Genomics 112(5):3021–3035

Matthews C, Arshad M, Hannoufa A (2019) Alfalfa response to heat stress is modulated by microRNA156. Physiologia Plantarum 165 (4):830–842

Ma JB, Ye K, Patel DJ (2004) Structural basis for overhang-specific small interfering RNA recognition by the PAZ domain. Nature 429:318–322

Medina C, da Rocha M, Magliano M, Ratpopoulo A, Revel B, Marteu N et al (2017) Characterization of microRNAs from *Arabidopsis* galls highlights a role for miR159 in the plant response to the root-knot nematode *Meloidogyne incognita*. New Phytologist 216:882–896

Millar AA, Waterhouse PM (2005) Plant and animal microRNAs: Similarities and differences. Functional & Integrative Genomics 5(3):129

Mishra R, Mohanty JN, Chand SK et al (2018) Can-miRn37a mediated suppression of ethylene response factors enhances the resistance of chilli against anthracnose pathogen *Colletotrichum truncatum* L. Plant Science 267:135–147

Mittler R (2006) Abiotic stress, the field environment and stress combination. Trends in Plant Science 11 (1):15–19

Miura K, Ikeda M, Matsubara A, Song XJ, Ito M, Asano K, Matsuoka M, Kitano H, Ashikari M (2010) OsSPL14 promotes panicle branching and higher grain productivity in rice. Nature Genetics 42 (6):545–549

Morkunas I, Mai VC, Gabrys B (2011) Phytohormonal signaling in plant responses to aphid feeding. Acta Physiologiae Plantarum 33:2057–2073

Naseem M, Srivastava M, Tehseen M, Ahmed N (2015) Auxin crosstalk to plant immune networks: A plant-pathogen interaction perspective. Current Protein & Peptide Science 16(5):389–394

Navarro L, Dunoyer P, Jay F, Arnold B, Dharmasiri N, Estelle M, Voinnet O, Jones JD (2006) A plant miRNA contributes to antibacterial resistance by repressing auxin signaling. Science 312(5772):436–439

Nelson PT, Baldwin DA, Scearce LM, Oberholtzer JC, Tobias JW, Mourelatos Z (2004) Microarray-based, high-throughput gene expression profiling of microRNAs. Nature Methods 1 (2):155–161

Ni Z, Hu Z, Jiang Q, Zhang H (2013) GmNFYA3, a target gene of miR169, is a positive regulator of plant tolerance to drought stress. *Plant Molecular Biology* 82:113–129

Niu QW, Lin SS, Reyes JL, Chen KC, Wu HW, Yeh SD, Chua NH (2006) Expression of artificial microRNAs in transgenic *Arabidopsis thaliana* confers virus resistance. Nature Biotechnology 24(11):1420–1428

Niu D, Xia J, Jiang C, Qi B, Ling X, Lin S, Zhang W, Guo J, Jin H, Zhao H (2016) *Bacillus cereus* AR156 primes induced systemic resistance by suppressing miR825/825* and activating defense-related genes in *Arabidopsis*. Journal of Integrative Plant Biology 58 (4):426–439

Ouyang T, Liu Z, Han Z, Ge Q (2019) MicroRNA detection specificity: Recent advances and future perspective. Analytical Chemistry 91 (5):3179–3186

Ouyang S, Park G, Atamian HS, Han CS, Stajich JE, Kaloshian I et al (2014) MicroRNAs suppress NB domain genes in tomato that confer resistance to *Fusarium oxysporum*. PLoS Pathogens 10(10):e1004464

Pacheco R, Garcıa-Marcos A, Barajas D, Martianez J, Tenllado F (2012) PVX-potyvirus synergistic infections differentially alter microRNA accumulation in *Nicotiana benthamiana*. Virus Research 165:231–235

Palatnik JF, Allen E, Wu X, Schommer C, Schwab R, Carrington JC, Weigel D (2003) Control of leaf morphogenesis by microRNAs. Nature 425 (6955):257–263

Pan X, Nichols RL, Li C, Zhang B (2019) MicroRNA-target gene responses to root knot nematode (*Meloidogyne incognita*) infection in cotton (*Gossypium hirsutum* L.). Genomics 111:383–390

Pantaleo V, Szittya G, Moxon S, Miozzi L, Moulton V, Dalmay T et al (2010) Identification of grapevine microRNAs and their targets using high-throughput sequencing and degradome analysis. The Plant Journal 62:960–976

Parker JS, Roe SM, Barford D (2004) Crystal structure of a PIWI protein suggests mechanisms for siRNA recognition and slicer activity. EMBO Journal 23:4727–4737

Patwa N, Nithin C, Bahadur RP, Basak J (2019) Identification and characterization of differentially expressed *Phaseolus vulgaris* miRNAs and their targets during mungbean yellow mosaic India virus infection reveals new insight into *Phaseolus*-MYMIV interaction. Genomics 111(6):1333–1342

Peace RJ, Hassani MS, Green JR (2019) miPIE: NGS-based prediction of miRNA using integrated evidence. Scientific Reports 9 (1):1–10

Pegler JL, Oultram JMJ, Grof CPL, Eamens AL (2019) DRB1, DRB2, and DRB4 are required for appropriate regulation of the microRNA399/PHOSPHATE2 expression module in *Arabidopsis thaliana*. Plants 8:124

Perez-Quintero AL, Quintero A, Urrego O, Vanegas P, Lopez C (2012) Bioinformatic identification of cassava miRNAs differentially expressed in response to infection by *Xanthomonas axonopodis* pv *manihotis*. BMC Plant Biology 12:29

Piya S, Kihm C, Rice JH, Baum TJ, Hewezi T (2017) Cooperative regulatory functions of miR858 and MYB83 during cyst nematode parasitism. Plant Physiology 174(3):1897–1912

Pradhan B, Naqvi AR, Saraf S, Mukherjee SK, Dey N (2015) Prediction and characterization of tomato leaf curl new Delhi virus (ToLCNDV) responsive novel microRNAs in *Solanum lycopersicum*. Virus Research 195:183–195

Pritchard CC, Cheng HH, Tewari M (2012) MicroRNA profiling: Approaches and considerations. Nature Reviews Genetics 13 (5):358–369

Rajwanshi R, Devi KJ, Sharma GR, Lal B (2019) Role of miRNAs in plant-microbe interaction. *In:* Kumar M., Muthusamy A., Kumar V., Bhalla-Sarin N. (eds) In vitro Plant Breeding towards Novel Agronomic Traits. Springer, Singapore. https://doi.org/10.1007/978-981-32-9824-8_10

Rambani A, Hu Y, Piya S, Long M, Rice JH, Pantalone V, Hewezi T (2020) Identification of differentially methylated miRNA genes during compatible and incompatible interactions between soybean and soybean cyst nematode. Molecular Plant-Microbe Interactions 33(11):1340–1352

Ramesh SV, Chouhan BS, Kumar G, Praveen S, Chand S (2017a) Expression dynamics of *Glycine max* (L) Merrill microRNAs (miRNAs) and their targets during Mungbean yellow mosaic India virus (MYMIV) infection. Physiological and Molecular Plant Pathology 100:13–22

Ramesh SV, Ratnaparkhe MB, Kumawat G, Gupta GK, Husain SM (2014) Plant miRNAome and antiviral resistance: A retrospective view and prospective challenges. Virus Genes 48:1–14

Ramesh SV, Sahu PP, Prasad M, Praveen S, Pappu HR (2017b) Geminiviruses and plant hosts: A closer examination of the molecular arms race. Viruses 9:256

Reyes CA, Ocolotobiche EE, Marmisolle FE, Robles Luna G, Borniego MB, Bazzini AA et al (2016) Citrus *Psorosis* virus 24 K protein interacts with citrus miRNA precursors affects their processing and subsequent miRNA accumulation and target expression. Molecular Plant Pathology 17:317–329

Rhoades MW, Bartel DP, Bartel B (2006) MicroRNAs and their regulatory roles in plants. Annual Review of Plant Biology 57:19–53

Robert-Seilaniantz A, Grant M, Jones JDG (2011) Hormone crosstalk in plant disease and defense: More than just jasmonate-salicylate antagonism. The Annual Review of Phytopathology 49:317–343

Rogers K, Chen X (2013) Biogenesis turnover and mode of action of plant microRNAs. Plant Cell 25(7):2383–99

Saini RP, Raman V, Dhandapani G, Malhotra EV, Sreevathsa R, Kumar PA, Sharma TR, Pattanayak D (2018) Silencing of HaAce1 gene by host-delivered artificial microRNA disrupts growth and development of *Helicoverpa armigera*. PLoS One 13(3):e0194150.

Salvador-Guirao R, Baldrich P, Weigel D, Rubio-Somoza I, San Segundo B (2018) The microRNA miR773 is involved in the *Arabidopsis* immune response to fungal pathogens. Molecular Plant-Microbe Interactions 31(2):249–259

Sattar S, Addo-Quaye C, Thompson GA (2016) miRNA-mediated auxin signalling repression during vat-mediated aphid resistance in *Cucumis melo*. Plant, Cell and Environment 39 (6):1216–1227

Schwarz DS, Hutvagner G, Du T, Xu ZS, Aronin N, Zamore PD (2003) Asymmetry in the assembly of the RNAi enzyme complex. Cell 115:199–208

Schwarzkopf M, Pierce NA (2016) Multiplexed miRNA northern blots via hybridization chain reaction. Nucleic Acids Research. 44 (15):e129–e129

Shahbaz M, Pilon M (2019) Conserved Cu-microRNAs in *Arabidopsis thaliana* function in copper economy under deficiency. Plants 8:141

Shivaprasad PV, Chen HM, Patel K, Bond DM et al (2012) A microRNA superfamily regulates nucleotide binding site leucine-rich repeats and other mRNAs. Plant Cell 24:859–874

Shivaraj S, Dhakate P, Mayee P, Negi M, Singh A (2014) Natural genetic variation in MIR172 isolated from *Brassica* species. Biologia Plantarum 58 (4):627–640

Shriram V, Kumar V, Devarumath RM, Khare TS, Wani SH (2016) MicroRNAs as potential targets for abiotic stress tolerance in plants. Frontiers in Plant Science 7:817

Simon-Mateo C, Garcia JA (2006) MicroRNA-guided processing impairs plum pox virus replication but the virus readily evolves to escape this silencing mechanism. The Journal of Virology 80:2429–2436

Singh PK, Campbell MJ (2013) The interactions of microRNA and epigenetic modifications in prostate cancer. Cancer 5:998–1019

Snyman MC, Solofoharivelo MC, Souza-Richards R, Stephan D, Murray S, Burger JT (2017) The use of high-throughput small RNA sequencing reveals differentially expressed microRNAs in response to aster yellows phytoplasma-infection in *Vitis vinifera* cv ‘Chardonnay’. PLoS One 12 (8):e0182629

Song JB, Gao S, Wang Y, Li BW, Zhang YL, Yang ZM (2016) miR394 and its target gene LCR are involved in cold stress response in *Arabidopsis*. Plant Gene 5:56–64

Song JJ, Smith SK, Hannon GJ, Joshua-Tor L (2004) Crystal structure of Argonaute and its implications for RISC slicer activity. Science 305:1434–1437

Song C, Yu M, Han J, Wang C, Liu H, Zhang Y et al (2012) Validation and characterization of *Citrus sinensis* microRNAs and their target genes. BMC Research Notes 5:235

Soto-Suarez M, Baldrich P, Weigel D, Rubio-Somoza I, San Segundo B (2017) The *Arabidopsis* miR396 mediates pathogen-associated molecular pattern-triggered immune responses against fungal pathogens. Scientific Reports 7:44898

Stief A, Altmann S, Hoffmann K, Pant BD, Scheible W-R, Bäurle I (2014) *Arabidopsis* miR156 regulates tolerance to recurring environmental stress through SPL transcription factors. The Plant Cell 26 (4):1792–1807

Sun G (2012) MicroRNAs and their diverse functions in plants. Plant Molecular Biology 80:17–36

Sunkar R, Girke T, Jain PK, Zhu J-K (2005) Cloning and characterization of microRNAs from rice. The Plant Cell 17 (5):1397–1411

Sunkar R, Li YF, Jagadeeswaran G (2012) Functions of microRNAs in plant stress responses. Trends in Plant Science 17(4):196–203

Sunkar R, Zhu JK (2004) Novel and stress-regulated microRNAs and other small RNAs from *Arabidopsis*. The Plant Cell 16 (8):2001–2019

Tan Gana NH, Victoriano AF, Okamoto T (2012) Evaluation of online miRNA resources for biomedical applications. Genes to Cells 17 (1):11–27

Tian B, Wang S, Todd TC, Johnson C, Tang G, Trick HN (2017) Genome-wide identification of soybean microRNA responsive to soybean cyst nematodes infection by deep sequencing. BMC Genomics 18:572

Trda L, Boutrot F, Claverie J, Brule D, Dorey S, Poinssot B (2015) Perception of pathogenic or beneficial bacteria and their evasion of host immunity: Pattern recognition receptors in the frontline. Frontier Plant Science 6:219

Ueno T, Funatsu T (2014) Label-free quantification of microRNAs using ligase-assisted sandwich hybridization on a DNA microarray. PLoS One 9 (3):e90920

Usha S, Jyothi M, Suchithra B, Dixit R, Rai D (2017) Computational identification of microRNAs and their targets from finger millet (*Eleusine coracana*). Interdisciplinary Sciences: Computational Life Sciences 9 (1):72–79

Valoczi A, Hornyik C, Varga N, Burgyan J, Kauppinen S, Havelda Z (2004) Sensitive and specific detection of microRNAs by northern blot analysis using LNA-modified oligonucleotide probes. Nucleic Acids Research 32 (22):e175–e175

Wamiq G, Khan JA (2018) Overexpression of ghr-miR166b generates resistance against *Bemisia tabaci* infestation in *Gossypium hirsutum* plants. Planta 247 (5):1175–1189

Wang H, Jiao X, Kong X, Hamera S, Wu Y, Chen X, Fang R, Yan Y (2016) A signaling Cascade from miR444 to RDR1 in rice antiviral RNA silencing pathway. Plant Physiology 170 (4):2365–2377

Wang ST, Sun XL, Hoshino Y, Yu Y, Jia B, Sun ZW, Sun MZ, Duan XB, Zhu YM (2014) MicroRNA319 positively regulates cold tolerance by targeting OsPCF6 and OsTCP21 in rice (*Oryza sativa* L.). PLoS one 9(3):e91357

Wang WC, Lin FM, Chang WC, Lin KY, Huang HD, Lin NS (2009) miRExpress: Analyzing high-throughput sequencing data for profiling microRNA expression. BMC Bioinformatics 10 (1):1–13

Wang W, Liu D, Chen D, Cheng Y, Zhang X, Song L, Hu M, Dong J, Shen F (2019) MicroRNA414c affects salt tolerance of cotton by regulating reactive oxygen species metabolism under salinity stress. RNA Biology 16 (3):362–375

Wang Y, Sun F, Cao H, Peng H, Ni Z, Sun Q, Yao Y (2012) TamiR159 directed wheat TaGAMYB cleavage and its involvement in anther development and heat response. PLoS One 7 (11):e48445

Warthmann N, Chen H, Ossowski S, Weigel D, Herve P (2008) Highly specific gene silencing by artificial miRNAs in rice. PLoS One 3(3):e1829

Werner S, Wollmann H, Schneeberger K, Weigel D (2010) Structure determinants for accurate processing of miR172a in *Arabidopsis thaliana*. Current Biology 20:42–48

Wightman B, Ha I, Ruvkun G (1993) Posttranscriptional regulation of the heterochronic gene lin-14 by lin-4 mediates temporal pattern formation in *C. elegans* Cell 75(5):855–862

Wong J, Gao L, Yang Y, Zhai J, Arikit S, Yu Y, Duan S, Chan V, Xiong Q, Yan J (2014) Roles of small RNA s in soybean defense against *Phytophthora sojae* infection The Plant Journal 79 (6):928–940

Wu FL, Qi JY, Meng X, Jin WB (2020) miR319c acts as a positive regulator of tomato against *Botrytis cinerea* infection by targeting TCP29. Plant Science 300:110610

Wu J, Yang Z, Wang Y, Zheng L, Ye R, Ji Y et al (2015) Viral-inducible Argonaute18 confers broad-spectrum virus resistance in rice by sequestering a host microRNA. Elife 4:e05733

Wu J, Yang R, Yang Z, Yao S, Zhao S, Wang Y, Li P, Song X, Jin L, Zhou T, Lan Y et al (2017) ROS accumulation and antiviral defence control by microRNA528 in rice. Nature Plants 3 (1):16203

Wu L, Zhou H, Zhang Q, Zhang J, Ni F, Liu C, Qi Y (2010) DNA methylation mediated by a microRNA pathway. Molecular Cell 38:465–475

Xia K, Wang R, Ou X, Fang Z, Tian C, Duan J, Wang Y, Zhang M (2012) OsTIR1 and OsAFB2 downregulation via OsmiR393 overexpression leads to more tillers, early flowering and less tolerance to salt and drought in rice. PLoS One 7(1):e30039

Xia Z, Zhao Z, Li M, Chen L, Jiao Z, Wu Y, Zhou T, Yu W, Fan Z (2018) Identification of miRNAs and their targets in maize in response to sugarcane mosaic virus infection. Plant Physiology and Biochemistry 125:143–152

Xie F, Frazier TP, Zhang B (2011) Identification, characterization and expression analysis of microRNAs and their targets in the potato (*Solanum tuberosum*). Gene 473(1):8–22

Xie M, Zhang S, Yu B (2015) microRNA biogenesis, degradation and activity in plants. Cellular and Molecular Life Sciences 72:87–99

Xie Z, Allen E, Fahlgren N, Calamar A, Givan SA, Carrington JC (2005) Expression of *Arabidopsis* MIRNA genes. Plant Physiology 38(4):2145–2154

Yang ZY, Ebright YW, Yu B, Chen XM (2006) HEN1 recognizes 21-24 nt small RNA duplexes and deposits a methyl group onto the 2′ OH of the 3′ terminal nucleotide. Nucleic Acids Research 34:667–675

Yang L, Jue D, Li W, Zhang R, Chen M, Yang Q (2013) Identification of miRNA from eggplant (*Solanum melongena* L) by small RNA deep sequencing and their response to *Verticillium dahliae* infection. PLoS One 8(8):e72840

Yang C, Li D, Mao D, Liu X, Ji C, Li X, Zhao X, Cheng Z, Chen C, Zhu L (2013) Overexpression of microRNA 319 impacts leaf morphogenesis and leads to enhanced cold tolerance in rice (*Oryza sativa* L.). Plant, Cell and Environment 36 (12):2207–2218

Yang L, Mu X, Liu C, Cai J, Shi K, Zhu W et al (2015) Overexpression of potato miR482e enhanced plant sensitivity to *Verticillium dahliae* infection. Journal of Integrative Plant Biology 57:1078–1088

Yan Z, Hossain MS, Valdes-Lopez O, Hoang NT, Zhai J et al (2016) Identification and functional characterization of soybean root hair microRNA s expressed in response to *Bradyrhizobium japonicum* infection. Plant Biotechnology Journal 14(1) 332–341

Yao Y, Guo G, Ni Z, Sunkar R, Du J, Zhu J-K, Sun Q (2007) Cloning and characterization of microRNAs from wheat (*Triticum aestivum* L.). Genome Biology 8 (6):1–13

Ye W, Shen CH, Lin Y, Chen PJ, Xu X, Oelmüller R, Yeh KW, Lai Z (2014) Growth promotion-related miRNAs in *Oncidium orchid* roots colonized by the endophytic fungus *Piriformospora indica*. PLoS One 9(1):e84920

Yin Z, Murawska Z, Xie F, Pawełkowicz M, Michalak K, Zhang B, Lebecka R (2019) microRNA response in potato virus Y infected tobacco shows strain-specificity depending on host and symptom severity. Virus Research 260:20–32

Yin JQ, Zhao RC, Morris KV (2008) Profiling microRNA expression with microarrays. Trends in Biotechnology 26 (2):70–76

Yogindran S, Rajam MV (2021) Host-derived artificial miRNA-mediated silencing of ecdysone receptor gene provides enhanced resistance to *Helicoverpa armigera* in tomato. Genomics 113(1 Pt 2):736–747

Yu X, Hou Y, Chen W, Wang S, Wang P, Qu S (2017) *Malus hupehensis* miR168 targets to ARGONAUTE1 and contributes to the resistance against *Botryosphaeria dothidea* infection by altering defense responses. Plant and Cell Physiology 58(9):1541–1557

Yu B, Yang ZY, Li JJ, Minakhina S, Yang MC, Padgett RW et al (2005) Methylation as a crucial step in plant microRNA biogenesis. Science 307: 932–935

Zang W, Gao S, Zhou X, Chellappan P, Chen Z, Zhou X, Zhang X, Fromuth N, Coutino G, Coffey M, Jin H (2011) Bacteria-responsive microRNAs regulate plant innate immunity by modulating plant hormone networks. Plant Molecular Biology 75(1–2): 93–105

Zhang B (2015) MicroRNA: A new target for improving plant tolerance to abiotic stress. Journal of Experimental Botany 66 (7):1749–1761

Zhang L, Chia JM, Kumari S, Stein JC, Liu Z, Narechania A, Maher CA, Guill K, McMullen MD, Ware D (2009) A genome-wide characterization of microRNA genes in maize. PLoS Genetics 5 (11):e1000716

Zhang P, Liu Y, Zhang Y, Liu C, Wang Z, Li Z (2013) Multiplex ligation-dependent probe amplification (MLPA) for ultrasensitive multiplexed microRNA detection using ribonucleotide-modified DNA probes. Chemical Communications 49 (85):10013–10015

Zhang J, Li Z, Wang H, Wang Y, Jia H, Yan J (2011a) Ultrasensitive quantification of mature microRNAs by real-time PCR based on ligation of a ribonucleotide-modified DNA probe. Chemical Communications 47 (33):9465–9467

Zhang Q, Li Y, Zhang Y, Wu C, Wang S, Hao L, Wang S, Li T (2017) md-miR156ab and md-miR395 target WRKY transcription factors to influence apple resistance to leaf spot disease. Frontiers of Plant Science 8:526

Zhang B, Wang Q, Wang K, Pan X, Liu F, Guo T, Cobb GP, Anderson TA (2007) Identification of cotton microRNAs and their targets. Gene 397(1–2):26–37

Zhang B, Wang Q (2015) MicroRNA-based biotechnology for plant improvement. Journal of Cellular Physiology 230(1):1–5

Zhang W, Zhang H, Qi F, Jian G (2016) Generation of transcriptome profiling and gene functional analysis in *Gossypium hirsutum* upon *Verticillium dahliae* infection. Biochemical and Biophysical Research Communications 473(4):879–885

Zhang T, Zhao YL, Zhao JH, Wang S, Jin Y, Chen ZQ, Fang YY, Hua CL, Ding SW, Guo HS (2016) Cotton plants export microRNAs to inhibit virulence gene expression in a fungal pathogen. Nature Plants 2(10):16153

Zhang X, Zou Z, Gong P, Zhang J, Ziaf K, Li H, Xiao F, Ye Z (2011b) Over-expression of microRNA169 confers enhanced drought tolerance to tomato. Biotechnology Letters 33 (2):403–409

Zhao B, Liang R, Ge L, Li W, Xiao H, Lin H, Ruan K, Jin Y (2007) Identification of drought-induced microRNAs in rice. Biochemical and Biophysical Research Communications 354 (2):585–590

Zhao Y, Zhang C, Liu W, Gao W, Liu C, Song G, Li WX, Mao L, Chen B, Xu Y (2016) An alternative strategy for targeted gene replacement in plants using a dual-sgRNA/Cas9 design. Scientific Reports 6 (1):1–11

Zhou J, Deng K, Cheng Y, Zhong Z, Tian L, Tang X, Tang A, Zheng X, Zhang T, Qi Y (2017) CRISPR-Cas9 based genome editing reveals new insights into microRNA function and regulation in rice. Frontiers in Plant Science 8:1598

Zhou X, Khare T, Kumar V (2020) Recent trends and advances in identification and functional characterization of plant miRNAs. Acta Physiologiae Plantarum 42 (2):1–21

Zhou X, Wang G, Sutoh K, Zhu JK, Zhang W (2008) Identification of cold-inducible microRNAs in plants by transcriptome analysis. Biochimica et Biophysica Acta 1779 (11):780–788

Zhou Q, Wang Z, Zhang J, Meng H, Huang B (2012) Genome-wide identification and profiling of microRNA-like RNAs from *Metarhizium anisopliae* during development. Fungal Biology 116(11):1156–1162

Zhu QH, Fan L, Liu Y, Xu H, Llewellyn D, Wilson I (2013) miR482 regulation of NBS-LRR defense genes during fungal pathogen infection in cotton. PLoS ONE 8(12):e84390

6 Identification and Characterization of Plant ncRNA Responsive to Climate Change

Suraj Patil, Shrusthti Joshi and Vinay Kumar
Department of Biotechnology, Savitribai Phule Pune University, Pune, India

6.1 INTRODUCTION

One of the greatest challenges faced by agriculturists in this century are the changes in climatic conditions, which include varied weather patterns, amplified temperature ranges, acidification of oceans resulting in a decline in corals, imperiled ecosystems and, ultimately, greatly impacted food security. The Intergovernmental Panel on Climate Change (IPCC) defines climate changes as "a change in the state of the climate that can be identified (e.g., using statistical tests) by changes in the mean and/or the variability of its properties, and that persists for an extended period, typically decades or longer" (Cubasch et al. 2021). It is often observed that the terms *weather* and *climate* are used interchangeably, which is a misconception; weather pertains to short-term changes in local or regional atmospheric conditions, while climate refers to long-term atmospheric conditions.

Agricultural practices and agroecosystems are impacted severely by climate change. The change in temperature and/or precipitation pattern affects the growth and yield of crops, making their maintenance more difficult. Land degradation due to heavy rainfall, drought severity, heat stress, dry spells, wind, sea-level rise and permafrost thaw are other effects of climatic change (Shukla et al. 2019). Food security, food safety and sustainable agriculture development have been on focus and new research is necessary to reduce the impact of climatic change, which causes loss of yield, plant growth and development. This is supported by various studies published between 1980 and 2015 that illustrate the global decline in the yield of wheat, maize and cereal crops by 21%, 40% and 10%, respectively, under drought conditions due to climate change. Further, a decline of yield by 6 ± 2.9%, 7.4 ± 4.5%, 3.2 ± 3.7% and 3.1% in wheat, maize, rice and soybean, respectively, was reported with every °C rise in global temperature (Asseng et al. 2015; Daryanto et al. 2016; Zhoa et al. 2017; Kim et al. 2019).

Plants being sessile in nature have developed stress-responsive pathways that inherently help them overcome the challenges posed due to changes in their optimal environmental condition due to climate. One of the mechanisms involves extensive use of post-transcriptional regulation by non-coding RNAs (ncRNAs). NcRNAs refer to transcripts that no longer code for proteins and instead play vital regulatory roles that are post-transcriptional in nature. Untranslated transcripts of RNA molecules are marked as ncRNA. The first ncRNA was found in baker's yeast and was characterized as an alanine tRNA (Holley et al. 1965), leading to the discovery of several other regulatory ncRNAs of different shapes and sizes. These controlled RNA interferences (RNAi) play a major role in the adaptation of plants to the abiotic stresses occurring due to climatic changes, conferring them with tolerance for their survival. These ncRNAs are often observed in the gene silencing phenomenon that involves sequence-specific gene legislation encapsulated via double-stranded RNA, resulting in inhibition of protein formation and expression of the gene (Zheng and Qu 2015; Brant and Budak 2018;

DOI: 10.1201/9781003369288-6

Yan et al. 2020, 2021). A large number of ncRNAs have been identified and characterized in plants such as *Arabidopsis thaliana, Zea mays, Vitis vinifera, Manihot esculenta* and *Medicago truncatula* (Kapusta et al. 2013; Li et al. 2014; Liu et al. 2015; Gao et al. 2019; Zhang et al. 2019b; Li et al. 2021).

ncRNAs are categorized based on their size into small and long non-coding RNAs (lncRNA). Furthermore, functionally vital RNAs (transfer RNA [tRNA], ribosomal RNA [rRNA]), small ncRNAs (micro-RNAs [miRNAs]), small interfering RNAs (siRNAs), piwi-interacting RNAs (piRNAs), circular RNA (circRNA), and the lncRNAs (X-inactivation-specific transcript [XIST] and the HOX-antisense-intergenic RNA [HOTAIR] etc.) (Renyi and Jiankang 2014; Santosh et al. 2014; Brant and Budak 2018) have been created. These ncRNAs work by post-transcriptional gene silencing (PTGS) as discovered in many organisms, such as fungi, animals, and plants (Brant and Budak 2018). The mRNA formed is cleaved by the DICER or Dicer-like enzyme. Small ncRNA consists of an RNA-induced silencing complex (RISC) and Argonaute proteins (AGOs). Drosha and Pasha are part of the microprocessor protein complex wherein Drosha and Dicer are RNase III enzymes, Pasha is a dsRNA-binding protein and Argonautes are RNase H enzymes (Wilson and Doundna, 2013; Kamthan et al. 2015; Brant and Budak 2018).

The knowledge on ncRNAs and their importance in cell regulatory pathways has impressively increased, hence, much attention has been placed on the role of ncRNAs in regulatory processes and establishing data on newly discovered classes of ncRNAs like piRNAs (Siomi and Siomi, 2009) and tRNA-derived fragments (tRFs) (Thompson and Parker 2009; Garcia-Silva et al. 2012; Martens-Uzunova et al. 2013; Kumar et al. 2014). To support the current research in this field, several dedicated computational sources have been established, allowing quick, easy and detailed access to ncRNA-related data. Along with the databases, various identification techniques have been developed to suit the needs of ncRNA characterization. Different classes of ncRNAs have specific databases where their identification and characterization can be conducted. These ncRNAs are often validated in the lab using RLM-RACE, blotting, PCR analysis, etc. In this chapter, we discuss the advancement of ncRNA research in plants and focus on identification and characterization techniques and their application in crop improvement under climatic changes.

6.2 IDENTIFICATION AND CHARACTERIZATION OF ncRNAs: TRENDS AND DEVELOPMENT

The current development and advancement in sequencing technologies have facilitated multitudinal research analysis. This resulted in the discovery of genes, transcriptional factors, gene families and genes associated with agronomic traits. Along with this, a new area of ncRNA-mediated gene regulation has also been identified, which is formed from ncRNA transcripts (Paul and Dhandapani 2019). In addition, the availability of sequence data in the form of expressed sequence tags (ESTs), complementary DNA (cDNA), microarrays, whole genome sequence (WGS), and RNA sequencing (RNA seq) have further enhanced the identification process. The advancement in sequencing technologies that has made the process more specific, fast and easily available has provided a new paradigm in ncRNA research. A brief description of modern tools and techniques currently in use for identifying and characterizing different classes of ncRNAs and their related work is given next (Figure 6.1).

6.2.1 Next-Generation Sequencing

The advent of next-generation sequencing (NGS) technologies has made specific sequencing possible for identification of ncRNA, enabling us to acquire data in this field. NGS, also called deep sequencing technologies, has revolutionized the discovery and identification of various novel and conserved ncRNAs (Si et al. 2017). The degradome sequencing technique is ncRNA-specific, wherein it provides a plant profile containing species of ncRNAs, such as miRNA, along with their target mRNAs, allowing the identification novel miRNAs and target mRNAs. It is utilized

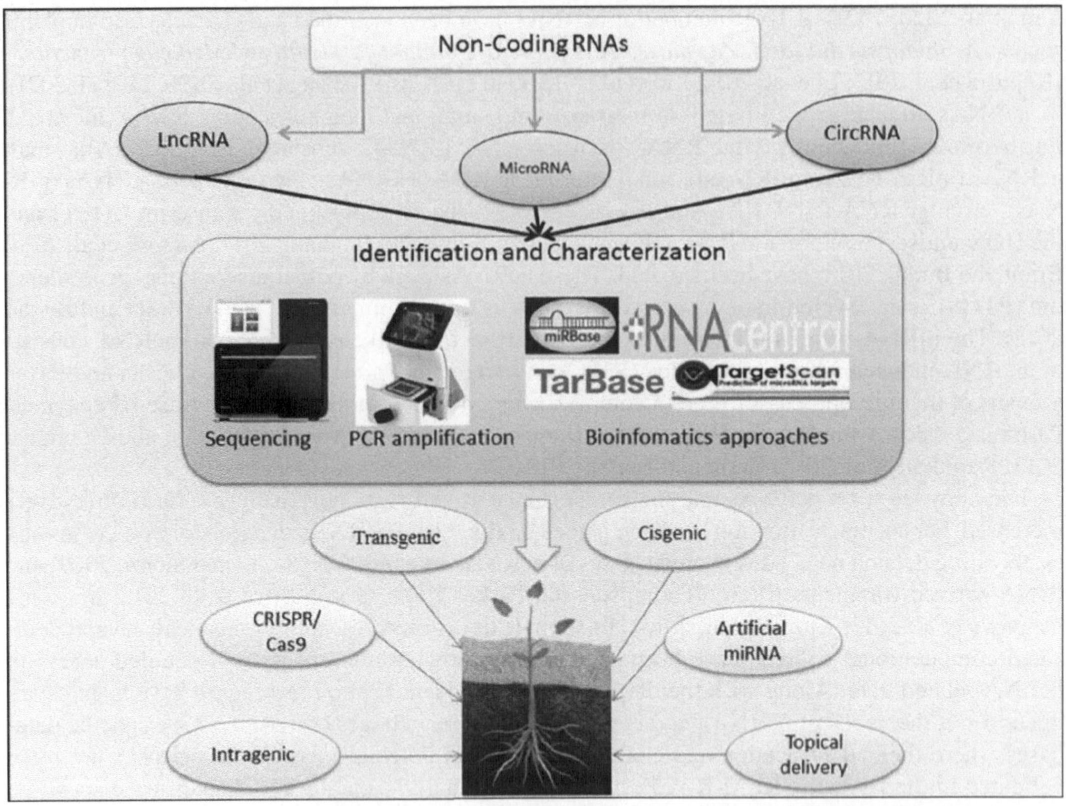

FIGURE 6.1 Approaches used for identification and characterization of ncRNA for crop improvement.

for degraded RNA (hence its name) under which sequencing and identification of microRNA (miRNA), short-interfering RNA (siRNA), turnover RNA fragments, and miRNA-/siRNA-cleaved target RNA fragments can be performed. It combines modified 5′ rapid amplification of cDNA ends (5′-RACE) and NGS for specific amplification and isolation of ncRNAs (Lin et al. 2019). Third-generation sequencing, like single-molecule real-time (SMRT) sequencing (Pacific Biosciences), Helicos sequencing (genetic analysis system) and Nanopore are also utilized for identification of ncRNAs and their targets (Patil et al. 2021). These methods have enabled several genome-wide identifications, which enriched our knowledge with growing numbers of ncRNAs publicly available.

This was reported in *Populus*, where degradome sequencing has been applied to identify cold responsive miRNAs and their targets using Illumina HiSeq 2000 (Bao et al. 2019). In fruits of *Solanum lycopersicum* L., genome-wide identification of circRNAs was performed by deep sequencing at different ripening stages to analyze the functional significance of Phytoene Synthase 1 (PSY1)-derived circRNA and its role in carotenoid biosynthesis (Tan et al. 2017). For this, the total RNA was extracted and filtered using a Ribo-Zero Magnetic Kit (Epicentre, Madison, WI, USA) for rRNA removal, and linear RNA was digested by RNase R (Epicentre, Madison, WI, USA). The RNA-Seq library was prepared using the TruSeq RNA LT Sample Prep Kit v2 (Illumina, San Diego, CA) and high-throughput sequencing was performed on the HiSeq 2500 instrument with 100 bp paired-end reads (Genenergy Biotechnology Co., Ltd, Shanghai, China). The sequencing techniques have allowed cloning and comparative analysis of small RNA and provide a baseline for identification of miRNAs. Using this method, identification of 16 *A. thaliana* miRNAs was conducted to show different expression patterns during different developmental stages (Reinhart et al. 2002). Further, based on this technique, ESTs in plants like *A. thaliana*, *Oryza sativa and Brassica rapa* were used for miRNA identification (Archak and Nagaraju 2007; Zhang et al. 2009; Dhandapani et al. 2011). Similarly, the HiSeq 2000 sequencing platform (Illumina, San Diego, CA, USA) was used in identification of high-temperature-responsive lncRNAs in an abiotic stress-tolerant poplar species (*Populus simonii*) using strand-specific RNA sequencing (ssRNA-seq) (Song et al. 2020a).

The procedure of ncRNA identification evolves with the advancement in technologies and techniques. A recently developed technique, miRNA sequencing (miRNA seq) and lncRNA sequencing (lncRNA seq) have been designed to focus on specific ncRNAs of interest. This resulted in identification of novel miRNAs and lncRNAs that were unidentifiable earlier. The drawback of these methods lies in their expense. Hence, RNA seq experiments became widely used due to their cost-effective identification of ncRNAs (Paul and Dhandapani 2019).

6.2.2 PCR Identification

The identification of lncRNA in plants is carried out by utilizing cDNA sequences, RNA sequencing and strand-specific RNA sequencing technologies in plants (Wen et al. 2020; Liu et al. 2015a). These are usually identified by PCR amplification. The potential target genes are divided into cis- and trans-acting of lncRNAs according to their regulatory methods. RNAplex is used to screen trans-target genes by calculating the binding free energy of lncRNA and its co-expressed protein-coding genes (Zhang et al. 2019). LncRNA973 was identified in *Gossypium* spp by conducting RT-PCR analysis. For this identification, the isolated RNA was used for first-strand cDNA synthesis, which was later used as a template for lncRNA and mRNAs qRT-PCR amplification. This was later used for miRNA expression detection using Mir-X™ miRNA First-Strand Synthesis and TB Green™ Premix Ex Taq™ II (Tli RNaseH Plus) (TaKaRa) (Zhang et al. 2019).

6.2.3 Bioinformatics Approaches

Along with lab approaches, *in silico* analysis using bioinformatics methods has enabled identification of ncRNAs. Various databases have been developed that collectively showcase known and conserved sequences according to a researcher's needs. From the datasets established by the software and its makers, the candidate genes along with their targets are often identified and later characterized structurally, functionally and physiologically by comparison with a homologous base. NONCODE (http://www.noncode.org/) is a comprehensive database harboring ncRNAs, especially lncRNAs. NONCODEV6 provides the full scope of lncRNAs across plants and animals. The number of plant lncRNAs in NONCODEV6 is about 94,697. It shows lncRNA annotations with tissue expression profiles and predicted functions in five common plants along with its conservation annotation at the transcript level for 23 plant species (Zhao et al. 2021a). The Plant ncRNA database (PNRD; http://structuralbiology.cau.edu.cn/PNRD) has been developed to provide a way to conduct fundamental research on plant ncRNAs. It allows the users access to public resources. A collection of 25,739 entries belonging to 11 different types of ncRNAs from 150 plant species is available. Its advantages include a user-friendly interface, multifunctional enhanced browsing options and more background data for users to download. Also, it has integrated currently prevalent technologies and toolkits to strengthen the capability of the database (Yi et al. 2015).

For miRNA detection, the most commonly used databases are Plant miRNA database (PMRD) (http://bioinformatics.cau.edu.cn/PMRD/) (Zhang et al. 2010), Tarbase (http://dianalab.e-ce.uth.gr/html/diana/web/index.php?r=tarbasev8) (Karagkoni et al. 2018), MiRDB (http://www.mirdb.org/) (Chen and Wang 2020) and miRbase (https://www.mirbase.org/) (Kozomoro et al. 2019) that contain all published miRNAs based on which can compare and identify novel miRNAs in plant species. Some other known miRNA databases include miRNEST 2.0 (http://mirnest.amu.edu.pl) (Szcześniak and Makałowska 2014, Target prediction for plant miRNAs (TAPIR) (http://bioinformatics.psb.ugent.be/webtools/tapir/) (Bonnet et al. 2010), PlanTE-MIR DB (http://bioinfo-tool.cp.utfpr.edu.br/plantemirdb/) (Lorenzetti et al. 2016), Plant miRNA Encyclopedia (PmiREN) (http://www.pmiren.com/)

(Guo et al. 2020), Degradome-Based Plant MiRNA–Target Interaction And Network Database (DPMIND) (https://cbi.njau.edu.cn/DPMIND/) (Fei et al. 2018) and Plant microRNA Knowledge Base (PmiRKB) (https://bis.zju.edu.cn/pmirkb/) (Meng et al. 2011). In addition to databases, various tools have been developed for the identification of miRNAs from sequencing data. Web-based and downloadable tools like RNA22 (https://cm.jefferson.edu/rna22/) (Loher and Rigoutsos, 2012), PITA (https://www.mybiosoftware.com/pita-6-microrna-prediction-tool.html), psRNATarget (https://www.zhaolab.org/psRNATarget/, miRNA-truncation and tailing analysis (miTRATA), and Iwa-miRNA (Zhang et al. 2022) are commonly used for miRNA characterization.

PlantCircNet, a database for plant circRNA–miRNA–mRNA regulatory networks, (http://bis.zju.edu.cn/plantcircnet/ and GreenCircRNA, for plant circRNAs that act as miRNA decoys, are databases specific for circRNA identification and characterization also used to study miRNA-circRNA interaction. CIRI is an de novo algorithm that uses NGS to specifically detect circRNA (Gao et al. 2015). It is an unbiased, annotation-independent approach that accurately detects novel circRNAs based on paired chiastic clipping (PCC) signals combined with a filtering system able to remove false positives. It is able to specifically identify intronic/intergenic circRNAs as well as fragments. Similarly, lncRScan is used to specifically detect lncRNA by overcoming the technical problems encountered in RNA seq. It eliminates partial transcripts and artifacts that occur due to the specificity of RNA seq in the assembled transcriptome. Along with this, the identification of lncRNA from the protein-coding mRNAs is achieved (Sun et al. 2015). Short Uncharacterized RNA Finder (SURFr) is another recent algorithm for identification and analysis of ncRNA-derived RNAs (Kasukurthi et al. 2021). For tRNA, tsRNAsearch is a Nextflow pipeline for the identification of tRNA and ncRNA fragments from small RNA seq data (https://github.com/GiantSpaceRobot/tsRNAsearch) (Donovan et al. 2021).

6.3 PIPELINES USED FOR CLIMATE RESILIENCE CROP PRODUCTION

During stress conditions, numerous genes are differentially expressed as part of a plant's response to stress tolerance. These genes usually include transcription factors, ROS scavengers, protein stabilizer and ncRNAs such as miRNAs, circRNAs, LncRNAs, etc. Manipulating these differentially expressed genes can aid abiotic stress-susceptible crops. Developing a climatic smart-crop using biotechnological tools can be achieved by using various strategies such as gene overexpression, downexpression, knockdown and target mimicry. Each of these approaches has been utilized to develop abiotic stress-tolerant crops that can be better suited for climate resilience. Among clusters of ncRNAs, miRNA, circRNA and lncRNA manipulations have been extensively investigated for developing abiotic stress-tolerant crops.

6.3.1 circRNA

CircRNAs are endogenous and non-coding, generated by back splicing in which 3′ and 5′ ends are covalently closed by a 3′,5′-phosphodiester bond or a 2′,5′-phosphodiester bond at the junction sites. CircRNA size varies according to species and tissue that cannot be confined in the range of 100 nt to 4 kb (Zhao et al. 2019). CircRNA could be generated from any genomic location, such as introns, exons and intergenic regions (Chen et al. 2015). Due to its circular loop structure, circRNAs are highly stable in nature and are not affected by exoribonuclease, such as RNase R that has 3′–5′ exoribonuclease activity. Due to this nature, circRNA can be isolated and purified from total cellular RNAs by treatment of RNase B. With detection and identification of abundant circRNAs in plant systems, it has also revealed their probable role in biological processes, especially abiotic stress tolerance, protein binding, transcription regulation and miRNA binding (Chen et al. 2015).

Transcriptomics, sequencing and genome-wide studies have shown numerous differentially expressed circRNA in response to abiotic stress, especially in drought (Zhang et al. 2019), calcium deficiency-induced tip-burn (Wang et al. 2019), heat (Yang et al. 2019), multiple salt and drought stressors (Medina et al. 2021) and multiple drought and heat stressors (Zhou et al. 2020).

This transcriptomic study has opened up the possibilities of exploring circRNA in crop improvement programs. Encouraged by the enormous data generated through transcriptomic analysis, transgenic studies were carried out recently in efforts to develop stress-resilient crops. Accordingly, gene constructs targeting overexpression of circRNA, RNAi-mediated circRNA knockdown and CRISPR-mediated gene knockouts were developed. In one such study, Gao et al. (2019) carried out overexpression of Vv-circATS1, a circRNA from *V. vinifera* derived from glycerol-3-P acyltransferase, which showed improved cold tolerance in *A. thaliana*. Authors achieved this by amplifying the genomic region of circRNA_1218, circRNA_1975, circRNA_4328, circRNA_4363, circRNA_4473, circRNA_5664 and circRNA_7172 with or without flanking introns from grapes, and then cloned them into a pHB vector. Similarly, a study was carried out on tomato fruit by Tan et al. (2017), where circRNA derived from PSY1 and Phytoene desaturase (PDS), responsible in fruit ripening, were overexpressed. Here, overexpression vectors were created in which genomic fragments consisting of exons and flanking sequences were cloned and, later, these fragments were introduced into the expression vector pCAMBIA1301 in the opposite orientation under the CaMV 35 S promoter. Subsequently, these vectors were introduced into *Agrobacterium tumefaciens* EHA105, which was further used to transform the tomato.

In addition to overexpression studies, circRNA knockdown strategies also reveal the behavior of crops under stress conditions. For instance, Song et al. (2021) created a transgene-containing overexpression and knockdown of circRNAs Circ_0003418 and Circ_0000408. For knockdown analysis, siRNAs were designed at the junction sequences of candidate circRNAs and 3′-ends of parental genes (PtoXBAT32.3 and PtoXBAT32.5). Partial complementarity of siRNA overlap with cognate linear RNA was avoided by limiting the overlap sequence to 10 nts. Further Agrobacterium-mediated transformation into the poplar callus was created with these knockdown vectors to generate stable circRNA and parental gene-dominant repression. Transformed knockdown Circ_0003418 showed decreased transcript abundance of its parental gene in contrast to an overexpressed transgene. This experiment also demonstrated that PtoXBAT32.5 expression was induced with the up-regulation of Circ_0003418 under heat stress, thereby indicating Circ_0003418 as a negative regulator for heat tolerance.

6.3.2 microRNA (miRNA)

MiRNAs are most extensively investigated amongst ncRNA entities. They are single-stranded, 19–25 nt long and present in all eukaryotes. It is now well established that miRNAs are known to play a role in abiotic stress tolerance. During stress conditions, miRNAs up-regulate or down-regulate their respective target genes via multiple pathways, such as mRNA cleavage, chromatin remodeling, translational repression and/or methylation of DNA (Patil et al. 2021). For crop improvement programs involving miRNA, transgenics can be achieved by transgenesis, cisgenesis, intragenesis, artificial miRNA (amiRNA), target mimicry and MIR gene editing.

MiRNA-targeted transgenes confirming drought, heat, salinity, heavy metals, light and combined multiple stress conditions were studied. In drought stress tolerance, Hang et al. (2021) developed a transgenic *Lolium perenne* (perennial ryegrass) line to confer tolerance against drought by the heterologous expression of miR408 from *O. sativa* (Os-miR408). Transgenic *L. perenne* exhibited higher leaf-relative water content (RWC), lower electrolyte leakage (EL) and less lipid peroxidation. Further, these results suggested the improved drought tolerance expressing Os-miR408 transgenic plants could be a result of changes in leaf morphology, thereby maintaining cellular water availability and increased antioxidative capacity. While in salinity stress, Liu et al. (2019) created a transgenic *A. thaliana* with constitutively expressed ghr-miR414c from *Gossypium herbaceum* along with ectopic expression of GhFSD1 that showed enhanced salinity tolerance. Authors used the pRI201-AN vector with overexpressed GhFSD1 and ghr-miR414c under control of cauliflower mosaic virus 35S promoter (CaMV 35S). These developed plasmids were further transformed into *A. tumefaciens* strain GV3101 and subsequently used for *Arabidopsis* transformation by the floral dip transformation method, as ghr-miR414c targets the coding sequence region of iron superoxide dismutase gene (GhFSD1). Overexpression of ghr-miR414c showed increased sensitivity to salinity stress, yielding a phenotype similar to that of GhFSD1-silenced cotton.

Artificial miRNA, synonymous mutation and miRNA knockdown strategies have also been explored for generating abiotic-stress-tolerant crops containing reduced levels of negative regulating miRNAs. Abiotic tolerance can be possible by knocking down miRNAs that target and negatively regulate genes responsible for maintaining cellular homeostasis during stress conditions, such as osmoprotectants, ROS scavengers, HSP and transcription factors of such genes. In such an attempt, Zhao et al. (2021b) created a synonymous mutation at the target site of miR396a in Regulating Factor (GRF) 15. The authors created a transgenic poplar with *Pag*GRF15 overexpression containing a six nucleotide mutation at the target site of miR396. The developed transgenic plants showed enhanced expression of the GRF15 gene, thereby further conferring greater heat-stress tolerance than wild type crops. In another study, Liu et al. (2018) developed a transgenic *Arabidopsis* with amiRNA MIM160 that mimicked the inhibitor of miR160, T-DNA insertion mutant of miR160 as well as miR160 overexpressing transgenic lines. The authors found that the MIM160 and mutant lines showed reduced tolerance to heat while the overexpressing miR160 transgenic *Arabidopsis* showed enhanced heat-stress tolerance and seedling survival. In addition to these studies, various other success stories of transgenic crops targeting miRNAs are highlighted in Table 6.1.

6.3.3 Long Non-Coding RNAs (lncRNA)

LncRNAs are 200 nt longer with lower protein coding potential and are present in eukaryotes (Li et al. 2021). Unlike miRNAs which are transcribed by RNA polymerase II or III, lncRNA in plants can be transcribed mostly by either polymerase II along with other polymerase including RNA Pol III, IV, and V (Li et al. 2021). LncRNA can be classified based on localization as long intergenic non-coding (linc) RNAs, long non-coding natural antisense transcripts (lncNATs), long intronic ncRNAs and overlapping lncRNAs (Derrien et al. 2012). In plants, lncRNAs are known to play key roles in various plant-growth and developmental processes such as biotic and abiotic stress, seedling and flower development, root organogenesis and gene silencing through multiple pathways (Heo and Sung 2011; Ariel et al. 2014; Bardou et al. 2014; Matzke and Mosher, 2014; Wang et al. 2014; Zhang et al. 2014; Berry et al. 2015; Liu et al. 2015; Wang et al. 2015; Fedak et al. 2016; Seo et al. 2017).

The transgenic approach has been recently explored targeting lncRNA for improving crop tolerance to abiotic stress. In such efforts, Zhang et al. (2019) used transgenics to functionally characterize the role and localization of lncRNA973 under salinity stress. Accordingly, the authors developed overexpressed and knockdown lines of *Arabidopsis* in which overexpressed lines were created by inserting lncRNA973 transcripts under the Ubi promoter in the pCAMBIA1300-GFP vector, while knockdown lines were created using virus-induced gene silencing (VIGS) where lncRNA sequenced were cloned into the TRV2 vector. For transformation, developed constructs were transferred in the *A. tumefaciens* strain GV3101 and transformed into WT *Arabidopsis* using the floral dip method. LncRNA might exhibit potential as a competing endogenous RNA (ceRNA), wherein lncRNA harbors miRNA-responsive elements (MREs) that compete with mRNAs for miRNA target sites, thus, subsequently acting as miRNA gene silencing. In an attempt to characterize such ceRNAs, Chen et al. (2021) carried out transcriptome-wide identification of drought-responsive lncRNAs and subsequently constructed an lncRNA-mediated ceRNA network by analyzing competing relationships between mRNAs and lncRNAs. Furthermore, they created a transgenic rice with overexpression of shortlisted lncRNA TCONS_00021861 and its miRNA miR528-3p. This was achieved by amplifying the sequences of lncRNA TCONS_00021861 and miR528-3p, subsequently inserting it into the pEASY-Blunt vector and p1301-35 SNOS expression vector, respectively. Further, these vectors were transferred into *A. tumefaciens* and transformed using the floral dip method. Authors found that lncRNA TCONS_00021861 could act as ceRNA while sponging miR528-3p, which further activates the IAA biosynthetic pathway, subsequently conferring drought stress tolerance. Similar examples related to transgenic studies that are focused on miRNAs, circRNAs and lncRNAS are mentioned in Table 6.1.

TABLE 6.1
Plant ncRNAs and their Exploration for Engineering Crops for Conferring Environmental Stress Tolerance

Non-coding RNA	Stress	Strategy	Transgenic Plant	Responses	Reference
			MicroRNA		
miR408	Drought	↑	*Cicer arietinum* L	Indirect regulation of DREB and other drought-responsive genes conferring tolerance against drought.	Hajyzadeh et al. (2015)
	Salt	↑	*Nicotiana benthamiana*	Promoted seed germination and reduced the accumulation of reactive oxygen species under salt stress.	Guo et al. (2018)
	CO_2 fixation	↑	*Oryza sativa*	Cleavage of OsUCL8 by miR408 affects copper homeostasis in the plant cell, which, in turn, affects the abundance of plastocyanin proteins and photosynthesis in rice, increases in the number of panicle branches and had slightly longer grains.	Zhang et al. (2017)
miRNA414c	Salt	↓	*Arabidopsis thaliana*	Overexpressing gh-miR414c decreased the expression of GhFSD1 and increased sensitivity to salinity stress, yielding a phenotype similar to GhFSD1-silenced cotton. Acts as negative regulator.	Wang et al. (2019)
miR156	Drought	↑	*Solanum lycopersicum*	The post-trauma of drought, which causes late reopening of stomatal opening, was enhanced by miR156 in a strigolactones-dependent manner. Enhances plant revival after drought stress application.	Visentin et al. (2020)
	Salt	↑&↓	*Malus domestica*	Weakened salt resistance by cleaving the target. Overexpression of MdSPL13 with knockdown of miR156 conferred tolerance in the transgenic line. Acts as a negative regulator.	Ma et al. (2021)
miR827	Drought	↑	*Hordeum vulgare* L.	Enhanced leaf-level stomatal conductance and photosynthetic assimilation.	Ferdous et al. (2017)
miR1508a	Drought	↑	*Glycine max* L.	Dwarfing, thick cell walls, lower survival rates and greater leaf water loss was observed. Negative regulator.	Sun et al. (2020)
	Cold			Exhibited cold tolerance at the germination and young seedling stages along with higher soluble sugar content. Positive regulator.	
miR163	Light	↑	*A. thaliana*	Enhanced primary root elongation without hampering the lateral root growth in a light-dependent manner. Contributes positively to regulation of root photomorphogenesis mediated by the HY5-miR163-PXMT1 network.	Li et al. (2021)
miR828	Light	↓	*Brassica rapa*	Light-induced down-regulation of BrmiR828 can negatively regulate the target transcript levels leading to the accumulation of MYB transcription factors that positively regulate anthocyanin biosynthesis in light-exposed seedlings of *B. rapa*.	Zhou et al. (2020)

(*Continued*)

TABLE 6.1 (*Continued*)
Plant ncRNAs and their Exploration for Engineering Crops for Conferring Environmental Stress Tolerance

Non-coding RNA	Stress	Strategy	Transgenic Plant	Responses	Reference
OsmiR156k	Cold	↑	*O. Sativa*	Decreased plant cold tolerance at the young seedling growth stage, as evidenced by lower survival rates, chlorophyll contents proline contents and inhibit the seedling growth at the very early seedling stage under cold stress.	Cui et al. (2015)
miR398	Heat	↑&↓	*A. thaliana*	Knockdown of NAT398b and NAT398c promotes miR398 processing, resulting in stronger plant thermotolerance owing to the silencing of miR398-targeted genes; in contrast, their overexpression activates NAT398b and NAT398c, causing poorer thermotolerance due to the up-regulation of miR398-targeted genes.	Li et al. (2020)
			Circular RNA		
Vv-circATS1	Cold	↑	*A. thaliana*	Overexpression of Vv-circATS1, a circRNA derived from glycerol-3-P acyltransferase, improved cold tolerance.	Gao et al. (2019)
Circ_0003418	Heat	↑	*Populus tomentosa*	PtoXBAT32.5 expression was induced with up-regulation of Circ_0003418, resulting in increased production of ethylene and peroxidation of membrane lipids, indicating that Circ_0003418 is a negative regulator of heat tolerance via the ubiquitin-mediated protein modification pathway.	Song et al. (2020)
circGORK (Guard cell outward-rectifying K+-channel)	Drought	↑	*A. thaliana*	Plants overexpressing circGORK were hypersensitive to abscisic acid, but insensitive to drought, suggesting a positive role of circGORK in drought tolerance.	Zhang et al. (2018)
CRIR1	Cold	↑	*Manihot esculenta*	CRIR1 regulates a range of cold stress-related genes in a CBF-independent pathway and, hence, shows enhanced cold tolerance in transgenic plants.	Li et al. (2021)
			Long Non-Coding RNA		
TCONS_00021861	Drought	↑	*O. sativa*	Overexpression of TCONS_00021861 attenuated the repression of miR528-3p on YUCCA7, leading to increased IAA (Indole-3-acetic acid) content and auxin overproduction phenotypes and conferring drought tolerance.	Chen et al. (2021)
LncRNA973	Salt	↑&↓	*A. thaliana*	Inceased expression showed salt tolerance and enhanced the seed germination rate, fresh weights and root lengths of the transgenic plants increased, while lncRNA973 knockdown plants wilted, and the leaves became yellowed and dropped under salt-stress conditions, indicating that the tolerance to salt stress had decreased compared with wild type.	Zhang et al. (2019)

6.4 CONCLUSION

Climate change is inevitable and will cause drastic effects on all living organisms. Amongst these, plants are highly prone to damage due to their sessile nature. Drastic climate change has led to the experience of a consortium of abiotic stress on plants and other organisms. These climatic stresses give rise to episodes of enhanced heat stress, drought stress, elevated CO_2 levels, salinity stress as well as their deleterious combined consequences. With increasing population in addition to such climatic changes, a critical imbalance of supply and demand of food will occur in the near future. Thus, it is an immediate need to upgrade and develop crops that can better sustain climatic change with enhanced yield. Genetic engineering can be the most rapid, promising and target-specific tool that can be used to create climate-smart crops. At the molecular level, plants respond to stress conditions in many ways. Stress conditions can cause differential expression of genes and proteins. Targeting such genes and expressing proteins responsible for cellular homeostasis and stress tolerance was a previously used approach. In addition to structural genes, plants also express non-coding genes that are now known to play pivotal roles in various cellular mechanisms. NcRNAs include various classes such as miRNA, siRNAs, circRNAs, piRNAs and lncRNAs. These ncRNAs are now known to be involved in stress conditions in which they modulate the functions of their target genes. Research in plant ncRNA is still ongoing, but some studies have demonstrated their potential in crop improvement. Therefore, identifying and characterizing unexplored ncRNAs can aid in understanding plant response. In support of this, advanced techniques such as NGS, third-generation and degradome sequencing have been developed in identifying ncRNAs. In addition, recent developments in bioinformatics, artificial intelligence and machine learning-based tools have enabled identifications, characterization and validation of data for stress-responsive ncRNAs. Utilizing data generated from high-throughput techniques can further generate climate-smart crops and aid sustainable development.

REFERENCES

Archak S, Nagaraju J (2007) Computational prediction of rice (*Oryza sativa*) miRNA targets. Genom Proteom Bioinform; 5(3–4): 196–206. https://doi.org/10.1016/S1672-0229(08)60007-8

Ariel F, Jegu T, Latrasse D, Romero-Barrios N, Christ A, Benhamed M, Crespi M (2014) Noncoding transcription by alternative RNA polymerases dynamically regulates an auxin-driven chromatin loop. Mol Cell; 55(3): 383–396. https://doi.org/10.1016/j.molcel.2014.06.011

Asseng S, Ewert F, Martre P, et al. (2015) Rising temperatures reduce global wheat production. Nat Clim Chang; 5: 143–147. https://doi.org/10.1038/nclimate2470

Bao H, Chen M, Chen H, et al. (2019) Transcriptome-wide identification of miRNA targets and a TAS3-homologous gene in *Populus* by degradome sequencing. Genes Genomics; 41: 849–861. https://doi.org/10.1007/s13258-019-00797-8

Bardou F, Ariel F, Simpson CG, et al. (2014) Long noncoding RNA modulates alternative splicing regulators in *Arabidopsis*. Dev Cell; 30(2): 166–176. https://doi.org/10.1016/j.devcel.2014.06.017

Berry S, Dean C (2015) Environmental perception and epigenetic memory: mechanistic insight through FLC. Plant J; 83(1): 133–148. https://doi.org/10.1111/tpj.12869

Bonnet E, He Y, Billiau K, et al. (2010) TAPIR, a web server for the prediction of plant microRNA targets, including target mimics. Bioinformatics; 26: 1566–1568. https://doi.org/10.1093/bioinformatics/btq233

Brant EJ, Budak H (2018) Plant small non-coding RNAs and their roles in biotic stresses. Front Plant Sci; 9:1038. https://doi.org/10.3389/fpls.2018.01038

Cui N, Sun X, Sun M, Jia B, Duanmu H, et al. (2015) Overexpression of OsmiR156k leads to reduced tolerance to cold stress in rice (*Oryza sativa*). Mol Breed; 35: 1–11. https://doi.org/10.1007/s11032-015-0402-6

Chen Y, Wang X (2020) miRDB: an online database for prediction of functional microRNA targets. Nucleic Acids Res; 48: D127–D131. https://doi.org/10.1093/nar/gkz757

Chen I, Chen C, Chuang T (2015) Biogenesis, identification, and function of exonic circular RNAs. Wiley Interdiscip Rev RNA; 6(5):563–579. https://doi.org/10.1002/wrna.1294

Chen J, Zhong Y, Qi X (2021) LncRNA TCONS_00021861 is functionally associated with drought tolerance in rice (*Oryza sativa* L.) via competing endogenous RNA regulation. BMC Plant Biol; 21(1): 1–12. https://doi.org/10.1186/s12870-021-03195-z

Cubasch U, Wuebbles D, Chen D, Facchini MC, Frame D, Mahowald N, J-GW. (2021). AR5 Climate Change 2013: The Physical Science Basis — IPCC. Cambridge University Press, Cambridge, United Kingdom and New York; 1585.

Dhandapani V, Ramchiary N, Paul P, et al. (2011) Identification of potential microRNAs and their targets in *Brassica rapa* L. Mol Cells; 32: 21–37. https://doi.org/10.1007/s10059-011-2313-7

Donovan P, McHale N, Venø M, Prehn J (2021) tsRNAsearch: a pipeline for the identification of tRNA and ncRNA fragments from small RNA-sequencing data. Bioinformatics; 37(23): 4424–4430. https://doi.org/10.1093/bioinformatics/btab515

Daryanto S, Wang L, Jacinthe P-A (2016) Global synthesis of drought effects on maize and wheat production. PLoS One; 11: e0156362. https://doi.org/10.1371/journal.pone.0156362

Derrien T, Johnson R, Bussotti G, Tanzer A, et al. (2012) The GENCODE v7 catalog of human long noncoding RNAs: analysis of their gene structure, evolution, and expression. Genome Res; 22(9): 1775–1789. https://doi.org/10.1101/gr.132159.111

Fedak H, Palusinska M, Krzyczmonik K, et al. (2016) Control of seed dormancy in Arabidopsis by a cis-acting noncoding antisense transcript. Proc Natl Acad Sci USA; 113: E7846–E7855. https://doi.org/10.1073/pnas.1608827113

Fei Y, Wang R, Li H, et al. (2018) DPMIND: degradome-based plant miRNA-target interaction and network database. Bioinformatics; 34: 1618–1620. https://doi.org/10.1093/bioinformatics/btx824

Gao Z, Li J, Luo M, et al. (2019) Characterization and cloning of grape circular RNAs identified the cold resistance-related Vv-circATS1. Plant Physiol; 180(2): 966–985. https://doi.org/10.1104/pp.18.01331

Gao Y, Wang J, Zhao F (2015) CIRI: an efficient and unbiased algorithm for denovo circularRNA identification. Genome Biol; 16: 4. https://doi.org/10.1186/s13059-014-0571-3

Guo Z, Kuang Z, Wang Y, et al. (2020) PmiREN: a comprehensive encyclopedia of plant miRNAs. Nucleic Acids Res; 48(D1): D1114–D1121. https://doi.org/10.1093/nar/gkz894

Garcia-Silva MR, Cabrera-Cabrera F, Guida MC, Cayota A (2012) Hints of tRNA-derived small RNAs role in RNA silencing mechanisms. Genes (Basel); 3: 603–614. https://doi.org/10.3390/genes3040603

Hajyzadeh M, Turktas M, Khawar KM, Unver T (2015) miR408 overexpression causes increased drought tolerance in chickpea. Gene; 555: 186–193. https://doi.org/10.1016/j.gene.2014.11.002

Hang N, Shi T, Liu Y, et al. (2021) Overexpression of os-microRNA408 enhances drought tolerance in *perennial ryegrass*. Physiol Plant; 172: 733–747. https://doi.org/10.1111/ppl.13276

Heo JB, Sung S (2011) Vernalization-mediated epigenetic silencing by a long intronic noncoding RNA. Science; 331: 76–79. https://doi.org/10.1126/science.1197349

Holley RW, Apgar J, Everett GA, Madison JT, Marquisee M, Merrill SH, Penswick JR, Zamir A (1965) Structure of a ribonucleic acid. Science; 147(3664): 1462–1465. http://www.jstor.org/stable/1715055

Kamthan A, Chaudhuri A, Kamthan M, Datt A (2015) Small RNAs in plants: recent development and application for crop improvement. Front Plant Sci; 6: 208. https://doi.org/10.3389/fpls.2015.00208

Kang T, Yu C-Y, Liu Y, et al. (2020) Subtly manipulated expression of ZmmiR156 in *tobacco* improves drought and salt tolerance without changing the architecture of transgenic plants. Front Plant Sci; 10. https://doi.org/10.3389/fpls.2019.01664

Kapusta A, Kronenberg Z, Lynch J, et al. (2013) Transposable elements are major contributors to the origin, diversification, and regulation of vertebrate long noncoding RNAs. PLoS Genetics; 9(4): e1003470. https://doi.org/10.1371/journal.pgen.1003470

Karagkouni D, Paraskevopoulou MD, Chatzopoulos S, et al. (2018) DIANA-TarBase v8: a decade-long collection of experimentally supported miRNA–gene interactions. Nucleic Acids Res; 46: D239–D245. https://doi.org/10.1093/nar/gkx1141

Kasukurthi MV, Houserova D, Huang Y, et al. (2021) SURFR: a real-time platform for non-coding RNA fragmentation analysis using wavelets. 2021 IEEE International Conference on Bioinformatics and Biomedicine (BIBM). Retrieved from https://par.nsf.gov/biblio/10321133. https://doi.org/10.1109/BIBM52615.2021.9669696

Kim W, Iizumi T, Nishimori M (2019) Global patterns of crop production losses associated with droughts from 1983 to 2009. J Appl Meteorol Climatol; 58: 1233–1244. https://doi.org/10.1175/JAMC-D-18-0174.1

Kozomara A, Birgaoanu M, Griffiths-Jones S (2019) miRBase: from microRNA sequences to function. Nucleic Acids Res; 47: D155–D162. https://doi.org/10.1093/nar/gky1141

Kumar P, Anaya J, Mudunuri SB, Dutta A (2014) Meta-analysis of tRNA derived RNA fragments reveals that they are evolutionarily conserved and associate with AGO proteins to recognize specific RNA targets. BMC Biol; 12: 78. https://doi.org/10.1186/PREACCEPT-5867533061403216

Li S, Cheng Z, Dong S, Li Z, Zou L, Zhao P, Guo X, Bao Y, Wang W, Peng M (2021) Global identification of full-length cassava lncRNAs unveils the role of cold-responsive intergenic lncRNA 1 in cold stress response. Plant Cell Environ. Advance online publication. https://doi.org/10.1111/pce.14236

Li T, Lian H, Li H, et al. (2021) HY5 regulates light-responsive transcription of microRNA163 to promote primary root elongation in Arabidopsis seedlings. J Integr Plant Biol; 63: 1437–1450. https://doi.org/10.1111/jipb.13099

Li Y, Li X, Yang J, He Y (2020) Natural antisense transcripts of MIR398 genes suppress microR398 processing and attenuate plant thermotolerance. Nat Commun; 11: 5351. https://doi.org/10.1038/s41467-020-19186-x

Li X, Wu Z, Fu X, Han W (2014) lncRNAs: insights into their function and mechanics in underlying disorders. Mutat Res Rev Mutat Res; 762: 1–21. https://doi.org/10.1016/j.mrrev.2014.04.002

Lin SS, Chen Y, Lu MYJ (2019) Degradome Sequencing in Plants. In: de Folter S. (eds) Plant MicroRNAs. Methods in Molecular Biology, vol 1932. Humana Press, New York, NY. https://doi.org/10.1007/978-1-4939-9042-9_15

Lin J-S, Kuo C-C, Yang I-C, Tsai W-A, Shen Y-H, Lin C-C, Liang Y-C, Li Y-C, Kuo Y-W, King Y-C, Lai H-M, Jeng S-T (2018) MicroRNA160 modulates plant development and heat shock protein Gene expression to mediate heat tolerance in *Arabidopsis*. Front Plant Sci; 9: 68. https://doi.org/10.3389/fpls.2018.00068

Liu D, Chen D, Cheng Y, et al. (2019) MicroRNA414c affects salt tolerance of cotton by regulating reactive oxygen species metabolism under salinity stress, RNA Biology; 16(3): 362–375. https://doi.org/10.1080/15476286.2019.1574163

Liu J., Wang H, Chua N-H (2015) Long noncoding RNA transcriptome of plants. Plant Biotechnol J; 13: 319–328. https://doi.org/10.1111/pbi.12336

Liu X, Hao LL, Li DY, Zhu LH, Hu SN (2015b) Long non-coding RNAs and their biological roles in plants. Genom Proteom Bioinform; 13: 137–147. https://doi.org/10.1016/j.gpb.2015.02.003

Li W, Wang T, Zhang Y, Li Y (2016) Overexpression of soybean miR172c confers tolerance to water deficit and salt stress, but increases ABA sensitivity in transgenic Arabidopsis thaliana. J Exp Bot; 67: 175–194. https://doi.org/10.1093/jxb/erv450

Lorenzetti A, Paschoal D (2016) PlanTE-MIR DB: a database for transposable element-related microRNAs in plant genomes. Funct Integr Genomics; 16: 235–242. https://doi.org/10.1007/s10142-016-0480-5

Loher P, Rigoutsos I (2012) Interactive exploration of RNA22 microRNA target predictions. Bioinformatics; 28(24): 3322–3323. https://doi.org/10.1093/bioinformatics/bts615

Martens-Uzunova ES, Olvedy M, Jenster G (2013) Beyond microRNA—novel RNAs derived from small non-coding RNA and their implication in cancer. Cancer Lett; 340: 201–211. https://doi.org/10.1016/j.canlet.2012.11.058

Matzke MA, Mosher RA (2014) RNA-directed DNA methylation: an epigenetic pathway of increasing complexity. Nat Rev Genet; 15(6): 394–408. https://doi.org/10.1038/nrg3683

Ma Y, Xue H, Zhang F, et al. (2021) The miR156/SPL module regulates apple salt stress tolerance by activating MdWRKY100 expression. Plant Biotechnol J; 19: 311–323. https://doi.org/10.1111/pbi.13464

Medina CA, Samac DA, Yu LX (2021) Pan-transcriptome identifying master genes and regulation network in response to drought and salt stresses in alfalfa (*medicago sativa* l.). Sci Rep; 11: 17203. https://doi.org/10.1038/s41598-021-96712-x

Meng Y, Gou L, Chen D, et al. (2011) PmiRKB: a plant microRNA knowledge base. Nucleic Acids Res; 39: D181–D187. https://doi.org/10.1093/nar/gkq721

Patil S, Joshi S, Jamla M, Zhou X, Taherzadeh MJ, Suprasanna P, Kumar V (2021) MicroRNA-mediated bioengineering for climate-resilience in crops. Bioengineered; 12(2): 10430–10456. https://doi.org/10.1080/21655979.2021.1997244

Patil S, Joshi S, Jamla M, Zhou X, Taherzadeh MJ, Suprasanna P, Kumar V (2021) MicroRNA-mediated bioengineering for climate-resilience in crops. Bioengineered; 12(2): 10430–10456. https://doi.org/10.1080/21655979.2021.1997244.

Paul P, Dhandapani V (2019) Plant noncoding RNAs identification and characterization - a short review. Ann Adv Biomed Sci; 2(4): 000133. http://dx.doi.org/10.23880/aabsc-16000134

Reinhart B, Weinstein E, Rhoades M, Bartel B, Bartel D (2002) MicroRNAs in plants. Genes Dev; 16(13): 1616–1626. http://doi.org/10.1101/gad.1004402

Renyi L, Jiankang Z (2014). Non-coding RNAs as potent tools for crop improvement. Natl Sci Rev; 1(2): 186–189. https://doi.org/10.1093/nsr/nwu006

Santosh B, Akhil V, Pramod KY (2014). Non-coding RNAs: biological functions and applications. Cell Biochem Funct; 33(1): 14–22. https://doi.org/10.1002/cbf.3079

Seo JS, Sun HX, Park BS, et al. (2017) ELF18-INDUCED LONG-NONCODING RNA associates with mediator to enhance expression of innate immune response genes in Arabidopsis. Plant Cell; 29: 1024–1038. https://doi.org/10.1105/tpc.16.00886

Song Y, Chen P, Liu P, Bu C, Zhang D (2020) High-temperature-responsive poplar lncRNAs modulate target gene expression via RNA interference and act as RNA scaffolds to enhance heat tolerance. Int J Mol Sci; 21(18): 6808. https://doi.org/10.3390/ijms21186808

Song Y, Bu C, Chen P, Liu P, Zhang D (2021) Miniature inverted repeat transposable elements cis-regulate circular RNA expression and promote ethylene biosynthesis, reducing heat tolerance in *Populus tomentosa*. J Exp Bot; 72(5): 1978–1994. https://doi.org/10.1093/jxb/eraa570

Shukla PR, Skea J, Buendia EC, et al. IPCC, 2019: Climate change and land: an IPCC special report on climate change, desertification, land degradation, sustainable land management, food security, and greenhouse gas fluxes in terrestrial ecosystems. https://www.ipcc.ch/site/assets/uploads/2019/11/SRCCL-Full-Report-Compiled-191128.pdf

Si X, Wang Q, Zhang L, et al. (2017) Survey of gene splicing algorithms based on reads. Bioengineered; 8: 750–758. https://doi.org/10.1080/21655979.2017.1373538

Siomi H, Siomi M (2009). On the road to reading the RNA-interference code. Nature; 457(7228): 396–404. https://doi.org/10.1038/nature07754

Song Y, Bu C, Chen P, Liu P, Zhang D (2021) Miniature inverted repeat transposable elements cis-regulate circular RNA expression and promote ethylene biosynthesis, reducing heat tolerance in *Populus tomentosa*. J Exp Bot; 72(5): 1978–1994. https://doi.org/10.1093/jxb/eraa570

Sun M, Jing Y, Wang X, et al. (2020) Gma-miR1508a confers dwarfing, cold tolerance, and drought sensitivity in soybean. Mol Breed; 40. https://doi.org/10.1007/s11032-020-01116-w

Sun L, Liu H, Zhang L, Meng J (2015). lncRScan-SVM: a tool for predicting long non-coding RNAs using support vector machine. PLoS One; 10(10): e0139654. https://doi.org/10.1371/journal.pone.0139654

Szcześniak MW, Makałowska I (2014) miRNEST 2.0: a database of plant and animal microRNAs. Nucleic Acids Res; 42: D74–D77. https://doi.org/10.1093/nar/gkt1156

Tan J, Zhou Z, Niu Y, Sun X, Deng Z (2017) Identification and functional characterization of tomato CircRNAs derived from genes involved in fruit pigment accumulation. Sci Rep; 7(1); 8594. https://doi.org/10.1038/s41598-017-08806-0

Thompson DM, Parker R (2009) The RNase Rny1p cleaves tRNAs and promotes cell death during oxidative stress in *Saccharomyces cerevisiae*. J Cell Biol; 185: 43–50. https://doi.org/10.1083/jcb.200811119

Visentin I, Pagliarani C, Deva E, et al (2020) A novel strigolactone-miR156 module controls stomatal behaviour during drought recovery. Plant Cell Environ; 43: 1613–1624. https://doi.org/10.1111/pce.13758

Wang Y, Fan X, Lin F, He G, Terzaghi W, et al. (2014) *Arabidopsis* noncoding RNA mediates control of photomorphogenesis by red light. Proc Natl Acad Sci U S A; 111(28): 10359–10364. https://doi.org/10.1073/pnas.1409457111

Wang W, Liu D, Chen D, et al. (2019) MicroRNA414c affects salt tolerance of cotton by regulating reactive oxygen species metabolism under salinity stress. RNA Biol; 16: 362–375. https://doi.org/10.1080/15476286.2019.1574163

Wang K, Liu Y, Teng F, et al. (2021) Heterogeneous expression of osa-MIR156bc increases abiotic stress resistance and forage quality of alfalfa. Crop J; 9; 1135–1144. https://doi.org/10.1016/j.cj.2020.11.009

Wang W, Wang J, Wei Q, et al. (2019) Transcriptome-wide identification and characterization of circular RNAs in leaves of Chinese cabbage (*Brassica rapa* l. Ssp. *pekinensis*) in response to calcium deficiency-induced tip-burn. Sci Rep; 9: 14544. https://doi.org/10.1038/s41598-019-51190-0

Wang M, Yuan D, Tu L, et al. (2015). Long noncoding RNAs and their proposed functions in fibre development of cotton (*Gossypium spp.*). New Phytol; 207(4): 1181–1197. https://doi.org/10.1111/nph.13429

Wen X, Ding Y, Tan Z, Wang J, Zhang D, Wang Y (2020). Identification and characterization of cadmium stress-related LncRNAs from *Betula platyphylla*. Plant Sci; 299: 110601. https://doi.org/10.1016/j.plantsci.2020.110601

Wilson RC, Doudna JA (2013) Molecular mechanisms of RNA interference. Annu Rev Biophys; 42: 217–239. https://doi.org/10.1146/annurev-biophys-083012-130404

Yang Z, Li W, Su X, et al. (2019) Early response of radish to heat stress by strand-specific transcriptome and miRNA analysis. Int J Mol Sci; 20(13): 3321. https://doi.org/10.3390/ijms20133321

Yan S, Qian J, Cai C, Ma Z, Li J, Yin M, Shen J (2020). Spray method application of transdermal dsRNA delivery system for efficient gene silencing and pest control on soybean aphid *aphis glycines*. J Pest Sci; 93(1): 449–459. https://doi.org/10.1007/s10340-019-01157-x

Yan S, Ren BY, Shen J (2021). Nanoparticle-mediated double-stranded RNA delivery system: a promising approach for sustainable pest management. Insect Sci; 28(1): 21–34. https://doi.org/10.1111/1744-7917.12822

Yi X, Zhang Z, Ling Y, Xu W, Su Z (2015). PNRD: a plant non-coding RNA database. Nucleic Acids Res; 43(D1): D982–D989. https://doi.org/10.1093/nar/gku1162

Zhang T, Zhai J, Zhang X, Ling L, et al. (2022). Interactive web-based annotation of plant MicroRNAs with iwa-miRNA. Genom Proteom Bioinform; 20(3): 557–567. https://doi.org/10.1016/j.gpb.2021.02.010
Zhang X, Dong J, Deng F et al. (2019a) The long non-coding RNA lncRNA973 is involved in cotton response to salt stress. BMC Plant Biol; 19: 459. https://doi.org/10.1186/s12870-019-2088-0
Zhang P, Fan Y, Sun X, et al. (2019b) A large-scale circular RNA profiling reveals universal molecular mechanisms responsive to drought stress in maize and *Arabidopsis*. Plant J; 98: 697–713. https://doi.org/10.1111/tpj.14267
Zhang YC, Liao JY, Li ZY, et al. (2014) Genome-wide screening and functional analysis identify a large number of long noncoding RNAs involved in the sexual reproduction of rice. Genome Biol; 15(12): 512. https://doi.org/10.1186/s13059-014-0512-1
Zhang J-P, Yu Y, Feng Y-Z, et al. (2017) MiR408 regulates grain yield and photosynthesis via a phytocyanin protein. Plant Physiol; 175: 1175–1185. https://doi.org/10.1104/pp.17.01169
Zhang Z, Yu J, Li D, et al. (2010) PMRD: plant microRNA database. Nucleic Acids Res; 38: D806–D813. https://doi.org/10.1093/nar/gkp818
Zhao C, Liu B, Piao S, et al. (2017) Temperature increase reduces global yields of major crops in four independent estimates. Proc Natl Acad Sci; 114: 9326–9331. https://doi.org/10.1073/pnas.1701762114
Zhao J, Qin B, Nikolay R, Spahn C, Zhang, G (2019) Translatomics: the global view of translation. Int J Mol Sci; 20(1): 212. https://doi.org/10.3390/ijms20010212
Zhao L, Wang J, Li Y, et al. (2021a). NONCODEV6: an updated database dedicated to long non-coding RNA annotation in both animals and plants. Nucleic Acids Res; 49(D1): D165–D171.
Zhao Y, Xie J, Wang S, et al. (2021b) Synonymous mutation in growth regulating factor 15 of miR396a target sites enhances photosynthetic efficiency and heat tolerance in poplar. J Exp Bot; 2: 4502–4519. https://doi.org/10.1093/jxb/erab120
Zheng LL, Qu LH (2015) Application of microRNA gene resources in the improvement of agronomic traits in rice. Plant Biotechnol J; 13: 329–336.
Zhou R, Yu X, Ottosen C, Zhao T (2020) High throughput sequencing of CircRNAs in tomato leaves responding to multiple stresses of drought and heat. Hortic Plant J; 6(1): 34–38. https://doi.org/10.1016/j.hpj.2019.12.004

7 Plant Non-Coding RNAs and Their Regulation in Salinity and Heavy Metal Stress

Arunima Bhattacharya[1] and Aryadeep Roychoudhury[2]

[1]Centre National de la Recherche Scientifique, Institut National de la Santé et de la Recherche Médicale, Institut Européen de Chimie et Biologie, University of Bordeaux, Pessac, France

[2]Discipline of Life Sciences, School of Sciences, Indira Gandhi National Open University, New Delhi, India

7.1 INTRODUCTION

Messenger RNAs (mRNAs) that act as templates for protein synthesis have been the principal focus of various research work for the major part of the past half century, while paradoxically, the majority of genomes transcribe to non-coding RNAs (ncRNAs). Often arbitrarily named as "junk RNA" or "intergenic RNA", ncRNAs were often considered to be mere by-products of mass transcription events, with little or no biological meaning (Zhang et al. 2019a). The gradual discoveries of ribosomal RNA (rRNA) and transfer RNA (tRNA) opened the door for further research into functional implications of ncRNAs in the second half of the 20th century, which evolved into high-throughput sequencing-aided profiling of the ncRNAs in present times, using a number of public databases such as Rfam (Griffiths-Jones et al. 2003), NONCODE (Liu et al. 2005), miRbase (Griffiths-Jones et al. 2006) and circBase (Glažar et al. 2014).

When facing a stress situation of any kind, the responses of a living organism to minimize the influence of that stress generally include options like tolerance, resistance, avoidance and ultimately escape. Plants, being sessile organisms, can resort to only the first three and, hence, regulations at the transcriptional and post-transcriptional levels are essential for survival (Tyczewska et al. 2016). In plants, endogenous ncRNAs act as key regulatory RNA molecules in gene expression during stress responses, forming complex interconnected networks, and this interrelationship has been in the purview of recent research, especially since the discovery of the phenomenon of RNA interference (RNAi) (Fire et al. 1998), which formed the stepping stone towards understanding the mechanism of microRNA (miRNA)- and small interfering RNA (siRNA)-mediated signalling cascades.

This chapter outlines the various functional ncRNAs in plants observed in response to salinity and heavy metal (HM) stress, and attempts to explore the holistic mechanism of signal transduction during such stress responses in plants.

7.2 ncRNAs: GENERAL MODE OF ACTION

7.2.1 Various ncRNA Species in Eukaryotes

A vast repertoire of ncRNAs is produced in eukaryotes due to transcription from different genomic regions and processing of RNA. Depending on their regulatory roles, ncRNAs can be divided into two broad categories: housekeeping ncRNAs, which abundantly and ubiquitously regulate generic cellular functions, and regulatory ncRNAs, which play key roles in the regulation of gene expression

DOI: 10.1201/9781003369288-7

at the epigenetic, transcriptional and post-transcriptional levels (Ponjavic et al. 2007; Cech and Steitz 2014; Peschansky and Wahlestedt 2014; Zhang et al. 2019a).

The housekeeping ncRNAs broadly consist of rRNAs, tRNAs, small nuclear RNAs (snRNAs), small nucleolar RNAs (snoRNAs), telomerase RNAs (Cerutti et al. 1968; Zieve and Penman 1976), translation interfering RNAs (tiRNAs) (Schimmel 2018), sno-derived RNAs (Taft et al. 2009), sno-miRNAs (Ono et al. 2011), sno-piRNAs (He et al. 2015), etc. (Figure 7.1). They usually range from 50 to 500 nucleotides (nt), are constitutively expressed in all cell types and are essential for viability of cells as they play major roles in protein synthesis, splicing, and also inhibition of translation by recruiting packed aggregates of proteins and RNAs under stressful situations (Schaefer et al. 2010; Venkatesh et al. 2016; Chen et al. 2019).

The major regulatory RNAs, on the other hand, can be further subdivided as per their average size into small ncRNAs (sncRNAs, <200 nucleotides) including miRNAs, siRNAs and piwi-interacting RNAs (piRNAs), and long ncRNAs (lncRNAs, >200 nucleotides) (Zhang et al. 2019a).

The major ncRNAs in plants are the miRNAs, siRNAs and lncRNAs, which are discussed in the following sections.

7.2.2 Major ncRNAs in Plants

7.2.2.1 miRNAs

Generated from transcribed hairpin loop structures, miRNAs are the most abundant sncRNAs, 20–22 nt in length, with roles in both nuclear and cytoplasmic gene expression regulation and

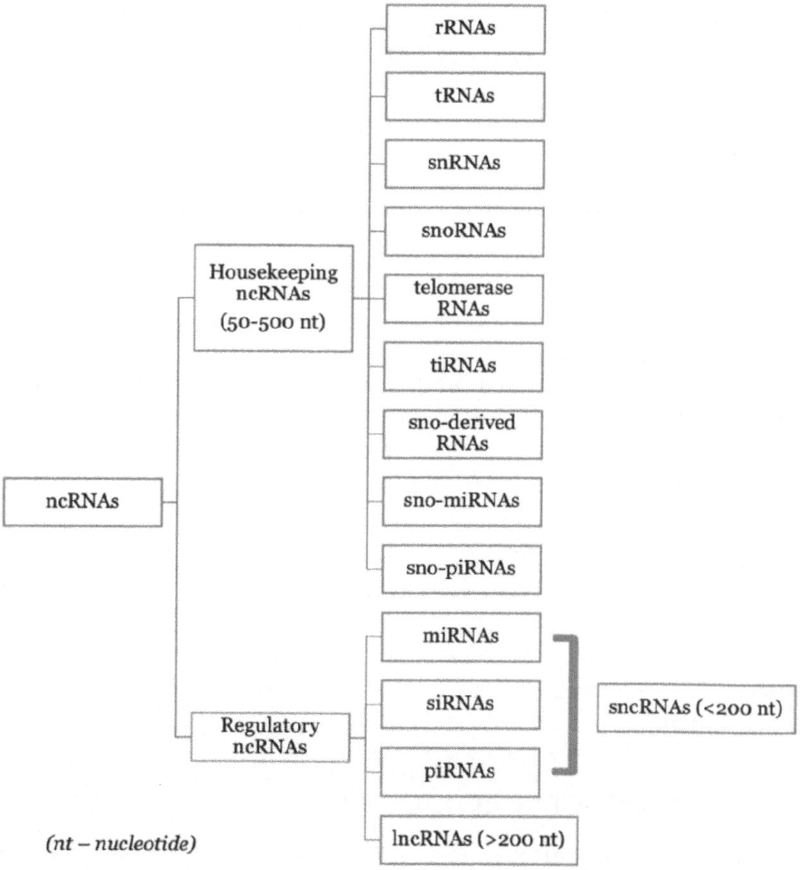

FIGURE 7.1 Classification of ncRNAs. [Adapted and modified from Waititu et al. (2020)].

post-transcriptional gene silencing (PTGS) (Lee et al. 1993; Reinhart et al. 2000; Lagos-Quintana et al. 2001; Waititu et al. 2020). miRNAs are, in fact, synthesized from miRNA genes (*MIR* genes), which are sequentially processed by RNA polymerase II (RNA pol II) and RNase III Dicer-like protein 1 (DCL1) (Rogers and Chen 2012; Rogers and Chen 2013; Axtell and Meyers 2018), resulting in an miRNA–miRNA* duplex, which is transported to the cytoplasm by HASTY (HST) and integrated into the RNA-induced silencing complex (RISC) containing ARGONAUTE 1 (AGO1) proteins which cleave mRNA and repress translation, thereby controlling gene silencing (Guleria et al. 2011; Kwak and Tomari 2012; Pareek et al. 2015; Waititu et al. 2020).

7.2.2.2 siRNAs

Massively important in RNAi, siRNAs are double-stranded ncRNAs, which, based on their origin, may be divided into two broad groups: exogenous siRNA (exo-siRNA), which are produced mostly from exogenous nucleic acids during artificial insertions or viral infections, and endogenous siRNA (endo-siRNA), which, in most cases, are either transcribed from transposon elements or arise from endogenous genomic loci (Waterhouse et al. 1998; Hamilton and Baulcombe 1999; Zhang et al. 2019a).

The major siRNAs in plants include trans-acting siRNAs (ta-siRNA), heterochromatic siRNAs (hc-siRNA) and natural antisense siRNAs (nat-siRNAs), each with distinct biogenesis cascades and modes of action. In general, biogenesis of any siRNA involves double-stranded RNAs (dsRNAs), DCL enzymes and RISC-containing AGO protein: ta-siRNAs need SUPPRESSOR OF GENE SILENCING 3 (SGS3), RNA-dependent RNA polymerases 6 (RDR6) and DCL4 (Peragine et al. 2004; Gasciolli et al. 2005); hc-siRNAs require Pol IV, DCL3, RDR2, 5S rRNA genes and a plant-specific RNA polymerase, NRPD1A (Zilberman et al. 2004; Mallory and Vaucheret 2006), while nat-siRNAs are generated using SGS3, RDR6, DCL1, DCL2 and NRPD1A (Borsani et al. 2005; Guleria et al. 2011; Waititu et al. 2020).

7.2.2.3 lncRNAs

These are non-coding transcripts longer than 200 nt in length that can be divided into subcategories in a number of manners:

- Cis-natural antisense (cis-NATs), trans-natural antisense transcripts (trans-NATs) and pseudogenes (Nie et al. 2012).
- Sense or antisense lncRNAs, based on the original strand from which they were transcribed (Nam and Bartel 2012).
- Divergent or convergent lncRNAs, based on the orientation of transcription.
- Intronic or intergenic lncRNAs, based on their location with respect to coding transcripts (Derrien et al. 2012).

Generally, transcribed by RNA polymerases II, III, IV and V, lncRNAs participate in myriad gene expression regulation pathways, including chromatin remodelling, transcriptional enhancer and/or repression and RNA splicing (Waititu et al. 2020). Some lncRNAs, in turn, could also be further processed to generate sncRNAs like miRNAs, piRNAs and snoRNAs (Röther and Meister 2011; Zhang et al. 2019a).

7.3 ncRNAs AND PLANT GENE REGULATION

sncRNAs, like the miRNAs and siRNAs, and lncRNAs regulate gene expression in plants at the transcriptional and post-transcriptional levels in a few major ways (Waititu et al. 2020), a brief overview of each is discussed in this section. These mechanisms are ultimately important in elucidating their roles in abiotic stress regulation in plants.

7.3.1 Ribosomal Translation Inhibition and Cleavage of Target mRNAs by miRNAs and siRNAs

The sncRNAs, like miRNAs and siRNAs, ultimately inhibit translation either by affecting the ribosomal machinery from proper functioning, or directly by cleavage of the target mRNA. Generally, they pair with target mRNAs at specific complementary binding sites and stimulate inhibition of translation via the Argonaute proteins; in plants, the major proteins being AGO1 and AGO10 (Rhoades et al. 2002; Aukerman and Sakai 2003; Cuellar and McManus 2005; Kim 2005).

An illustrative example would be that of the yet incompletely elucidated mechanism of restriction of ribosomal movement and recruitment due to binding of the RISC-AGO1 complex with the miRNA, to the 5′ untranslated regions (UTRs) of target mRNAs (Iwakawa and Tomari 2013), aided by a number of other proteins like ALTERED MERISTEM PROGRAM 1 (AMP1), VARICOSE (VCS), GW-repeat protein, KATANIN (KTN1) and also AGO10 (Yang et al. 2012; Li et al. 2013).

On the other hand, in the case of siRNAs integrated with RISC-AGO1 or AGO7 complexes, the target mRNA is cleaved around 10 nucleotides upstream of the 5′ end of the antisense strand (Liu et al. 2017), after which an exonuclease XRN4 (EXONUCLEASE 4) degrades both the 3′ and 5′ cleaved fragments (Souret et al. 2004; Ren et al. 2014). Additional AGO proteins involved in similar mechanisms include AGO2, AGO4 and AGO10, on which further research is in progress (Maunoury and Vaucheret 2011; Zhu et al. 2011).

7.3.2 Gene Silencing via DNA and Histone Methylation Induced by miRNAs and siRNAs

Hc-siRNAs, produced by the action of the Dicer-like enzyme DCL3, are generally transcribed at heterochromatic regions of the genome, and via cytosine methylation in sequence contexts of CG, CHG and CHH in cis (Zilberman et al. 2003; Xie et al. 2004), they silence gene expression through the RNA-directed DNA methylation (RdDM) pathway (Matzke et al. 2009; Creasey et al. 2014; Erdmann and Picard 2020). Further, given DNA methylation is considered an important epigenetic mark, histone methylation can also be linked to this event as H3K9 methyltransferases too are guided by the hc-siRNA/AGO4-RISC complex, ultimately leading to transcriptional gene silencing of the target sequence (Zilberman et al. 2003; Xie et al. 2004).

7.3.3 lncRNAs and Regulation of Gene Expression

The regulatory mechanisms of lncRNAs are more diversified than the sncRNAs in that they can regulate target mRNAs, both in cis and trans (Wu et al. 2020). While cis-acting lncRNAs mostly function in the vicinity of their sites of synthesis, operating on local sequences and/or chromosome regions on contiguous genes, the trans-acting lncRNAs function at sites distant from those of their origin, even on different chromosomes (Lee 2012). The role of lncRNAs as precursors of sncRNAs is also important (Lauressergues et al. 2015) and can be elucidated by the formation of dsRNA duplexes between lncRNAs and natural antisense transcripts (NATs) to give rise to sncRNAs involved in regulatory functions (Waititu et al. 2020).

Broadly, the regulatory mechanisms of lncRNAs can be classified (Figure 7.2) into four archetypes (Wang and Chang 2011; Wu et al. 2020):

i. **Signal:** This includes functions of lncRNAs as signalling molecules that can bind to transcription factors, thereby regulating the spatio-temporal expression of coding mRNAs. For example, the *Arabidopsis* lncRNACOLDAIR is responsible for recruitment of the protein complex Polycomb Repressive Complex 2 (PRC2) to the FLOWERING LOCUS C (FLC), by direct physical association. PRC2, in turn, promotes histone H3K27 methylation, changing the chromatin landscape of the FLC locus, ultimately inhibiting its expression and regulating vernalization (Heo and Sung 2011).

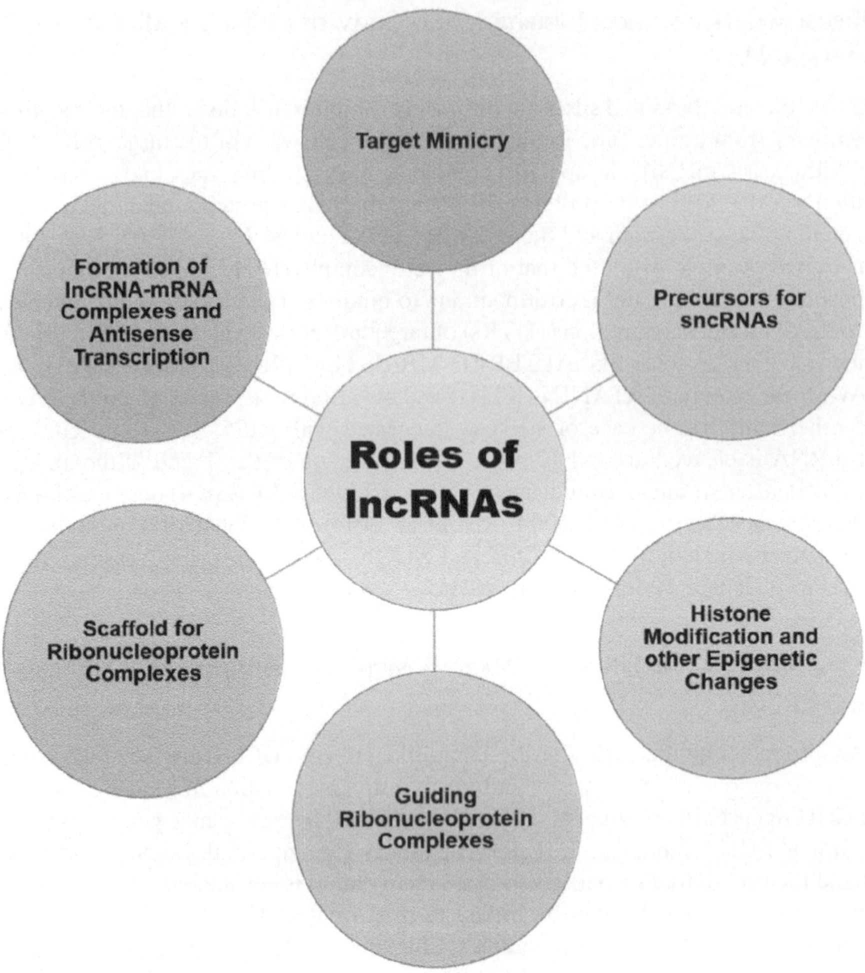

FIGURE 7.2 Schematic diagram showing regulatory roles of lncRNAs.

ii. **Decoy:** This refers to the indirect regulatory ability of lncRNAs via recruitment of RNA-binding proteins that, in turn, act as transcription factors, chromosome modifiers and other regulatory molecules (Wu et al. 2020). Additionally, lncRNAs also have the ability to competitively adsorb complementary miRNAs, thereby indirectly regulating the expression of the target mRNAs of these miRNAs (commonly called "target mimicry"); due to such a "sponge-like" function and the competitive nature of these lncRNAs, they are often called competitive endogenous RNAs or ceRNAs (Salmena et al. 2011). For example, under phosphate deficiency conditions, the *Arabidopsis* lncRNAs INDUCED BY PHOSPHATE STARVATION1 (IPS1) and At4 can competitively bind to the miRNA, miR399, and in turn derepress and/or upregulate the expression of its target gene, PHOSPHATE2 (PHO2), thereby regulating phosphate homeostasis (Chiou et al. 2006; Shin et al. 2006; Franco-Zorrilla et al. 2007).

iii. **Guide:** This includes the function of lncRNAs, in cis or trans, to guide ribonucleoprotein complexes and/or recruit chromatin modifiers to target genes. For example, the *Medicago truncatula* lncRNA Enod40 directly binds with RNA-binding proteins and participates in root nodule formation (Campalans et al. 2004).

iv. **Scaffold:** This includes the function of lncRNAs to combine multiple proteins to form ribonucleoprotein complexes using their protein-binding domains. Such complexes can, in turn, inhibit or activate transcription from target sequences. This function of lncRNAs as a platform, integrating myriad components via different binding sites, has been exemplified in epigenetic regulation in eukaryotes in general, but plant-specific instances are putative and under the domain of ongoing research (Wu et al. 2020).

7.4 REGULATION OF SALINITY STRESS RESPONSE IN PLANTS BY ncRNAs

7.4.1 Salinity Stress and General Responses in Plants

A number of constraints are faced by plants subjected to salinity or salt stress, as has been concisely summarized by Zhao et al. (2020). The foremost constraint is osmotic stress, hampering the ability of the plant to take up water from the soil. To counteract the immediate arrest of root and shoot cell expansion due to the resultant reduced turgor pressure, plant cells accumulate few compatible solutes or osmolytes to maintain homeostasis (Bohnert and Jensen 1996; Jogawat 2019). These include sugars, amino acids (especially proline), sugar alcohols, polyamines and other secondary metabolites, neutralizing the detrimental effects of the salinity stress and inducing tolerance in the plant cells (Roychoudhury and Chakraborty 2013; Roychoudhury and Das 2014; Roychoudhury and Banerjee 2016). Phytohormones, particularly abscisic acid (ABA), are also involved in abiotic stresses related to water deficit and osmotic stress (Bray 2002; Umezawa et al. 2010; Roychoudhury et al. 2013). ABA can trigger stomatal closure along with the activation and inhibition of expression of myriad genes, thereby enabling the plant to manage water loss, and in turn, salinity stress (Zhu 2002). ABA is also related to the regulation of stress responses at the post-transcriptional and epigenetic levels (Chinnusamy et al. 2008), indicating further roles in tolerance mechanisms (Contreras-Cubas et al. 2012).

Ion imbalance and/or ionic toxicity is the second constraint, mostly associated with overaccumulation of Na^+ and Cl^- in intracellular compartments. Combating the Na^+-mediated stress is related with the Salt Overly Sensitive (SOS) signalling pathway, whereby three major processes occur: (i) entry of Na^+ is minimized into the cells, (ii) compartmentation of Na^+ is maximized in the vacuole and (iii) efflux of Na^+ from the cell is maximized (Zhu et al. 1998; Ji et al. 2013). The Cl^- toxicity is met with tolerance mechanisms that cause sequestration into vacuoles using tonoplast ion exchangers and electrogenic proton pumps (Flowers and Colmer 2008; Flowers et al. 2015).

Salinity stress is also invariably related with the onset of oxidative stress (Zhao et al. 2020). Reactive oxygen species (ROS) like hydrogen peroxide (H_2O_2), superoxide anions ($O_2^{\bullet -}$), singlet oxygen ($^1O_2$) and hydroxyl radical ($OH^\bullet$) are produced in the apoplast, chloroplast, mitochondria and peroxisomes, generally by the membrane respiratory burst oxidase homologs AtRbohD and AtRbohF, whose genes are upregulated under salt stress (Miller et al. 2010). ROS at low concentrations can generally combat salt stress by reduction of Na^+/K^+ ratios by an influx of K^+ ions into the cytosol (Ma et al. 2012); however, excess ROS accumulation in saline conditions can lead to further detrimental effects including but not limited to lipid peroxidation in membranes, DNA damage, denaturation of proteins and an overall impairment of enzymatic activities (Noctor and Foyer 1998). Therefore, this concentration-dependent mechanism acts such that before reaching detrimental levels, the ROS activate a number of ROS-regulated ion channels in the cell membrane and organellar membranes, triggering a number of downstream cascades. Once the ROS cross the permissible limits, plant cells activate the ROS-scavenging machineries that include non-enzymatic antioxidants like ascorbate, glutathione and tocopherols (Das and Roychoudhury 2014), and enzymes like catalases, superoxide dismutases (SODs), ascorbate peroxidase and glutathione reductase involved in ascorbate-glutathione cycle (Roychoudhury and Basu 2012; Hanin et al. 2016). In fact, salt-tolerant plants have been shown to overexpress ascorbate peroxidase (Badawi et al. 2004), mitochondrial alternative oxidase AOX1 and Mn-SOD (Mittova et al. 2003; Giraud et al. 2008; Zhao et al. 2020). The various responses shown by a plant during water deficit are shown in Figure 7.3.

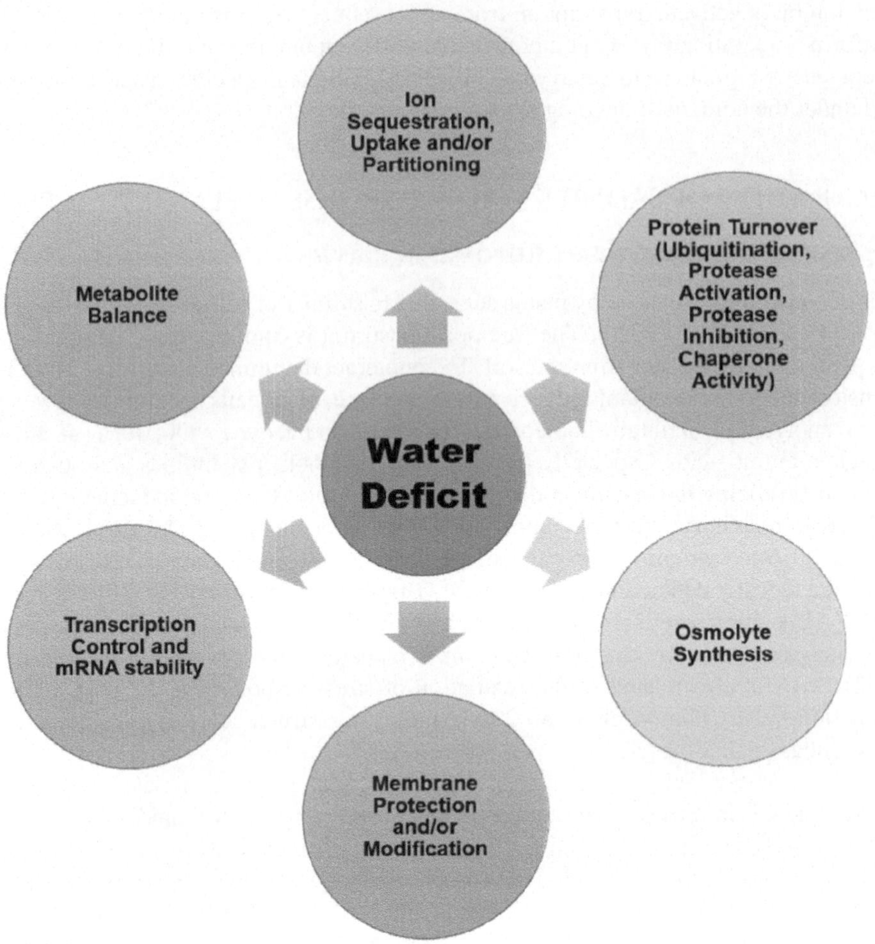

FIGURE 7.3 Various facets of plant response during water deficit. [Adapted and modified from Bohnert and Jensen (1996) and Bray (2002)].

7.4.2 ncRNAs and Salinity Stress Responses

Numerous experiments in the recent past have shown the importance of stress-responsive upregulation or downregulation of ncRNAs. In this section, a few such instances have been explored with respect to salinity stress responsive miRNAs, siRNAs and lncRNAs.

7.4.2.1 miRNAs

One of the major microarray-based analyses of stress-responsive miRNAs in *Arabidopsis thaliana* by Liu et al. (2008) revealed that miR156, miR158, miR159, miR165, miR167, miR168, miR169, miR171, miR319, miR393, miR394, miR396 and miR397 are upregulated in response to salt stress, while miR398 is downregulated. Similar studies by Ding et al. (2009) in maize showed that miR162, miR168, miR395 and miR474 are upregulated, while miR156, miR164, miR167 and miR396 are downregulated in response to salinity stress (Banerjee et al. 2017). Salinity and drought-responsive upregulation of miR156 (miR156_4), miR160 and miR164 in *Brassica juncea* and of miR37 and miR396a in *Brassica oleracea*, along with downregulation of miR156 (miR156_3) in *B. juncea* and of miR3, miR34, miR393 and miR855 in *B. oleracea* have been observed in studies by Bhardwaj et al. (2014) and Tian et al. (2014) and reviewed by Ahmed et al. (2020). Similarly, in rice,

TABLE 7.1
miRNAs Upregulated and Downregulated in Different Plant Species during Salinity Stress

Plant Species	miRNAs		References
	Upregulated	**Downregulated**	
Arabidopsis thaliana	miR156, miR158, miR159, miR165, miR167, miR168, miR169, miR171, miR319, miR393, miR394, miR396, miR397	miR398	Liu et al. (2008)
Zea mays	miR162, miR168, miR395, miR474	miR156, miR164, miR167, miR396	Ding et al. (2009)
Brassica juncea	miR156_4, miR160 miR164	miR156_3	Bhardwaj et al. (2014)
Brassica oleracea	miR37, miR396a	miR3, miR34, miR393, miR855	Tian et al. (2014)
Oryza sativa	miR395, miR474, miR845, miR851, miR854, miR901, miR903, miR1125	miR170, miR172, miR397, miR408, miR529, miR896, miR1030, miR1035, miR1050, miR1088, miR1126	Zhou et al. (2010), Contreras-Cubas et al. (2012)
Solanum lycopersicum H-2274	miR156aF, miR170-5p, miR171a5p, miR395a, miR398a-5p	N/A	Çakır et al. (2021)

upregulation of miR395, miR474, miR845, miR851, miR854, miR901, miR903 and miR1125 is observed, along with the downregulation of miR170, miR172, miR397, miR408, miR529, miR896, miR1030, miR1035, miR1050, miR1088 and miR1126 in response to drought and salinity stress (Zhou et al. 2010; Contreras-Cubas et al. 2012). Quantitative polymerase chain reaction (qPCR)-based expression profile studies in *Solanum lycopersicum* H-2274 plants under salt stress have also very recently shown upregulation of miR156aF, miR170-5p, miR171a5p, miR395a and miR398a-5p with respective tissue specificities (Çakır et al. 2021). Comprehensive studies on the different targets of these miRNAs will eventually further enhance our understanding of the roles of miRNAs in salinity stress response in plants. Table 7.1 attempts to summarize a list of differentially expressed miRNAs.

7.4.2.2 siRNAs

One of the very few major examples of salinity stress-responsive production of siRNA is that of nat-siRNA in *Arabidopsis*. As mentioned previously, plants upregulate production of proline (Pro) as an osmolyte in response to salinity stress and this is achieved by both activation of the biosynthesis of Pro and by inactivation of the degradation of Pro (Yoshiba et al. 1997). While L-Pro is synthesized by the action of delta(1)-pyrroline-5-carboxylate (P5C) synthetase (P5CS) and P5C reductase (P5CR), L-Pro is broken down by proline dehydrogenase (ProDH) and P5C dehydrogenase (P5CDH). The gene encoding P5CDH with constitutive expression overlaps with that encoding the salt stress-inducible SRO5 (similar to rcd one 5) in the anti-sense orientation. Hence, under salt stress, SRO5 produces a 24 nt nat-siRNA that mediates the cleavage of the P5CDH mRNA, in turn downregulating its expression and reducing the degradation of Pro and enhancing the accumulation of the osmolyte (Borsani et al. 2005; Wang et al. 2017).

7.4.2.3 lncRNAs

lncRNAs constitute a substantial part of ongoing research with respect to abiotic stress tolerance in plants (Urquiaga et al. 2021). Qin et al. (2017) reported that Drought Induced lncRNA (DRIR) in *Arabidopsis* acts as a positive regulator of drought and salt stress responses in plants such that there is a positively correlated facilitation of water transport, regulation of transcription of downstream genes and an overall enhancement of plant sensitivity to ABA, resulting in collective stress tolerance. Studies by Rajagopalan et al. (2006) and Ben Amor et al. (2009) identified upregulation of lncRNAs in *Arabidopsis* under NaCl treatment, including miRNA precursors like that of miR822, miR839 and miR869A. Several salt-responsive lncRNAs have also been reported in duckweed, *Spirodela polyrhiza,* by Fu et al. (2020). LncRNAsTCONS_00033722, TCONS_00044328, and TCONS_00059333 have been specifically targeted by a number of salt-responsive miRNAs, re-instating the concurrent regulation of salt stress responses by both miRNAs and lncRNAs. Specifically, TCONS_00033722 is commonly targeted by salt-responsive miRNAs, miR156, miR169, miR171, miR393, miR397 and miR398 (Sunkar et al. 2012), of which all but miR397 can regulate auxin signalling (Navarro et al. 2006; Huang et al. 2017; Shi et al. 2017). For example, PIN1, an auxin efflux carrier, is targeted by miR398, thus alleviating PIN1-induced growth inhibition by the miRNA-lncRNA interaction (Fu et al. 2020). In cotton, lncRNA973 overexpression can be correlated with increased salt tolerance (Zhang et al. 2019b). Generally localized in the nucleus, lncRNA973 is involved in stress alleviation by ROS-scavenging gene expression regulation, affecting the activation and/or inhibition of downstream transcription factors and genes, contributing to the overall salt stress response in cotton. Current research also aims to elucidate functional mechanisms of lncRNAs in salinity stress response in other agricultural crops like alfalfa (Wang et al. 2015), barley (Karlik and Gozukirmizi 2018) and sorghum (Sun et al. 2020), amongst others.

7.5 REGULATION OF HM STRESS RESPONSE IN PLANTS BY ncRNAs

7.5.1 HM Stress and General Responses in Plants

The principal and primary effect of HM stress in plants at the cellular level is the generation of oxidative stress by the metal ions. Under such conditions, ROS like H_2O_2, $O_2^{\bullet-}$, $^1O_2$ and $OH^{\bullet}$ are produced in cell organelles, such as mitochondria, peroxisomes and chloroplasts, which damage DNA and/or impair DNA repair mechanisms, impede the function and integrity of the phospholipid membranes, disrupt nutrient homeostasis and perturb the activity of proteins (Møller et al. 2007; Manara 2012; Li et al. 2016a). Plant cells have thereby developed a plethora of adaptive mechanisms for managing such stresses and preventing debilitating toxic reactions. The first line of defence involves prevention or reduction in uptake of restricting metal ions to the apoplast by binding them to the cell wall or cell exudates, thereby inhibiting long-distance transport (Zhao and Chengcai 2011; Manara 2012) and/or increased synthesis of amino acids, citric acid, malic acid, etc. (Thurman and Rankin 1982). At elevated concentrations, however, a complex network of storage and detoxification strategies become activated, involving chelation of metal ions in the cytosol and subsequent sequestration into the vacuole by vacuolar transporters (Zhao and Chengcai 2011; UlHaq et al. 2019). These chelating and metal-binding proteins mostly include naturally occurring intracellular cysteine-rich oligomeric phytochelatins and metallothioneins; they protect against stress-induced oxidative damage, maintain the homeostasis of essential transition metal ions like copper (Cu), iron (Fe), manganese (M) and zinc (Zn), and immobilize and sequester toxic metal ions like Cd (cadmium) as high molecular weight (HMW) complexes in the vacuole (Jabeen et al. 2009; Ashraf et al. 2010; Capdevila and Atrian 2011; Furini 2012; Hossain et al. 2012; Liu et al. 2015; Pochodylo and Aristilde 2017).

HM ions also profoundly disrupt cellular protein homeostasis by affecting their folding, stimulating aggregation of nascent, non-native and/or misfolded proteins, in turn leading to stress in the endoplasmic reticulum (ER) and a sharp drop in cell viability (Amm et al. 2014; Viehweger 2014; Clemens and Ma 2016; Liu and Howell 2016; Hasan et al. 2017). Proteins are the most vulnerable targets of HMs. Whether by formation of a complex with side-chain functional groups of proteins or displacement of essential ions from metalloproteins, the final results are impairment of physiological functions, interference in the folding process and a resultant quantitative deficiency of soluble proteins with an accumulation of proteotoxic aggregates. The response to this involves a set of "stress genes" which are induced to synthesize a group of proteins called heat shock proteins (HSPs) that act as molecular chaperones and assist correct folding of nascent and stress-accumulated misfolded proteins; this, in turn, prevents further aggregation and promotes selective degradation of misfolded or denatured proteins (Bierkens 2000; Gupta et al. 2010; Hüttner et al. 2012; Tamás et al. 2014; Park and Seo 2015). However, exclusive mechanisms for individual metal ions exist in many plants, such that the exposure of toxic levels of HM to the sensitive components of the cell can be avoided (Ghori et al. 2019).

7.5.2 ncRNAs and HM Stress Responses

The roles of ncRNAs in response to HM stress vary not only from one plant to another, but also based on which HM is responsible for eliciting the stress response. This section attempts to summarize the roles of miRNAs and lncRNAs in HM stress responses caused by Cd, mercury (Hg), arsenic (As), Mn and aluminium (Al) toxicity in different plants (Shriram et al. 2016; Banerjee et al. 2017; Ahmed et al. 2020; Jha et al. 2020).

7.5.2.1 Cd Toxicity

The main physiological system harmed due to Cd toxicity is the photosynthetic machinery, resulting in loss of chlorophyll content and carbon assimilation, leading to wilting, chlorosis, stunted growth and, ultimately, death of the cell (Rodríguez-Serrano et al. 2009, Banerjee et al. 2017). A number of studies have shown the following miRNA and lncRNA expression profiles in response to Cd toxicity in different plants:

7.5.2.1.1 miRNAs

Ding et al. (2016) reported downregulation of miR390 in response to Cd stress in rice with a consequent upregulation of *Oryza sativa* stress-responsive leucine-rich repeat receptor-like kinase (OsSRK) expression, the target gene of miR390. Further, overexpression of miR390 in transgenic plants showed a higher degree of Cd-induced oxidative stress with downregulation of OsSRK, underlining the importance of miR390-mediated Cd-toxicity stress response regulation. Xu et al. (2013) reported differential expression of 15 known and 8 new miRNA families by construction of small RNA libraries of *Raphanus sativus* L. (radish) in response to Cd toxicity; these miRNAs majorly targeted genes encoding for iron, ABC transporter proteins and phytochelatin-synthase-1 (PCS1). Overexpression of miR395 has been demonstrated in response to Cd-induced oxidative stress in *Brassica napus* (Zhang et al. 2013b). Since miR395 directly targets the low-affinity sulphate transporter BnSultr2;1 and ATP sulphurylases BnAPS3 and BnAPS4, which also play important roles in Cd and sulphur stress responses, miR395 finds special importance in Cd-related detoxification in *B. napus* (Huang et al. 2010). Genome-wide studies in *B. napus* have also shown the direct involvement of miR167 in post-transcriptional regulation of Natural resistance-associated macrophage proteins (NRAMP) metal ion transporter genes in response to Cd stress (Meng et al. 2017). Transcriptional regulation via targeting of the RNA helicase gene *BnaA04g26450D*, encoding BnRH24 by miR158, has also been reported in *B. napus*, along with differential expression profiles of 49 RNA helicases in response to Cd stress, further underlining the importance of miRNAs in HM stress response (Zhang et al. 2018).

7.5.2.1.2 lncRNAs

Deep RNA sequencing studies have shown the involvement of lncRNAs with Cd stress in rice, where upregulation of 69 lncRNAs and downregulation of 75 lncRNAs have been reported, along with additional gene expression regulation via lncRNA and miRNA associations (Chen et al. 2018).

Roles of lncRNAs in regulation of target genes involved in metabolism of cysteine and methionine under Cd stress have also been revealed by the functional analysis of differentially expressed lncRNAs in rice (Chen et al. 2018). qRT-PCR analyses by Feng et al. (2016) validated the functional roles of lncRNAsTCONS_00091906, TCONS_00033487 and TCONS_00097191 in *B. napus* under Cd stress and indicated roles in target mimicry for mRNAs EL628609, TC182597 and TC203372 involved in uptake and detoxification of Cd.

7.5.2.2 Hg Toxicity

Hg toxicity leads to ill-effects in plants in the form of stunted growth, abnormal vasculature, reduction in chlorophyll content and, most importantly, accumulation of ROS (Chen and Yang 2012a; Shriram et al. 2016). Various miRNAs have been reported in regulation of Hg stress responses in plants, for example, deep sequence analyses in *M. truncatula* have revealed upregulation of 15 conserved miRNA families and downregulation of 4 miRNA families in response to Hg stress, and 130 target genes corresponding to 58 known miRNA families have also been reported. Further, differential expression of miR160, miR166, miR171, miR319, miR393, miR395, miR398 and miR529 have been shown via qPCR and bioinformatics-based studies (Zhou et al. 2008, 2012). Also, downregulation of miR398 has been observed in the face of HM stress in *Arabidopsis* (Sunkar et al. 2006), with an increase in its target transcripts of Cu/Zn-SOD 2 (CSD2), which in turn combat HM-induced oxidative stress. Given miR398 also shows differential expression in the face of Hg toxicity in *M. truncatula*, its involvement in HM tolerance in plants is deemed to be important (Zhou et al. 2008).

7.5.2.3 As Toxicity

Despite chemically being a metalloid, As is included among the HMs based on its heaviness and toxicity, since even low levels of exposure to As can induce toxicity (Duffus 2002). Not only is it a Class I carcinogen, posing serious threat to human health via consumption of crops cultivated in As-contaminated soils, but arsenite toxicity is also harmful for plants, leading to a reduced photosynthetic rate, generation of ROS, peroxidation of lipids and perturbation of carbohydrate metabolism (Meharg and Hartley-Whitaker 2002; Stoeva et al. 2003; Requejo and Tena 2005; Srivastava et al. 2012; Shriram et al. 2016; Banerjee et al. 2017).

Studies on miRNA- and lncRNA-mediated regulation of As toxicity have been conducted mostly on rice and *B. juncea* and have been summarized as follows:

7.5.2.3.1 miRNAs

Yu et al. (2012) identified 36 As-responsive miRNAs in rice, 14 of which were involved in gene-expression regulation in signalling pathways related to transportation and metabolism. Upregulation of miR397b, miR408 and miR528 and downregulation of miR390 and miR1316 have been reported in indica rice roots by Liu and Zhang (2012). Transcription factors of the metabolic pathways of lipids and jasmonic acid (JA) were found to be targeted by the As-responsive miRNAs like miR319, in *B. juncea,* indicating that JA is related to the As-mediated stress response (Srivastava et al. 2009, 2013; Gupta et al. 2014).

7.5.2.3.2 lncRNAs

Studies based on target prediction and degradome-sequencing data-based validation by Tang et al. (2019) have shown the involvement of a number of lncRNAs in As toxicity in rice. However, most of them act as precursors of smaller ncRNAs with probable roles in target mimicry as well, all of which require further research (Jha et al. 2020).

7.5.2.4 Mn Toxicity

Although Mn is an important component for enzymatic reactions, acting as an inorganic catalyst, higher concentrations of Mn can induce HM-stress in plants, including differential expression of ncRNAs (Gupta et al. 2014). For example, miRNA macroarray hybridization and qRT-PCR-based studies by Valdés-López et al. (2010) identified 37 Mn-responsive miRNAs in *Phaseolus vulgaris* (common bean), of which the targets of miR1532, miR1510/2110, miR1515 and miR1508 were predicted to be receptor kinase proteins, nucleotide-binding site leucine-rich repeat (NBS-LRR) resistance-like proteins, HSPs and calcium-dependent protein kinases, respectively (Valdés-López et al. 2010; Shriram et al. 2016).

7.5.2.5 Al Toxicity

Aluminium can bind with carboxylate and phosphate groups in root cell walls, leading to structural changes, cell-wall expansion inhibition and decrease in root growth, which reduce mineral, nutrient and water uptake as well as affect cytoplasmic Ca^{2+} homeostasis and induce oxidative stress (Ma et al. 2004; Silva 2012). High-throughput genome sequencing in *M. truncatula* (Chen et al. 2012b) and soybean (Zeng et al. 2012) seedlings subjected to Al stress have revealed alteration in expression of 23 miRNAs in *M. truncatula* and 30 miRNAs with species-specific upregulation of miR396 and miR390 in soybean.

7.6 DATABASES OF ncRNAs

Essentially, all the works listed in this section are the outcome of results derived from extensive *in silico* research using a number of databases developed using intricate algorithms. A few such instances reflecting the advances of computational experimental biology have been listed in Table 7.2.

TABLE 7.2
Summary of Key Databases for Identification and Functional Annotation of ncRNAs in Plants

Database	Function	Latest URL	Latest Reference
miRbase	Database of published miRNA sequences and annotation	http://www.mirbase.org/	Kozomara et al. (2018)
miRPlant	Integrated tool for identification of plant miRNA	https://sourceforge.net/projects/mirplant/	An et al. (2014)
Rfam	Collection of RNA families, each represented by multiple sequence alignments, consensus secondary structures and covariance models	http://rfam.xfam.org/	Kalvari et al. (2021)
sRNAbench	Profiling of small RNAs and their sequence variants in single or multi-species high-throughput experiments	https://arn.ugr.es/srnatoolbox/srnabench/	Aparicio-Puerta et al. (2019)
PmiRKB (Plant miRNA Knowledge Base)	Includes miRNAs of *Arabidopsis thaliana* and *Oryza sativa*	http://bis.zju.edu.cn/pmirkb/	Meng et al. (2011)
miRTarBase	Database containing miRNA-target interactions	https://bio.tools/mirtarbase	Huang et al. (2020)
DeepLNC	Lnc-RNA prediction using Deep Neural Network	https://bioserver.iiita.ac.in/deeplnc/	Tripathi et al. (2016)
PNRD (Plant Non-coding RNA Database)	Comprehensive platform of plant ncRNA analysis	http://structuralbiology.cau.edu.cn/PNRD/	Yi et al. (2015)

Source: Adapted and Modified from Wang et al. (2017).

7.7 CONCLUSION AND FUTURE PERSPECTIVES

The "junk RNA" or "dark matter of the genome" is being gradually brought to light and, as outlined in various sections of this chapter, its roles are manifold in abiotic stress tolerance in plants. While it is true that expression studies of known ncRNAs and discovery of novel ncRNAs have become facilitated with the advent of better computational power and novel software and technology, their gap with functional annotation of these ncRNAs and subsequent verification of those functions *in vitro* and *in vivo* is ever-increasing. Practical applications of this continuously growing bank of discoveries are yet to be explored in detail. Initial works targeting the stress-responsive miRNAs and lncRNAs using the advanced technologies of Clustered Regularly Interspaced Short Palindromic Repeats associated nucleases (CRISPR)/Cas9-based gene editing and subsequent development of stress-tolerant transgenics and GMOs (genetically modified organisms) promise alternative approaches for crop breeding (Li et al. 2016b; Zhou et al. 2017, Zhang and Zhang 2020). Thus, with further unravelling of the intricate and detailed functions of ncRNAs in different signalling cascades and target mRNAs, it is hoped that a more holistic view of how sessile organisms like plants survive the vagaries of environmental stresses will gain clarity and accuracy.

ACKNOWLEDGEMENTS

Financial assistance from Science and Engineering Research Board, Government of India, through the grant [EMR/2016/004799] and Department of Higher Education, Science and Technology and Biotechnology, Government of West Bengal, through the grant [264(Sanc.)/ST/P/S&T/1G-80/2017] to Prof. Aryadeep Roychoudhury is gratefully acknowledged.

REFERENCES

Ahmed, W., Xia, Y., Li, R., Bai, G., Siddique, K., & Guo, P. (2020). Non-coding RNAs: Functional roles in the regulation of stress response in *Brassica* crops. Genomics. 112(2), 1419–1424.

Amm, I., Sommer, T., & Wolf, D. H. (2014). Protein quality control and elimination of protein waste: The role of the ubiquitin-proteasome system. Biochim Biophys Acta. 1843(1), 182–196.

An, J., Lai, J., Sajjanhar, A., Lehman, M. L., & Nelson, C. C. (2014). miRPlant: An integrated tool for identification of plant miRNA from RNA sequencing data. BMC Bioinf. 15(1), 275.

Aparicio-Puerta, E., Lebrón, R., Rueda, A., Gómez-Martín, C., Giannoukakos, S., Jaspez, D., Medina, J. M., Zubkovic, A., Jurak, I., Fromm, B., Marchal, J. A., Oliver, J., & Hackenberg, M. (2019). sRNAbench and sRNAtoolbox 2019: Intuitive fast small RNA profiling and differential expression. Nucl Acids Res. 47(W1), W530–W535.

Ashraf, M., Qztürk, M. A., & Ahmad, M. S. A. (2010). Plant Adaptation and Phytoremediation. Springer, New York, NY.

Aukerman, M. J., & Sakai, H. (2003). Regulation of flowering time and floral organ identity by a MicroRNA and its APETALA2-like target genes. The Plant Cell. 15(11), 2730–2741.

Axtell, M. J., & Meyers, B. C. (2018). Revisiting criteria for plant MicroRNA annotation in the era of big data. The Plant Cell. 30(2), 272–284.

Badawi, G. H., Yamauchi, Y., Kawano, N., Tanaka, K., & Tanaka, K. (2004). Over-expression of ascorbate peroxidase in tobacco chloroplasts enhances the tolerance to salt stress and water deficit. Physiol Plant. 121, 231–238.

Banerjee, S., Sirohi, A., Ansari, A. A., & Gill, S. S. (2017). Role of small RNAs in abiotic stress responses in plants. Plant Gene. 11, 180–189.

Ben Amor, B., Wirth, S., Merchan, F., Laporte, P., d'Aubenton-Carafa, Y., Hirsch, J., Maizel, A., Mallory, A., Lucas, A., Deragon, J. M., Vaucheret, H., Thermes, C., & Crespi, M. (2009). Novel long non-protein coding RNAs involved in Arabidopsis differentiation and stress responses. Genome Res. 19(1), 57–69.

Bhardwaj, A. R., Joshi, G., Pandey, R., Kukreja, B., Goel, S., Jagannath, A., Kumar, A., Katiyar-Agarwal, S., & Agarwal, M. (2014). A genome-wide perspective of miRNAome in response to high temperature, salinity and drought stresses in *Brassica juncea* (Czern) L. PloS One. 9(3), e92456.

Bierkens, J. G. (2000). Applications and pitfalls of stress-proteins in biomonitoring. Toxicology 153, 61–72.

Bohnert, H. J., & Jensen, R. G. (1996). Strategies for engineering water-stress tolerance inplants. Trends Biotechnol. 14, 89–97.

Borsani, O., Zhu, J., Verslues, P. E., Sunkar, R., & Zhu, J. K. (2005). Endogenous siRNAs derived from a pair of natural cis-antisense transcripts regulate salt tolerance in Arabidopsis. Cell. 123(7), 1279–1291.

Bray, E. A. (2002) Abscisic acid regulation of gene expression during water-deficit stress in the era of the Arabidopsis genome. Plant Cell Environ. 25:153–161.

Çakır, Z., Arıkan, B., Karpuz, B., & Turgut-Kara, N. (2021). Expression analysis of miRNAs and their targets related to salt stress in *Solanum lycopersicum* H-2274. Biotechnol Biotechnol Equip. 35(1), 283–290.

Campalans, A., Kondorosi, A., & Crespi, M. (2004). Enod40, a short open reading frame-containing mRNA, induces cytoplasmic localization of a nuclear RNA binding protein in *Medicago truncatula*. The Plant Cell. 16(4), 1047–1059.

Capdevila, M., & Atrian, S. (2011). Metallothionein protein evolution: A miniassay. J Biol Inorg Chem. 16, 977–989.

Cech, T. R., & Steitz, J. A. (2014). The noncoding RNA revolution-trashing old rules to forge new ones. Cell. 157(1), 77–94.

Cerutti, P., Holt, J. W., & Miller, N. (1968). Detection and determination of 5,6-dihydrouridine and 4-thiouridine in transfer ribonucleic acid from different sources. J Mol Biol. 34(3), 505–518.

Chen, Q., Meng, X., Liao, Q., & Chen, M. (2019). Versatile interactions and bioinformatics analysis of noncoding RNAs. Brief Bioinfor. 20(5), 1781–1794.

Chen, L., Shi, S., Jiang, N., Khanzada, H., Wassan, G. M., Zhu, C., Peng, X., Xu, J., Chen, Y., Yu, Q., He, X., Fu, J., Chen, X., Hu, L., Ouyang, L., Sun, X., He, H., & Bian, J. (2018). Genome-wide analysis of long non-coding RNAs affecting roots development at an early stage in the rice response to cadmium stress. BMC Genomics. 19(1), 460.

Chen, L., Wang, T., Zhao, M., Tian, Q., & Zhang, W. H. (2012b). Identification of aluminum-responsive microRNAs in *Medicago truncatula* by genome-wide high-throughput sequencing. Planta. 235, 375–386.

Chen, J., & Yang, Z. M. (2012a). Mercury toxicity, molecular response and tolerance in higher plants. Biometals. 25(5), 847–857.

Chinnusamy, V., Gong, Z., & Zhu, J. K. (2008) Abscisic acid-mediated epigenetic processes in plant development and stress responses. J Integr Plant Biol 50:1187–1195

Chiou, T. J., Aung, K., Lin, S. I., Wu, C. C., Chiang, S. F., & Su, C. L. (2006). Regulation of phosphate homeostasis by MicroRNA in *Arabidopsis*. Plant Cell. 18(2), 412–421.

Clemens, S., & Ma, J. F. (2016). Toxic heavy metal and metalloid accumulation in crop plants and foods. Annu Rev Plant Biol 67, 489–512.

Contreras-Cubas, C., Palomar, M., Arteaga-Vázquez, M., Reyes, J. L., & Covarrubias, A. A. (2012). Non-coding RNAs in the plant response to abiotic stress. Planta. 236(4), 943–958.

Creasey, K. M., Zhai, J., Borges, F., Van Ex, F., Regulski, M., Meyers, B. C., & Martienssen, R. A. (2014). miRNAs trigger widespread epigenetically activated siRNAs from transposons in *Arabidopsis*. Nature. 508(7496), 411–415.

Cuellar, T. L., & McManus, M. T. (2005). MicroRNAs and endocrine biology. J Endocrinol. 187(3), 327–332.

Das, K., & Roychoudhury, A. (2014). Reactive oxygen species (ROS) and response of antioxidants as ROS-scavengers during environmental stress in plants. Front Env Sci. 2, 53

Derrien, T., Johnson, R., Bussotti, G., Tanzer, A., Djebali, S., Tilgner, H., Guernec, G., Martin, D., Merkel, A., Knowles, D. G., Lagarde, J., Veeravalli, L., Ruan, X., Ruan, Y., Lassmann, T., Carninci, P., Brown, J. B., Lipovich, L., Gonzalez, J. M., Thomas, M., … Guigó, R. (2012). The GENCODE v7 catalog of human long noncoding RNAs: Analysis of their gene structure, evolution, and expression. Genome Res. 22(9), 1775–1789.

Ding, Y., Ye, Y., Jiang, Z., Wang, Y., & Zhu, C. (2016). MicroRNA390 is involved in cadmium tolerance and accumulation in rice. Front Plant Sci. 7, 235.

Ding, D., Zhang, L., Wang, H., Liu, Z., Zhang, Z., & Zheng, Y. (2009). Differential expression of miRNAs in response to salt stress in maize roots. Ann Bot. 103(1), 29–38.

Duffus, J. H. (2002) Heavy metal-a meaningless term. Pure Appl Chem. 74, 793–807.

Erdmann, R. M., & Picard, C. L. (2020). RNA-directed DNA methylation. PLoS Genet. 16(10), e1009034.

Feng, S. J., Zhang, X. D., Liu, X. S., Tan, S. K., Chu, S. S., Meng, J. G., Zhao, K. X., Zheng, J. F., & Yang, Z. M. (2016). Characterization of long non-coding RNAs involved in cadmium toxic response in *Brassica napus*. RSC Adv. 6(85), 82157–82173.

Fire, A., Xu, S., Montgomery, M. K., Kostas, S. A., Driver, S. E., & Mello, C. C. (1998). Potent and specific genetic interference by double-stranded RNA in *Caenorhabditis elegans*. Nature. 391(6669), 806–811.

Flowers, T. J., & Colmer, T. D. (2008). Salinity tolerance in halophytes. New Phytologist. 179(4), 945–963.
Flowers, T. J., Munns, R., & Colmer, T. D. (2015). Sodium chloride toxicity and the cellular basis of salt tolerance in halophytes. Ann Bot. 115(3), 419–431.
Franco-Zorrilla, J. M., Valli, A., Todesco, M., Mateos, I., Puga, M. I., Rubio-Somoza, I., Leyva, A., Weigel, D., García, J. A., & Paz-Ares, J. (2007). Target mimicry provides a new mechanism for regulation of microRNA activity. Nat Genet. 39(8), 1033–1037.
Fu, L., Ding, Z., Tan, D., Han, B., Sun, X., & Zhang, J. (2020). Genome-wide discovery and functional prediction of salt-responsive lncRNAs in duckweed. BMC Genomics. 21(1), 212.
Furini, A. (2012). Plants and Heavy Metals. Springer, Netherlands.
Gasciolli, V., Mallory, A. C., Bartel, D. P., & Vaucheret, H. (2005). Partially redundant functions of Arabidopsis DICER-like enzymes and a role for DCL4 in producing trans-acting siRNAs. Curr Biol. 15(16), 1494–1500.
Ghori, N. H., Ghori, T., Hayat, M. Q., Imadi, S. R., Gul, A., Altay, V., & Ozturk, M. (2019). Heavy metal stress and responses in plants. Int J Environ Sci Technol. 16(3), 1807–1828.
Giraud, E., Ho, L. H. M., Clifton, R., Carroll, A., Estavillo, G., Tan, Y. F., Howell, K. A., Ivanova, A., Pogson, B. J., & Millar, A. H., et al. (2008). The absence of Alternative Oxidase1a in *Arabidopsis* results in acute sensitivity to combined light and drought stress. Plant Physiol. 147, 595–610.
Glažar, P., Papavasileiou, P., & Rajewsky, N. (2014). circBase: A database for circular RNAs. RNA (New York, N.Y.). 20(11), 1666–1670.
Griffiths-Jones, S., Bateman, A., Marshall, M., Khanna, A., & Eddy, S. R. (2003). Rfam: An RNA family database. Nucl Acids Res. 31(1), 439–441.
Griffiths-Jones, S., Grocock, R. J., van Dongen, S., Bateman, A., & Enright, A. J. (2006). miRBase: microRNA sequences, targets and gene nomenclature. Nucl Acids Res. 34(Database issue), D140–D144.
Guleria, P., Mahajan, M., Bhardwaj, J., & Yadav, S. K. (2011). Plant small RNAs: Biogenesis, mode of action and their roles in abiotic stresses. Genomics, Proteomics Bioinf. 9(6), 183–199.
Gupta, O. P., Sharma, P., Gupta, R. K., & Sharma, I. (2014). MicroRNA mediated regulation of metal toxicity in plants: Present status and future perspectives. Plant Mol Biology. 84(1-2), 1–18.
Gupta, S. C., Sharma, A., Mishra, M., Mishra, R. K., & Chowdhuri, D. K. (2010). Heat shock proteins in toxicology: How close and How far? Life Sci. 86, 377–384.
Hamilton, A. J., & Baulcombe, D. C. (1999). A species of small antisense RNA in posttranscriptional gene silencing in plants. Science (New York, N.Y.). 286(5441), 950–952.
Hanin, M., Ebel, C., Ngom, M., Laplaze, L., & Masmoudi, K. (2016). New insights on plant salt tolerance mechanisms and their potential use for breeding. Front Plant Sci. 7, 1–17.
Hasan, M. K., Cheng, Y., Kanwar, M. K., Chu, X.-Y., Ahammed, G. J., & Qi, Z.-Y. (2017). Responses of plant proteins to heavy metal stress—A review. Front Plant Sci. 8:1492.
He, X., Chen, X., Zhang, X., Duan, X., Pan, T., Hu, Q., Zhang, Y., Zhong, F., Liu, J., Zhang, H., Luo, J., Wu, K., Peng, G., Luo, H., Zhang, L., Li, X., & Zhang, H. (2015). An lnc RNA (GAS5)/SnoRNA-derived piRNA induces activation of TRAIL Gene by site-specifically recruiting MLL/COMPASS-like complexes. Nucl Acids Res. 43(7), 3712–3725.
Heo, J. B., & Sung, S. (2011). Vernalization-mediated epigenetic silencing by a long intronic noncoding RNA. Science (New York, N.Y.). 331(6013), 76–79.
Hossain, M. A., Piyatida, P., da Silva, J. A. T., & Fujita, M. (2012). Molecular mechanism of heavy metal toxicity and tolerance in plants: Central role of glutathione in detoxification of reactive oxygen species and methylglyoxal and in heavy metal chelation. J Bot. 2012:872875.
Huang, H. Y., Lin, Y. C. D., Li, J., Huang, K. Y., Shrestha, S., Hong, H. C., Tang, Y., Chen, Y. G., Jin, C. N., Yu, Y., Xu, J. T., Li, Y. M., Cai, X. X., Zhou, Z. Y., Chen, X. H., Pei, Y. Y., Hu, L., Su, J. J., Cui, S. D., … Huang, H. D. (2020). miRTarBase 2020: Updates to the experimentally validated microRNA–target interaction database. Nucl Acids Res. 48, D148–D154.
Huang, W., Peng, S., Xian, Z., Lin, D., Hu, G., Yang, L., Ren, M., & Li, Z. (2017). Overexpression of a tomato miR171 target gene SlGRAS24 impacts multiple agronomical traits via regulating gibberellin and auxin homeostasis. Plant Biotechnol J. 15(4), 472–488.
Huang, S. Q., Xiang, A. L., Che, L. L., Chen, S., Li, H., Song, J. B., & Yang, Z. M. (2010), A set of miRNAs from *Brassica napus* in response to sulphate deficiency and cadmium stress. Plant Mol Biol. 8, 887–899.
Hüttner, S., Veit, C., Schoberer, J., Grass, J., & Strasser, R. (2012). Unraveling the function of *Arabidopsis thaliana* OS9 in the endoplasmic reticulum associated degradation of glycoproteins. Plant Mol Biol. 79, 21–33.
Iwakawa, H. O., & Tomari, Y. (2013). Molecular insights into microRNA-mediated translational repression in plants. Mol Cell. 52(4), 591–601.

Jabeen, R., Ahmad, A., & Iqbal, M. (2009). Phytoremediation of heavy metals: Physiological and molecular mechanisms. Bot Rev. 75, 339–364.

Jha, U. C., Nayyar, H., Jha, R., Khurshid, M., Zhou, M., Mantri, N., & Siddique, K. (2020). Long non-coding RNAs: Emerging players regulating plant abiotic stress response and adaptation. BMC Plant Biol. 20(1), 466.

Ji, H., Pardo, J. M., Batelli, G., Van Oosten, M. J., Bressan, R. A., & Li, X. (2013). The Salt Overly Sensitive (SOS) pathway: Established and emerging roles. Mol Plant. 6(2), 275–286.

Jogawat, A. (2019). Osmolytes and their role in abiotic stress tolerance in plants. In: Roychoudhury A, Tripathi D (Eds.), Molecular Plant Abiotic Stress. John Wiley & Sons, Ltd., UK, pp. 91–104.

Kalvari, I., Nawrocki, E. P., Ontiveros-Palacios, N., Argasinska, J., Lamkiewicz, K., Marz, M., Griffiths-Jones, S., Toffano-Nioche, C., Gautheret, D., Weinberg, Z., Rivas, E., Eddy, S. R., Finn, R. D., Bateman, A., & Petrov, A. I. (2021). Rfam 14: Expanded coverage of metagenomic, viral and microRNA families. Nucl Acids Res. 49(D1), D192–D200.

Karlik, E., & Gozukirmizi, N. (2018). Expression analysis of lncRNA AK370814 involved in the barley vitamin B6 salvage pathway under salinity. Mol Biol Rep. 45, 1597–1609.

Kim, V. N. (2005). Small RNAs: Classification, biogenesis, and function. Mol Cells. 19(1), 1–15.

Kozomara, A., Birgaoanu, M., & Griffiths-Jones, S. (2018). miRBase: From microRNA sequences to function. Nucl Acids Res. 47, D155–D162.

Kwak, P. B., & Tomari, Y. (2012). The n domain of argonaute drives duplex unwinding during RISC assembly. Nat Struct Mol Biol. 19(2), 145–151.

Lagos-Quintana, M., Rauhut, R., Lendeckel, W., & Tuschl, T. (2001). Identification of novel genes coding for small expressed RNAs. Science (New York, N.Y.). 294(5543), 853–858.

Lauressergues, D., Couzigou, J. M., Clemente, H. S., Martinez, Y., Dunand, C., Bécard, G., & Combier, J. P. (2015). Primary transcripts of microRNAs encode regulatory peptides. Nature. 520(7545), 90–93.

Lee, J. T. (2012). Epigenetic regulation by long noncoding RNAs. Science (New York, N.Y.). 338(6113), 1435–1439.

Lee, R. C., Feinbaum, R. L., & Ambros, V. (1993). The *C. elegans* heterochronic gene lin-4 encodes small RNAs with antisense complementarity to lin-14. Cell. 75(5), 843–854.

Li, M., Ahammed, G. J., Li, C., Bao, X., Yu, J., Huang, C., Yin, H., & Zhou, J. (2016a). Brassinosteroid ameliorates zinc oxide nanoparticles-induced oxidative stress by improving antioxidant potential and redox homeostasis in tomato seedling. Front Plant Sci. 7, 615.

Li, S., Liu, L., Zhuang, X., Yu, Y., Liu, X., Cui, X., Ji, L., Pan, Z., Cao, X., Mo, B., Zhang, F., Raikhel, N., Jiang, L., & Chen, X. (2013). MicroRNAs inhibit the translation of target mRNAs on the endoplasmic reticulum in *Arabidopsis*. Cell. 153(3), 562–574.

Li, M., Li, X., Zhou, Z., Wu, P., Fang, M., & Pan, X. et al. (2016b). Reassessment of the four yield-related genes Gn1a, DEP1, GS3, and IPA1 in rice using a CRISPR/Cas9 system. Front Plant Sci. 7, 377.

Liu, C., Bai, B., Skogerbø, G., Cai, L., Deng, W., Zhang, Y., Bu, D., Zhao, Y., & Chen, R. (2005). NONCODE: An integrated knowledge database of non-coding RNAs. Nucl Acids Res. 33(Database issue), D112–D115.

Liu, J. X., & Howell, S. H. (2016). Managing the protein folding demands in the endoplasmic reticulum of plants. New Phyto. 211, 418–428.

Liu, J., Shi, X., Qian, M., Zheng, L., Lian, C., & Xia, Y. et al. (2015). Copper induced hydrogen peroxide upregulation of a metallothionein gene, OsMT2c, from *Oryza sativa* L. confers copper tolerance in *Arabidopsis thaliana*. J. Hazard Mater. 294, 99–108.

Liu, H. H., Tian, X., Li, Y. J., Wu, C. A., & Zheng, C. C. (2008). Microarray-based analysis of stress-regulated microRNAs in *Arabidopsis thaliana*. RNA (New York, N.Y.). 14(5), 836–843.

Liu, Q., & Zhang, H. (2012). Molecular identification and analysis of arsenite stress-responsive miRNAs in rice. J Agric Food Chem. 60(26), 6524–6536.

Liu, T., Zhang, L., Chen, G., & Shi, T. (2017). Identifying and characterizing the circular RNAs during the lifespan of *Arabidopsis* leaves. Front Plant Sci. 8, 1278.

Mallory, A. C., & Vaucheret, H. (2006). Functions of microRNAs and related small RNAs in plants. Nat Genet. 38(Suppl), S31–S36.

Manara, A. (2012). "Plant responses to heavy metal toxicity," In: Furini A (Ed.), Plants and Heavy Metals. Springer, pp. 27–53.

Ma, J. F., Shen, R., Nagao, S., & Tanimoto, E. (2004). Aluminium targets elongating cells by reducing cell wall extensibility in wheat roots. Plant Cell Physiol. 45, 583–589.

Matzke, M., Kanno, T., Daxinger, L., Huettel, B., & Matzke, A. J. (2009). RNA-mediated chromatin-based silencing in plants. Curr Opin Cell Biol. 21(3), 367–376.

Maunoury, N., & Vaucheret, H. (2011). AGO1 and AGO2 act redundantly in miR408-mediated plantacyanin regulation. PLoS One. 6(12), e28729.

Ma, L., Zhang, H., Sun, L., Jiao, Y., Zhang, G., Miao, C., & Hao, F. (2012). NADPH oxidase AtrbohD and AtrbohF function in ROS-dependent regulation of Na^+/K^+ homeostasis in *Arabidopsis* under salt stress. J Exp Bot. 63, 305–317.

Meharg, A. A., & Hartley-Whitaker, J. (2002). Arsenic uptake and metabolism in arsenic resistant and nonresistant plant species. New Phytologist. 154(1), 29–43.

Meng, Y., Gou, L., Chen, D., Mao, C., Jin, Y., Wu, P., & Chen, M. (2011). PmiRKB: A plant microRNA knowledge base. Nucl Acids Res. 39(Database issue), D181–D187.

Meng, J. G., Zhang, X. D., Tan, S. K., Zhao, K. X., & Yang, Z. M. (2017). Genome-wide identification of Cd-responsive NRAMP transporter genes and analyzing expression of NRAMP 1 mediated by miR167 in *Brassica napus*. Biometals. 30(6), 917–931.

Miller, G., Suzuki, N., Ciftci-Yilmaz, S., & Mittler, R. (2010). Reactive oxygen species homeostasis and signalling during drought and salinity stresses. Plant Cell Environ. 33, 453–467.

Mittova, V., Tal, M., Volokita, M., & Guy, M. (2003). Up-regulation of the leaf mitochondrial and peroxisomal antioxidative systems in response to salt-induced oxidative stress in the wild salt-tolerant tomato species *Lycopersicon pennellii*. Plant Cell Environ. 26, 845–856.

Møller, I. M., Jensen, P. E., & Hansson, A. (2007). Oxidative modifications to cellular components in plants. Annu Rev Plant Biol. 58, 459–481.

Nam, J. W., & Bartel, D. P. (2012). Long noncoding RNAs in *C. elegans*. Genome Res. 22(12), 2529–2540.

Navarro, L., Dunoyer, P., Jay, F., Arnold, B., Dharmasiri, N., Estelle, M., Voinnet, O., & Jones, J. D. (2006). A plant miRNA contributes to antibacterial resistance by repressing auxin signaling. Science (New York, N.Y.). 312(5772), 436–439.

Nie, L., Wu, H. J., Hsu, J. M., Chang, S. S., Labaff, A. M., Li, C. W., Wang, Y., Hsu, J. L., & Hung, M. C. (2012). Long non-coding RNAs: versatile master regulators of gene expression and crucial players in cancer. Am J Transl Res. 4(2), 127–150.

Noctor, G., & Foyer, C. H. (1998). Ascorbate and glutathione: Keeping active oxygen under control. Annu Rev Plant Physiol Plant Mol Biol. 49, 249–279.

Ono, M., Scott, M., Yamada, K., Avolio, F., Barton, G., & Lamond, A. (2011). Identification of human miRNA precursors that resemble box C/D snoRNAs. Nucl Acids Res. 39. 3879–91.

Pareek, M., Yogindran, S., Mukherjee, S., & Rajam, M. (2015). Plant MicroRNAs: Biogenesis, Functions, and Applications. In: Bahadur B, Venkat Rajam M, Sahijram L, Krishnamurthy K (Eds.), Plant Biology and Biotechnology Volume II: Plant Genomics and Biotechnology. Springer, New Delhi, pp 639–661.

Park, C. J., & Seo, Y. S. (2015). Heat shock proteins: A review of the molecular chaperones for plant immunity. Plant Pathol J. 31, 323–333.

Peragine, A., Yoshikawa, M., Wu, G., Albrecht, H. L., & Poethig, R. S. (2004). SGS3 and SGS2/SDE1/RDR6 are required for juvenile development and the production of trans-acting siRNAs in Arabidopsis. Genes Dev. 18(19), 2368–2379.

Peschansky, V. J., & Wahlestedt, C. (2014). Non-coding RNAs as direct and indirect modulators of epigenetic regulation. Epigenetics. 9(1), 3–12.

Pochodylo, A. L., & Aristilde, L. (2017). Molecular dynamics of stability and structures in phytochelatin complexes with Zn, Cu, Fe, Mg, and Ca: Implications for metal detoxification. Environ Chem Lett. 15, 1–6.

Ponjavic, J., Ponting, C. P., & Lunter, G. (2007). Functionality or transcriptional noise? Evidence for selection within long noncoding RNAs. Genome Res. 17(5), 556–565.

Qin, T., Zhao, H., Cui, P., Albesher, N., & Xiong, L. (2017). A nucleus-localized long non-coding RNA enhances drought and salt stress tolerance. Plant Physiol. 175(3), 1321–1336.

Rajagopalan, R., Vaucheret, H., Trejo, J., & Bartel, D. P. (2006). A diverse and evolutionarily fluid set of microRNAs in *Arabidopsis thaliana*. Genes Dev. 20(24), 3407–3425.

Reinhart, B. J., Slack, F. J., Basson, M., Pasquinelli, A. E., Bettinger, J. C., Rougvie, A. E., Horvitz, H. R., & Ruvkun, G. (2000). The 21-nucleotide let-7 RNA regulates developmental timing in *Caenorhabditis elegans*. Nature. 403(6772), 901–906.

Ren, G., Xie, M., Zhang, S., Vinovskis, C., Chen, X., & Yu, B. (2014). Methylation protects microRNAs from an AGO1-associated activity that uridylates 5' RNA fragments generated by AGO1 cleavage. Proc Natl Acad Sci USA. 111(17), 6365–6370.

Requejo, R., & Tena, M. (2005). Proteome analysis of maize roots reveals that oxidative stress is a main contributing factor to plant arsenic toxicity. Phytochemistry. 66, 1519–1528.

Rhoades, M. W., Reinhart, B. J., Lim, L. P., Burge, C. B., Bartel, B., & Bartel, D. P. (2002). Prediction of plant microRNA targets. Cell. 110(4), 513–520.

Rodríguez-Serrano, M., Romero-Puertas, M. C., Pazmiño, D. M., Testillano, P. S., Risueño, M. C., Del Río, L. A., & Sandalio, L. M. (2009). Cellular response of pea plants to cadmium toxicity: Cross talk between reactive oxygen species, nitric oxide, and calcium. Plant Physiol. 150(1), 229–243.

Rogers, K., & Chen, X. (2012). microRNA biogenesis and turnover in plants. Cold Spring Harbor Symp Quant Biol, 77, 183–194.

Rogers, K., & Chen, X. (2013). Biogenesis, turnover, and mode of action of plant microRNAs. The Plant Cell. 25(7), 2383–2399.

Röther, S., & Meister, G. (2011). Small RNAs derived from longer non-coding RNAs. Biochimie. 93(11), 1905–1915.

Roychoudhury, A., & Banerjee, A. (2016). Endogenous glycine betaine accumulation mediates abiotic stress tolerance in plants. Trop Plant Res. 3: 105–111.

Roychoudhury, A., & Basu, S. (2012) Ascorbate-Glutathione and plant tolerance to various abiotic stresses. In: Anjum NA, Umar S, Ahmad A (Eds.), Oxidative stress in plants: Causes, Consequences and Tolerance. IK International Publishers, New Delhi, pp. 177–258.

Roychoudhury, A., & Chakraborty, M. (2013). Biochemical and molecular basis of varietal difference in plant salt tolerance. Annu Rev Res Biol. 3(4), 422–454.

Roychoudhury, A., & Das, K. (2014). Functional role of polyamines and polyamine-metabolizing enzymes during salinity, drought and cold stresses. In: Anjum N.A, Gill SS, Gill R (Eds.), Plant Adaptation to Environmental Change: Significance of Amino acids and their Derivatives. CAB International, Wallingford, pp. 141–156.

Roychoudhury, A., Paul, S., & Basu, S. (2013). Cross-talk between abscisic acid-dependent and abscisic acid-independent pathways during abiotic stress. Plant Cell Rep. 32, 985–1006

Salmena, L., Poliseno, L., Tay, Y., Kats, L., & Pandolfi, P. P. (2011). A ceRNA hypothesis: The Rosetta Stone of a hidden RNA language? Cell. 146(3), 353–358.

Schaefer, M., Pollex, T., Hanna, K., Tuorto, F., Meusburger, M., Helm, M., & Lyko, F. (2010). RNA methylation by Dnmt2 protects transfer RNAs against stress-induced cleavage. Genes Dev. 24(15), 1590–1595.

Schimmel, P. (2018). The emerging complexity of the tRNA world: Mammalian tRNAs beyond protein synthesis. Nat Rev Mol Cell Biol. 19(1), 45–58.

Shi, M., Hu, X., Wei, Y., Hou, X., Yuan, X., Liu, J., & Liu, Y. (2017). Genome-wide profiling of small RNAs and degradome revealed conserved regulations of miRNAs on auxin-responsive genes during fruit enlargement in peaches. Int J Mol Sci. 18(12), 2599.

Shin, H., Shin, H. S., Chen, R., & Harrison, M. J. (2006). Loss of At4 function impacts phosphate distribution between the roots and the shoots during phosphate starvation. Plant J: Cell Mol Biol. 45(5), 712–726.

Shriram, V., Kumar, V., Devarumath, R. M., Khare, T. S., & Wani, S. H. (2016). MicroRNAs as potential targets for abiotic stress tolerance in plants. Front Plant Sci. 7, 817.

Silva, S. (2012). Aluminium toxicity targets in plants. J Bot. 2012, 219462.

Souret, F. F., Kastenmayer, J. P., & Green, P. J. (2004). AtXRN4 degrades mRNA in *Arabidopsis* and its substrates include selected miRNA targets. Mol Cell. 15(2), 173–183.

Srivastava, S., Srivastava, A. K., Suprasanna, P., & D'Souza, S. F. (2009). Comparative biochemical and transcriptional profiling of two contrasting varieties of *Brassica juncea* L. in response to arsenic exposure reveals mechanisms of stress perception and tolerance. J Exp Bot. 60(12), 3419–3431.

Srivastava, S., Srivastava, A. K., Suprasanna, P., & D'Souza, S. F. (2013). Identification and profiling of arsenic stress-induced microRNAs in *Brassica juncea*. J Exp Bot. 64(1), 303–315.

Srivastava, S., Suprasanna, P., & D'Souza, S. F. (2012). Mechanisms of arsenic tolerance and detoxification in plants and their application in transgenic technology: A critical appraisal. Int J Phytorem. 14(5), 506–517.

Stoeva, N., Berova, M., & Zlatev, Z. (2003). Physiological response of maize to arsenic contamination. Biol Plant. 46(3), 449–452.

Sunkar, R., Kapoor, A., & Zhu, J. K. (2006). Posttranscriptional induction of two Cu/Zn superoxide dismutase genes in *Arabidopsis* is mediated by downregulation of miR398 and important for oxidative stress tolerance. Plant Cell. 18(8), 2051–2065.

Sunkar, R., Li, Y. F., & Jagadeeswaran, G. (2012). Functions of microRNAs in plant stress responses. Trends Plant Sci. 17(4), 196–203.

Sun, X., Zheng, H., Li, J., Liu, L., Zhang, X., & Sui, N. (2020). Comparative transcriptome analysis reveals new lncRNAs responding to salt stress in sweet sorghum. Front Bioeng Biotechnol. 8, 331.

Taft, R. J., Glazov, E. A., Lassmann, T., Hayashizaki, Y., Carninci, P., & Mattick, J. S. (2009). Small RNAs derived from snoRNAs. RNA (New York, N.Y.). 15(7), 1233–1240.

Tamás, M. J., Sharma, S. K., Ibstedt, S., Jacobson, T., & Christen, P. (2014). Heavy metals and metalloids as a cause for protein misfolding and aggregation. Biomolecules. 4, 252–267.

Tang, Z., Xu, M., Ito, H., Cai, J., Ma, X., Qin, J., Yu, D., & Meng, Y. (2019). Deciphering the non-coding RNA-level response to arsenic stress in rice (*Oryza sativa*). Plant Signaling Behav. 14(9), 1629268.

Thurman, D. A., & Rankin, J. L. (1982). The role of organic acids in zinc tolerance in *Deschampsia caespitosa*. New Phytologist. 91(4), 629–635.

Tian, Y., Tian, Y., Luo, X., Zhou, T., Huang, Z., Liu, Y., Qiu, Y., Hou, B., Sun, D., Deng, H., Qian, S., & Yao, K. (2014). Identification and characterization of microRNAs related to salt stress in broccoli, using high-throughput sequencing and bioinformatics analysis. BMC Plant Biol, 14, 226.

Tripathi, R., Patel, S., Kumari, V., Chakraborty, P., & Varadwaj, P. K. (2016). DeepLNC, a long non-coding RNA prediction tool using deep neural network. Netw Model Anal Health Inform Bioinform. 5:21.

Tyczewska, A., Bakowska-Zywicka, K., Gracz, J., & Twardowski, T. (2016). Stress responsive non-protein coding RNAs. Abiotic and Biotic Stress in Plants - Recent Advances and Future Perspectives. IntechOpen, pp. 153–181.

UlHaq, S., Khan, A., Ali, M., Khattak, A., Gai, M., Zhang, W. X., Wei, H. X., & Gong, M. (2019). Heat shock proteins: Dynamic biomolecules to counter plant biotic and abiotic stresses. Int J Mol Sci. 20(21), 5321.

Umezawa, T., Nakashima, K., Miyakawa, T., Kuromori, T., Tanokura, M., Shinozaki, K., & Yamaguchi-Shinozaki, K. (2010) Molecular basis ofthe core regulatory network in ABA responses: Sensing, signaling and transport. Plant Cell Physiol. 51, 1821–1839

Urquiaga, M. C. D. O., Thiebaut, F., Hemerly, A. S., & Ferreira, P. C. G. (2021). From trash to luxury: The potential role of plant LncRNA in DNA methylation during abiotic stress. Front Plant Sci. 11, 603246.

Valdés-López, O., Yang, S. S., Aparicio-Fabre, R., Graham, P. H., Reyes, J. L., Vance, C. P., et al. (2010). MicroRNA expression profile in common bean (*Phaseolus vulgaris*) under nutrient deficiency stresses and manganese toxicity. New Phytol. 187, 805–818.

Venkatesh, T., Suresh, P. S., & Tsutsumi, R. (2016). tRFs: miRNAs in disguise. Gene. 579(2), 133–138.

Viehweger, K. (2014). How plants cope with heavy metals. Bot Stud. 55, 35.

Waititu, J. K., Zhang, C., Liu, J., & Wang, H. (2020). Plant non-coding RNAs: Origin, biogenesis, mode of action and their roles in abiotic stress. Int J Mol Sci. 21(21), 8401.

Wang, K. C., & Chang, H. Y. (2011). Molecular mechanisms of long noncoding RNAs. Mol Cell. 43(6), 904–914.

Wang, T. Z., Liu, M., Zhao, M. G., Chen, R., & Zhang, W. H. (2015). Identification and characterization of long non-coding RNAs involved in osmotic and salt stress in *Medicago truncatula* using genome-wide high-throughput sequencing. BMC Plant Biol. 15, 131.

Wang, J., Meng, X., Dobrovolskaya, O. B., Orlov, Y. L., & Chen, M. (2017). Non-coding RNAs and their roles in stress response in plants. Genomics Proteomics Bioinf. 15(5), 301–312.

Waterhouse, P. M., Graham, M. W., & Wang, M. B. (1998). Virus resistance and gene silencing in plants can be induced by simultaneous expression of sense and antisense RNA. Proc Natl Acad Sci USA. 95(23), 13959–13964.

Wu, L., Liu, S., Qi, H., Cai, H., & Xu, M. (2020). Research progress on plant long non-coding RNA. Plants (Basel, Switzerland). 9(4), 408.

Xie, Z., Johansen, L. K., Gustafson, A. M., Kasschau, K. D., Lellis, A. D., Zilberman, D., Jacobsen, S. E., & Carrington, J. C. (2004). Genetic and functional diversification of small RNA pathways in plants. PLoS Biol. 2(5), E104.

Xu, L., Wang, Y., Zhai, L., Xu, Y., Wang, L., Zhu, X., Gong, Y., Yu, R., Limera, C., & Liu, L. (2013). Genome-wide identification and characterization of cadmium-responsive microRNAs and their target genes in radish (*Raphanus sativus* L.) Roots. J Exp Bot. 64(14), 4271–4287.

Yang, L., Wu, G., & Poethig, R. S. (2012). Mutations in the GW-repeat protein SUO reveal a developmental function for microRNA-mediated translational repression in Arabidopsis. Proc Natl Acad Sci USA. 109(1), 315–320.

Yi, X., Zhang, Z., Ling, Y., Xu, W., & Su, Z. (2015). PNRD: A plant non-coding RNA database. Nucl Acids Res. 43(Database issue), D982–D989.

Yoshiba, Y., Kiyosue, T., Nakashima, K., Yamaguchi-Shinozaki, K., & Shinozaki, K. (1997). Regulation of levels of proline as an osmolyte in plants under water stress. Plant Cell Physiol. 38(10), 1095–1102.

Yu, L. J., Luo, Y. F., Liao, B., Xie, L. J., Chen, L., Xiao, S., Li, J. T., Hu, S. N., & Shu, W. S. (2012). Comparative transcriptome analysis of transporters, phytohormone and lipid metabolism pathways in response to arsenic stress in rice (*Oryza sativa*). New Phytologist. 195(1), 97–112.

Zeng, Q. Y., Yang, C. Y., Ma, Q. B., Li, X. P., Dong, W. W., & Nian, H. (2012). Identification of wild soybean miRNAs and their target genes responsive to aluminum stress. BMC Plant Biol. 12:182.

Zhang, X., Dong, J., Deng, F., Wang, W., Cheng, Y., Song, L., Hu, M., Shen, J., Xu, Q., & Shen, F. (2019b). The long non-coding RNA lncRNA973 is involved in cotton response to salt stress. BMC Plant Biol. 19:459.

Zhang, L. W., Song, J. B., Shu, X. X., Zhang, Y., & Yang, Z. M. (2013b). miR395 is involved in detoxification of cadmium in *Brassica napus*. J Hazard Mater. 250–251, 204–211.

Zhang, X. D., Sun, J. Y., You, Y. Y., Song, J. B., & Yang, Z. M. (2018). Identification of Cd-responsive RNA helicase genes and expression of a putative BnRH24 mediated by miR158 in canola (*Brassica napus*). Ecotoxicol Environ Saf. 157, 159–168.

Zhang, P., Wu, W., Chen, Q., & Chen, M. (2019a). Non-coding RNAs and their integrated networks. J Integr Bioinfor. 16(3), 20190027.

Zhang, S., Yue, Y., Sheng, L., Wu, Y., Fan, G., Li, A., Hu, X., ShangGuan, M., & Wei, C. (2013a). PASmiR: A literature-curated database for miRNA molecular regulation in plant response to abiotic stress. BMC Plant Biol. 13(1), 33.

Zhang, D., & Zhang, B. (2020). SpRY: Engineered CRISPR/Cas9 harnesses newgenome-editing power. Trends Genet. 36, 546–548.

Zhao, Y., & Chengcai, C. (2011). Towards understanding plant response to heavy metal stress. In: Shanker A (Ed.). Abiotic Stress in Plants - Mechanisms and Adaptations. Tech Europe, Rijeka, pp. 59–78.

Zhao, C., Zhang, H., Song, C., Zhu, J. K., & Shabala, S. (2020). Mechanisms of plant responses and adaptation to soil salinity. The Innovation. 1(1), 100017.

Zhou, J., Deng, K., Cheng, Y., Zhong, Z., Tian, L., Tang, X., et al. (2017). CRISPR cas9 based genome editing reveals new insights into microRNA function and regulation in rice. Front Plant Sci. 8, 1598.

Zhou, L., Liu, Y., Liu, Z., Kong, D., Duan, M., & Luo, L. (2010). Genome-wide identification and analysis of drought-responsive microRNAs in *Oryza sativa*. J Exp Bot. 61(15), 4157–4168.

Zhou, Z. S., Wang, S. J., & Yang, Z. M. (2008). Biological detection and analysis of mercury toxicity to alfalfa (*Medicago sativa*) plants. Chemosphere. 70, 1500–1509.

Zhou, Z. S., Zeng, H. Q., Liu, Z. P., & Yang, Z. M. (2012). Genome-wide identification of *Medicago truncatula* microRNAs and their targets reveals their differential regulation by heavy metal. Plant Cell Environ. 35, 86–99.

Zhu, J. K. (2002) Salt and drought stress signal transduction in plants. Annu Rev Plant Biol 53:247–273

Zhu, H., Hu, F., Wang, R., Zhou, X., Sze, S. H., Liou, L. W., Barefoot, A., Dickman, M., & Zhang, X. (2011). *Arabidopsis* Argonaute10 specifically sequesters miR166/165 to regulate shoot apical meristem development. Cell. 145(2), 242–256.

Zhu, J. K., Liu, J., & Xiong, L. (1998). Genetic analysis of salt tolerance in Arabidopsis: Evidence for a critical role of potassium nutrition. The Plant Cell. 10(7), 1181–1191.

Zieve, G., & Penman, S. (1976). Small RNA species of the HeLa cell: Metabolism and subcellular localization. Cell. 8(1), 19–31.

Zilberman, D., Cao, X., & Jacobsen, S. E. (2003). ARGONAUTE4 control of locus-specific siRNA accumulation and DNA and histone methylation. Science (New York, N.Y.). 299(5607), 716–719.

Zilberman, D., Cao, X., Johansen, L. K., Xie, Z., Carrington, J. C., & Jacobsen, S. E. (2004). Role of *Arabidopsis* ARGONAUTE4 in RNA-directed DNA methylation triggered by inverted repeats. Curr Biol. 14(13), 1214–1220.

8 Ionizing Radiation-Induced Non-Coding RNA in Plants and Their Implication in Mutation Breeding

Suvendu Mondal[1,2*], *Subham Bhakta*[1,2], *Parmeshwar K. Sahu*[3], *and Richa Sao*[3]
[1]Nuclear Agriculture and Biotechnology Division, Bhabha Atomic Research Centre, Mumbai, India
[2]Homi Bhabha National Institute, Mumbai, India
[3]Department of Genetics and Plant Breeding, Indira Gandhi Krishi Vishwavidyalaya, Raipur, India

8.1 INTRODUCTION

The study on non-coding RNA (ncRNAs) has gained momentum in recent years since its discovery as an essential biomolecule that plays a considerable role in contributing to genetic and phenotypic diversities in living systems. In the eukaryotic living system, only 2% of the total transcribed eukaryotic genome generates protein products. The remaining transcriptome comprises ncRNAs, which were previously thought of as "noise" due to the absence of coded protein. However, high-throughput sequencing analysis and experimental validation have confirmed their involvement in the epigenetic, transcriptional, and post-transcriptional gene-regulatory levels. Further, the advancement of effective next-generation sequencing technologies in conjunction with homology-based and/or experimental approaches has been vital in discovering the role of ncRNAs in plants.

Ionizing radiation (IR) has been exploited for the creation of genetic variability and new types of crop cultivars, post-harvest storage of agricultural and horticultural commodities, and quarantine services. IR is a potent genotoxic agent and inducer of oxidative stress, which generates a wide range of genetic and epigenetic changes in the plant genome both directly, by energetic disruption of DNA integrity, and indirectly, through the formation of reactive oxygen species (ROS) via water radiolysis. Radiation injury either due to natural background radiation, or an accident, will affect multiple cellular pathways or produce heterogeneous effects across the plant population. In the instance of radiation therapy or high-background radiation exposure, the goal of biodosimetry is to aid in the prognosis and early treatment of normal tissue damage. Plant biodosimetry will allow for the rapid detection of background radiation in the fallout area and mark them for restricted human movement. Biodosimetry was born from the recognition that the physical dose of radiation may differ from the extent of biological perturbations, and examines chromosomal, metabolomic, proteomic, molecular, and other physiological changes. Additionally, biodosimetry can be more quantifiable and verifiable than the evaluation of clinical symptoms, which can have ranges in severity and onset. Early determination of the extent of radiation injury through ncRNA biomarker detection could provide a means to treat this damage early and potentially prevent late radiation-induced disease in an area of radiation fallout (Xiao et al., 2020). The research on IR-dependent DNA

DOI: 10.1201/9781003369288-8

damage response (DDR) is very limited in plants. More work on this field can open new avenues towards molecular understanding of IR dependent miRNAs in plants. A schematic figure represents the DDR response involving an unknown factor responsible for ncRNA (especially miRNA) induction in plants is depicted in Figure 8.1. This chapter will briefly describe physiology of radiation damage, different types of ncRNA in plants responsive to IR, and their implication in genetic improvement of plants.

8.2 PHYSIOLOGY OF RADIATION DAMAGE

Radiation induces direct and indirect forms of DNA damage. Of the different types of DNA damage, radiation-induced double-strand breaks are considered the most lethal. Direct DNA damage occurs due to the direct deposition of radiation energy onto DNA molecules, whereas indirect damage arises from deposition of energy in other molecules (mainly cellular water) and the generation of free radicals, which then react with DNA. The four principal steps of radiation damage to intracellular molecules are as follows: (1) physical deposition of energy, (2) generation of primary radicals in the target molecules, (3) reaction of primary radicals on adjacent molecules, and (4) production of chemically stable damage by unstable radicals (Reisz et al., 2014). Additionally, ROS produced in the mitochondria during irradiation is another driver of DNA damage. Historically, hydroxyl radicals have been considered the key species in DNA damage, however, recent research has determined the key species is a carbonate radical anion formed in the Fenton reaction (Lles et al., 2019). This carbonate radical anion is more abundant in physiological conditions than the hydroxyl radical and generates 8-oxo-7,8-dihydroguanine at GGG sites in DNA, which signals for DNA repair to begin (Lles et al., 2019). Effects of radiation are a function of dose with a threshold

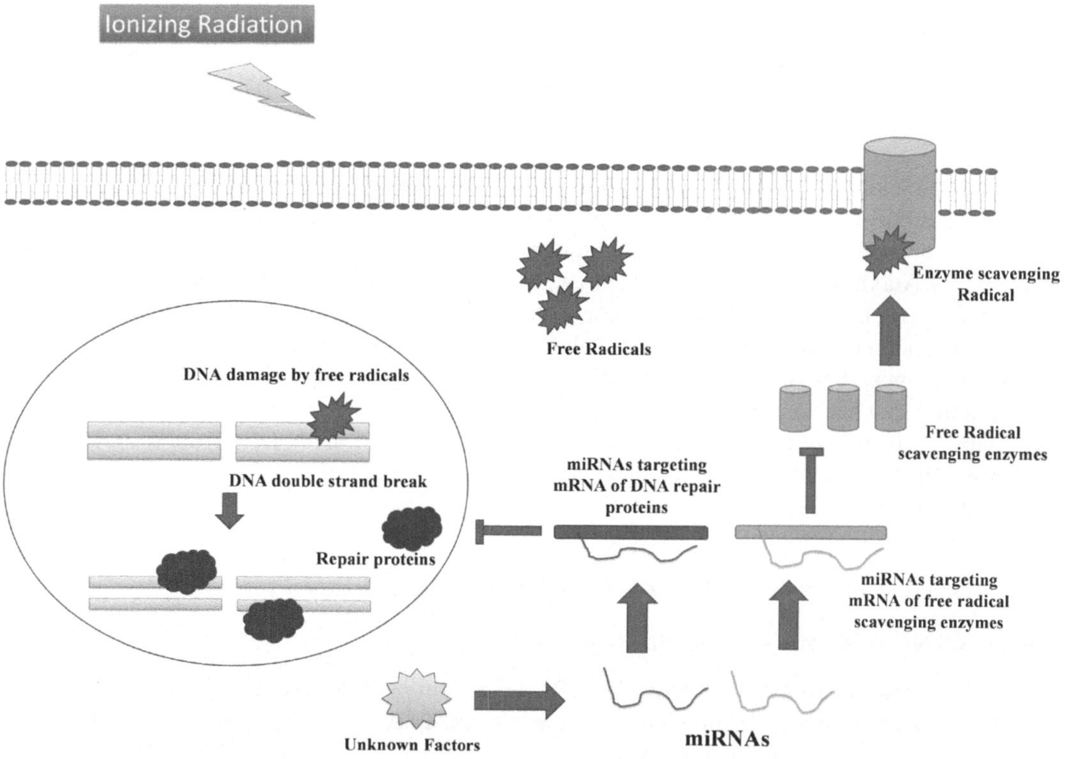

FIGURE 8.1 Schematic representation of the induction of miRNA in response to ionizing radiation.

above which injury is imminent and below which most people, cells, tissues, or animals experience little to no acute effects (Mathes and Alexander, 1996). Radiation response is also determined by the type of radiation source, volume of tissue irradiated, and post-recovery time. A cell's specific radiosensitivity is, in part, determined by DNA replication, recombination, and repair (Willers et al., 2004). Double-strand breaks are repaired via homologous recombination (HR) or non-homologous end joining (NHEJ) (Willers et al., 2004). DDR to double-strand breaks includes protein recruitment, modifications such as histone phosphorylation, and activation of signaling pathways such as p53-p21 (Li et al., 2018). IR can also induce senescence in normal cells, resulting in persistent cell-cycle arrest or mutation leading to cancer or cell death.

8.3 RADIATION RESPONSE IN PLANTS

Plants are comparably more radioresistant than other higher living organisms, including humans. Most of the plants tolerate >10 times higher doses than the lethal dose for human (5 Gy). There are some plant species which can tolerate very high doses up to 1,000 Gy (Sparrow and Miksche, 1961; Caplin and Willey, 2018; Gudkov et al., 2019). The evolution process acclimatized plant species to respond effectively to small changes in radiation levels through physiological adjustments (Vanhoudt et al., 2014). IR creates lipid peroxidation, oxidative modification of proteins, and nucleic acids during interaction with these cellular macromolecules inside cells (Reisz et al., 2014). Post-radiation death of plants results from the damage to DNA molecules inflicted by IR. A significant part (about 70–80%) of the IR-related DNA damage is caused by ROS formed during radiolysis of water and only 20–30% of the damage is due to the direct absorption of high-energy IR quanta by the target DNA molecules (Gudkov et al., 2019). Genotoxic stress like IR damages plant cells by producing different free radicals such as H_2O_2, $[O_2^-]$, and $OH^\bullet$. The H_2O_2, $[O_2^-]$ radicals are very unstable and can produce $OH^\bullet$ with the help of different enzymes. The $OH^\bullet$ radical is the main source of cellular toxicity — it can act upon all biological macromolecules including DNA and causes heavy damage in cells. Super oxide dismutase (SOD) is the key enzyme which maintains the cellular redox balance by neutralizing the free radicals. There are mainly three types of SODs present in plant copper/zinc (Cu/Zn) SOD, manganese (Mn) SOD, and iron (Fe) SOD. As a result of direct energy absorption from IR by the DNA molecules, single- and double-strand DNA breaks are formed. Although most of these breaks are repairable, the very load of such damage during IR exposure may lead to error-prone repair and can lead to chromosomal aberrations or mutations. In addition, double-strand DNA breaks trigger the signaling cascades mediated by Ataxia telangiectasia mutated (ATM) kinases (Culligan et al., 2006; Esnault et al., 2010). Such signaling event may affect morphology, cellular metabolism, respiration, photosynthesis, redox homeostasis and antioxidant balance, plant hormone modulation etc. (Gudkov et al., 2019). Sometimes, these effects of IR on plants show a transgenerational effect. Kovalchuk et al. (2004) stated that the long-term changes in the state of the genome are maintained by hypermethylation. The role of dynamic histone modifications (mainly, methylation and acetylation) in response to IR was found evident in maintaining the upregulation of DNA repair-related genes in plants (Mondal et al., 2016).

8.4 ROLE OF NON-CODING RNA IN GENOMIC STRESS

8.4.1 siRNA in Response to Ionizing Radiation

SiRNAs are heterochromatic siRNAs (hc-siRNAs) in plants mainly derived from transposons and other repetitive sequences in genomes. Single-stranded RNA transcripts are generated from DNA repeats by DNA-dependent RNA polymerase IV (Pol IV) (Herr et al., 2005; Onodera et al., 2005) and converted by RNA-dependent RNA polymerase 2 (RDR2) into double strand RNAs (dsRNAs) (Xie et al., 2004). These dsRNAs were further processed by DCL3 into 24 nucleotide (nt) hc-siRNAs (Xie et al., 2004). The newly generated hc-siRNAs are associated with the AGO4 subfamily members and direct DNA methylation by the *de novo* DNA methyltransferase DRM2 through a pathway known as RNA-directed DNA methylation (RdDM) (Matzke et al., 2009; Law and Jacobsen, 2010). The role of siRNAs was established in diverse biological roles including the different biotic and abiotic stresses of plants (Kharaiwesh et al., 2012). In addition to the key role, the different types of naturally occurring siRNAs in various molecular processes in plants, such as stress defense, epigenetic control, and transposon suppression, it was recently shown that they might be involved, directly or indirectly, in DDR and DNA recombination and repair. It was shown that *Arabidopsis* mutants, deficient in the Dicer-like (DCL) enzymes or RNA-dependent RNA polymerases, display differential response to methyl methane sulphonate (MMS) varying from higher sensitivity to higher tolerance compared to the wild type. In addition, *dcl2* and *dcl3* mutants have decreased the ability to repair UV-C-generated DNA lesions, suggesting a link between siRNA biogenesis and DNA repair (Yao et al., 2010). On the other hand, experiments in *Arabidopsis* showed that *ago* mutant plants were not impaired in the meiotic recombination, however, the sensitivity of *ago2* as well as of *ago9* mutants to gamma radiation and mitomycin C was confirmed (Oliver et al., 2014).

Wei et al. (2012) found significant roles of siRNA processing protein (Dicer-like protein: DCL2, DCL3, and DCL4) machinery in DNA double-strand break repair in *Arabidopsis.* The highest amount of reduction in double strand break (DSB) repair was found in mutant background of DCL3 (*dcl3*). This requirement of DCL proteins in DSB repair triggers the idea of involvement of siRNA. Further experiments proved that these DSB-induced new siRNAs were produced from adjacent genomic DNA areas of the DSB site. The siRNAs induced from such DSB events were termed as double-strand break-induced RNA (diRNA). In particular, the production of diRNAs was reduced by 98% in the *dcl3* mutant, consistent with its predominant effect on DSB repair efficiency. Further, mutations in RNA-dependent RNA polymerase (RDR2 and RDR6) and RNA polymerase IV and V significantly reduced diRNA production in DSB sites (Wei et al., 2012). Moreover, the production of diRNAs was found dependent on ATR (an ATM-associated protein that encodes a phosphatidyl inositol 3-kinase-like protein kinase [PI3 kinase]).The mechanism of involvement of diRNAs in DSB repair was later elucidated in animal cell culture models. Towards this, identification of an effector protein AGO2 that typically recruits diRNAs in the DSB site that, in collaboration with RAD51, repair the DSB (Gao et al., 2014). The diRNA binding capability and catalytic activity of AGO2, in turn, help the accumulation of Rad51 at the DSB sites and facilitate HR repair (Gao et al., 2014). In literature, upregulation of AGO2 is reported in response to gamma irradiation (Culligan et al., 2006; Mondal et al., 2016). Further, siRNA has a role in recognizing DNA-photoproducts induced by UV rays through the complexation with Damage Binding Protein 2 (DDB2) and Argonaute 1 (AGO1) (Schalk et al., 2017).

8.4.2 miRNA in Response to Ionizing Radiation

MiRNA in plants exhibit translation inhibition mechanisms without a notable effect on mRNA stability (Brodersen et al., 2008). MiRNAs are involved in DDR regulation by targeting most of the proteins involved at each step of the DDR. In general, a bulk of IR-responsive miRNAs also resorts to other DNA-damaging agents and H_2O_2. Further, these responsive miRNAs are associated with the DDR (Simone et al., 2009). Several miRNAs and their cognate targets, such as miR-24/miR-138 (H2AX), miR-182 (BRCA1), miR-421/miR-101 (ATM), and miR-504/miR-125b (p53), are known to be key regulators in the DDR pathway (Joly-Tonetti and Lamartine, 2012; Kim et al., 2016). All these miRNAs regulate HR and NHEJ by targeting Rad51 and Ku 70/80, respectively. Very limited studies have been published on the identification of radiation-responsive miRNAs in plants (Table 8.1). In response to UV-B radiation, miR156, miR159, miR160, miR164, miR167, miR169, miR395, and miR171 were identified as radiation-responsive miRNA genes in *Arabidopsis thaliana*, *Populus tremula*, and *Triticum aestivum* (Zhou et al., 2007; Jia et al.,

TABLE 8.1
List of Responsive miRNA in Plants against Ionizing and Non-Ionizing Radiation

Type of Radiation	Plant Species	Responsive miRNAs	References (available in Kim et al., 2016)
UV-B	*Arabidopsis* sp.	miR156, miR157, miR159, miR160, miR165, miR166, miR167, miR169, miR170, miR171, miR172, miR393, miR398, miR401	Zhou et al., 2007
UV-B	*Populus tremula*	miR156, miR160, miR164, miR165, miR166, miR167, miR168, miR398, miR408, miR159, miR169, miR390, miR393, miR395, miR399, miR472	Jia et al., 2009
UV-B	*Triticum aestivum*	miR156, miR159, miR164, miR167, miR171, miR395	Wang et al., 2013
UV-B	*Zea mays*	miR156, miR164, miR165, miR166, miR171, miR172, miR395, miR396, miR398, miR399, miR444, miR529, miR533a*, miR896, miR903, miR1427, miR1858a	Casati, 2013
H_2O_2	*Oryza sativa*	miR169, miR397, miR528, miR827, miR1425, miR319a.2, miR408-5p	Li et al., 2011
Gamma rays	*O. sativa*	*Osa*-miR414, *Osa*-miR164e, and *Osa*-miR408	Macovei and Tuteja, 2013
Gamma rays (with 200 Gy and 24 h recovery)	*Arabidopsis* sp.	miR164a, miR169d, miR172b*, miR403, miR827, miR840, miR850	Kim et al., 2016
Carbon ion	*O. sativa*	miR156, miR164, miR2097-5p	Zhang et al., 2011
Low energy N^+ ion (with 2×10^{17} N^+/cm^2)	*O. sativa*	Upregulated: mir1320-3p, miR1320-5p, miR156b-3p, miR156c-5p, miR156c-3p/g-3p, miR1561-5p, miR398b, miR6250 Downregulated: mir156a/e/i, miR156k, miR160f-5p, miR166j-5p, miR1846e, miR399d	Li et al., 2018

2009; Wang et al., 2013). These UV-B-responsive miRNA genes had various light-responsive and stress-related *cis*-elements in their upstream regions. IR-like carbon ion beam-induced miR164a, miR164c, miR164d, and miR156a-j were reported in rice seedlings (Zhang et al., 2011). Gamma rays induced miRNAs, Osa-miR408, Osa-miR414, and Osa-miR164e target specific helicases and influence the DNA repair mechanism in rice (Macovei and Tuteja, 2013). A novel miR69 of *Brachypodium distachyon* found targeting Rad 50 mRNA in response to H_2O_2 stress that is often associated with IR (Lv et al., 2016). Kim et al. (2016) noticed the high transcriptional induction of miR840 and miR850 in wild type and four other mutants (*cmt3-11t*, *drd1-6*, *ddm1-2*, *drm2*) of *Arabidopsis* regardless of any developmental stage in response to 200 Gy gamma rays at 24 hours post-recovery. Although the actual target genes and functions of miR840 and miR850 were not studied in detail, the consensus-induced expression in different mutant backgrounds of these two miRNAs was proposed as reproducible genetic markers in *Arabidopsis* plants exposed to gamma rays (Kim et al., 2016). In another way, direct involvement of miRNA in DDR was proven through the impairment of fertility and compromise in HR and meiosis in the mutant background of miRNA machinery genes such as *hyl1*, *hen1*, *dcl1*, *hst*, and *ago1* in *Arabidopsis*. Additionally, the *hen1-13* mutant was found to be hypersensitive to gamma irradiation, which mainly causes susceptibility to double-strand breaks to be repaired by HR (Oliver et al., 2017). Other than this experimental evidence, there are many theoretical studies which predicted that two DNA repair proteins, MRE11 and MUTL-homolog 1 (MLH1), are targeted by miR5261 and miR5761 in *Citrus sinensis* (Liang et al., 2017). These few reports show that miRNAs are an inevitable part of the IR-dependent double-stranded break repair system. The research on IR-dependent DDR response is very limited in plants. More work in this field can open new avenues towards molecular understanding of IR-dependent miRNAs in plants.

8.4.3 Long Non-Coding RNA (lncRNA) in Response to Ionizing Radiation

Long non-coding RNAs (lncRNAs) are a large and diverse group of polyadenylated or non-polyadenylated transcripts longer than 200 nt with no or minimal protein-coding potential, which originate from intergenic or intronic regions or from the antisense transcription of genes (Ma et al., 2013; Kung et al., 2013) and have a role in various developmental responses in plants (Datta and Paul, 2019). Based on their origin, lncRNAs can be further classified into four categories: long intronic non-coding RNAs, natural antisense transcript (NAT) region, overlapping lncRNAs (that partially overlap the coding gene), and long intergenic non-coding RNAs (lincRNAs; that lie between two protein coding genes) (Rinn and Chang, 2012; Datta and Paul, 2019). Many transposable elements (TEs) are intimately associated with the biogenesis of lncRNAs (Cho, 2018). *In silico* analyses of the total transcriptome data of human and *Arabidopsis* revealed more than 15,000 lncRNAs in cellular systems (Kapusta and Feschotte, 2014; Liu et al., 2015). TE sequences contribute 20–42% of all lncRNAs in vertebrates. Many lncRNAs overlap with TEs or other repeats in *Arabidopsis* (Kelley and Rinn, 2012; Liu et al., 2012; Kapusta et al., 2013; Kapusta and Feschotte, 2014; Liu et al., 2015). Wan et al. (2013a&b) reported significant transcriptional regulation of hundreds of lncRNAs in human and mouse cells in response to DNA DSBs. The model plant species *Arabidopsis* showed large number of lncRNA upregulation in response to various stresses including drought, high salt, chilling, and abscicic acid (ABA) treatment (Liu et al., 2012). Comparison of 80 Gy X-ray (150 kV, 12 mA, 88 minute)-induced transcripts in *Arabidopsis* with a plant lncRNA database, PLncDB, (Jin et al., 2013) revealed 86 transcriptionally upregulated (≥2-fold) lncRNAs in plants in response to IR (Wang et al., 2016). This induction of lncRNAs was found to be ATM dependent as only seven such lncRNAs were upregulated in *atm* mutant plants. Although the overall fraction of lncRNAs that are transcriptionally regulated early after DSB induction is low (0.58%), the fraction of ATM-dependently regulated lncRNAs is high (93%) among the transposons. The study revealed that ATM downstream factors, BRCA1, DRM1, JMJ30, AGO2, and the ATM-independent AGO4, participate in the regulation of individual TEs and lncRNAs (Wang et al., 2016).

8.5 IMPLICATION OF ncRNA IN MUTATION BREEDING

Due to the extensive complementarity of most of the plant miRNAs to the plant mRNAs, miRNA guides cleavage of the target mRNA or facilitates translational inhibition. These targets usually have perfect Watson-Crick complementarity at the six base pairs surrounding the cleavage site, supporting the idea that the pairing in this central region is important for cleavage (Mallory et al., 2004). Tang et al. (2003) proved that the dominant mutant allele *phv-1d* of the PHV (PHAVOLUTA gene) harbored a mutation in the complementary binding site of miR165/166. This mutation does not allow miR165/166 to bind properly and, thus, cleaves it less efficiently than the wild type sequence, which results in a semi-dominant phenotype. In the same line of observation, Mallory et al. (2004) revealed that the silent (no changes in amino acid) substitution in the miR165/166 complementary site of the PHB (PHABULOSA) gene can mimic the dominant mutation of *phv 1d*. Further study proved that the substitution in the 3′ region of the PHB-miRNA complementary site is more disruptive to RNA cleavage (Mallory et al., 2004). Ochando et al. (2006) also proved the same type of mutation resulted into semi-dominant, gain of function *icu4-1* and *icu4-2* dominant alleles of the INCURVATA 4 (ICU 4) gene. Such a mechanism may play in dominant/semi-dominant mutations derived from induced mutagenesis through IR due to the mutation in the complementary binding site of miRNA in the target gene. Conversely, mutation may appear in the dedicated miRNA gene and result in the gain of function mutation with some pleiotropic effects due to the other targets of the mutated miRNA. Such a mutation was created through nDart/aDart system in the 38 bp upstream transcription initiation site of the *miR156d* gene of rice (cv. Nipponbare), which resulted in a semi-dominant busy dwarf tiller 1 (*Bdt* 1) mutant (Hayashi-Tsugane et al., 2015). Conversely, Ji et al. (2019) isolated a semi-dominant shortened uppermost internode 4 (Sui-4) mutant from the Ac/Ds insertional mutant population of cv. Dongjin (Japonica rice) where a T-to-A nucleotide substitution in the supernumerary bract (*SNB*) gene resulted in the compromised interaction of miR172 and SUI4/SNB mRNA and, thus, expression of SUI4/SNB became enhanced. This enhancement of the SNB transcript resulted in a decrement of activity of cell expansion and cell-wall synthesis and consequently resulted in a shortened internode (Ji et al., 2019). The advancement of site-specific genome editing tools such as TALEN and CRISPR/Cas9 has helped to validate these findings of gene-expression regulation through mutation in the complementary binding sites of miRNA in the target gene in rice (Lin et al., 2021). In contrast, a site-directed mutation approach also demonstrated successful creation of mutations in *OsmiR528* toward the involvement of this miRNA in salt tolerance (Zhou et al., 2017) and the role of *OsmiR390* in shoot apical meristem development (Bi et al., 2020). CRISPR/Cas9-mediated mutagenesis also revealed that the deletion of more than two bases in miRNA genes OsmiR171f and OsmiR818b significantly reduced the level of corresponding mature miRNAs in rice and compromised the plant response to drought (Chung et al., 2020). Since the mechanism of creating mutations by IR through DSBs almost the same as in the case of CRISPR/Cas9, where the error-prone NHEJ repair takes a lead to incorporate mismatches or create small indels, similar kinds of dominant mutations may appear in induced mutagenesis. Such dominant/semi-dominant mutations were evident in gamma-ray-induced mutagenesis for seed-coat color (Mondal et al., 2007) and seedling height (Badigannavar and Mondal, 2010) in the groundnut. Deep sequencing along with the MutMap technique have the ability to pinpoint such mutations in crop plants.

REFERENCES

Badigannavar, A.M. and Mondal, S. 2010. Induction of mutations for plant height and inheritance of dwarf mutant in groundnut (*Arachis hypogaea* L.) Through gamma ray irradiation. *Electronic Journal of Plant Breeding*, 1:156–161.

Bi, H., Fei, Q., Li, R., Liu, B., Xia, R., Char, S.N., Meyer, B.C. and Yang, B. 2020. Disruption of miRNA sequences by TALENs and CRISPR/Cas9 induces varied lengths of miRNA production. *Plant Biotechnology Journal*, 18:1526–1536.

Brodersen, P., Sakvarelidze-Achard, L., Bruun-Rasmussen, M., Dunoyer, P., Yamamoto, Y.Y., Sieburth, L. and Voinnet, O. 2008. Widespread translational inhibition by plant miRNAs and siRNAs. *Science*, 320(5880):1185. doi: 10.1126/science.1159151.

Caplin, N. and Willey, N. 2018. Ionizing radiation, higher plants, and radioprotection: From acute high doses to chronic low doses. *Frontiers in Plant Science* 9:847. https://doi.org/10.3389/fpls.2018.00847.

Casati, P. 2013. Analysis of UV-B regulated miRNAs and their targets in maize leaves. *Plant Signal Behav.* 8(10): e26758. doi: 10.4161/psb.26758.

Cho, J. 2018. Transposon derived non-coding RNAs and their function in plants. *Frontiers in Plant Science*, 9:600. doi. 10.3389/fpls.2018.00600.

Chung, P.J., Chung, H., Oh, N., Choi, J., Bang, S.W., Jung, S.E., Jung, H., Shim, J.S. and Kim, J. 2020. Efficiency of recombinant CRISPR/rCas9-mediated miRNA Gene editing in rice. *International Journal of Molecular Sciences*, 21(24):9606. doi: 10.3390/ijms21249606.

Culligan, K.M., Robertson, C.E., Foreman, J., Doerner, P. and Britt, A.B. 2006. ATR and ATM play both distinct and additive roles in response to ionizing radiation. *The Plant Journal*, 48(6):947. doi: 10.1111/j.1365-313X.2006.02931.x.

Datta, R. and Paul, S. 2019. Long non-coding RNAs: Fine-tuning the developmental responses in plant. *Journal of Biosciences*, 44(4):77.

Esnault, M.A., Legue, F. and Chenal, C. 2010. Ionizing radiation: Advances in plant response. *Environmental and Experimental Botany*, 68(3):23–237.

Gao, M., Wei, W. and Li, M.M., et al. 2014. Ago2 facilitates Rad51 recruitment and DNA double-strand break repair by homologous recombination. *Cell Research*, 24: 532–541. https://doi.org/10.1038/cr.2014.36.

Gudkov, S.V., Grinberga, M.A., Sukhova, V. and Vodeneeva, V. 2019. Effect of ionizing radiation on physiological and molecular processes in plants. *Journal of Environmental Radioactivity*, 202:8–24.

Hayashi-Tsugane, M., Maekawa, M. and Tsugane, K. 2015. A gain-of-function bushy dwarf tiller 1 mutation in rice microRNA Gene miR156d caused by insertion of the DNA transposon nDart1. *Scientific Reports*,, 5:14357. doi: 10.1038/srep14357.

Herr, A.J., Jensen, M.B., Dalmay, T. and Baulcombe, D.C. 2005. RNA polymerase IV directs silencing of endogenous DNA. *Science*, 308:118–120.

Jia, X., Wang, W. X., Ren, L., Chen, Q. J., Mendu, V., Willcut, B., Dinkins, R., Tang, X. and Tang, G. 2009. Differential and dynamic regulation of miR398 in response to ABA and salt stress in *Populus tremula* and *Arabidopsis thaliana*. *Plant Molecular Biology*, 71:51–59.

Ji, H., Han, C., Lee, G.S., Jung, K.H., Kang, D.Y., Oh, J., Oh, H., Cheon, K.S., Kim, S.L., Choi, I., Baek, J. and Kim, K.H. 2019. Mutations in the microRNA172 binding site of SUPERNUMERARY BRACT (SNB) suppress internode elongation in rice. *Rice*, 12:62 https://doi.org/10.1186/s12284-019-0324-8.

Jin, J., Liu, J., Wang, H., Wong, L. and Chua, N.H. 2013. PLncDB: Plant long non-coding RNA database. *Bioinformatics*, 29:1068–1071

Joly-Tonetti, N. and Lamartine, J.. 2012. The Role of MicroRNAs in the Cellular Response to Ionizing Radiations. In Current Topics in Ionizing Radiation Research, edited by Mitsuru Nenoi, IntechOpen, 10.5772/34868.

Kapusta, A. and Feschotte, C. 2014. Volatile evolution of long non-coding RNA repertoires: Mechanisms and biological implications. *Trends in Genetic*, 30:439–452.

Kapusta, A., Kronenberg, Z., Lynch, V.J., Zhuo, X., Ramsay, L., Bourque, G., Yandell, M. and Feschotte, C. 2013. Transposable elements are major contributors to the origin, diversification, and regulation of vertebrate long non-coding RNAs. *PLoS Genetics*, 9:e1003470.

Kelley, D. and Rinn, J. 2012. Transposable elements reveal a stem cell-specific class of long non-coding RNAs. *Genome*, 13:R107.

Kharaiwesh, B., Zhu, J.K. and Zhu, J. 2012. Role of miRNAs and siRNAs in biotic and abiotic stress responses of plants. *Biochemicaet Biophysica Acta*, 1819:137–148

Kim, Y.K., Kim, B. and Kim, V.N. 2016. Re-evaluation of the roles of DROSHA, export in 5, and DICER in microRNA biogenesis. *PNAS* USA, 113(13):E1881. doi: 10.1073/pnas.1602532113.

Kovalchuk, I., Abramov, V., Pogribny, I. and Kovalchuk, O. 2004. Molecular aspects of plant adaptation to life in the Chernobyl zone. *Plant Physiology*, 135:357–363. https://doi.org/10.1104/pp.104.040477.

Kung, J.T., Colognori, D. and Lee, J. 2013. Long non-coding RNAs: Past, present, and future. *Genetics*, 193(3):651. doi: 10.1534/genetics.112.146704.

Law, J.A. and Jacobsen, S.E. 2010. Establishing, maintaining and modifying DNA methylation patterns in plants and animals. *Nature Review Genetics*, 11:204–220.

Liang, W.W., Huang, J.H., Li, C.P., Yang, L.T., Ye, X., Lin, D. and Chen, L.S. 2017. MicroRNA-mediated responses to long-term magnesium-deficiency in *Citrus sinensis* roots revealed by illumina sequencing. *BMC Genomics*, 18(1):657. doi: 10.1186/s12864-017-3999-5.

Li, T., Li, H., Zhang, Y.X., Liu, J.Y. 2011. Identification and analysis of seven H_2O_2-responsive miRNAs and 32 new miRNAs in the seedlings of rice (*Oryza sativa* L. ssp. *indica*). *Nucleic Acids Res.* 39(7):2821–2833. doi: 10.1093/nar/gkq1047.

Lin, Y., Zhu, Y., Cui, Y., Chen, R., Chen, Z., Li, G., Fan, M., Chen, J., Li, Y., Guo, X., Zheng, X., Chen, L. and Wang, F. 2021. Derepression of specific miRNA-target genes in rice using CRISPR/Cas9. *Journal of Experimental Botany*, 72:7067.7077

Liu, J., Jung, C., Xu, J., Wang, H., Deng, S., Bernad, L., Arenas-Huertero, C. and Chua, N.H. 2012. Genome-wide analysis uncovers regulation of long intergenic non-coding RNAs in *Arabidopsis*. *Plant Cell*, 24:4333.4345.

Liu, J., Wang, H. and Chua, N.H. 2015. Long noncoding RNA transcriptome of plants. *Journal of Plant Biotechnology*, 13:319.328

Li, M., You, L., Xue, J. and Lu, Y. 2018. Ionizing radiation-induced cellular senescence in normal, non-transformed cells and the involved DNA damage response: A mini review. *Frontiers in Pharmacology*, 9:522.

Lles, E., Mizrahi, A., Marks, V. and Meyerstein, D. 2019. Carbonate-radical-anions, and not hydroxyl radicals, are the products of the fenton reaction in neutral solutions containing bicarbonate. *Free Radicals in Biology and Medicine*, 131:1. doi: 10.1016/j.freeradbiomed.2018.11.015. doi: 10.1016/j.freeradbiomed.2018.11.015.

Lv, D.W., Zhen, S., Zhu, G.R., Bian, Y.W., Chen, G.X., Han, C.X., Yu, Z.T. and Yan, Y. 2016. High-throughput sequencing reveals H_2O_2 stress-associated MicroRNAs and a potential regulatory network in *Brachypodium distachyon* seedlings. *Frontiers in Plant Science*, 7:1567. doi: 10.3389/fpls.2016.01567.

Ma, L., Bajic, V.B. and Zhang, Z. 2013. On the classification of long non-coding RNAs. *RNA Biology*, 10(6):925–933. https://doi.org/10.4161/rna.24604.

Macovei, A. and Tuteja, N. 2013. Different expression of miRNAs targeting helicases in rice in response to low and high dose rate γ-ray treatments. *Plant Signaling and Behaviour*, 8(8): e25128. https://doi.org/10.4161/psb.25128.

Mallory, A.C., Reinhart, B.J., Jones-Rhoades, B.J., Tang, M.W., Zamore, G., Barton, P.D. and Bartel, M.K. 2004. MicroRNA control of PHABULOSA in leaf development: Importance of pairing to the microRNA 5' region. *EMBO Journal*, 23:3356.3364.

Mathes, S.J. and Alexander, J. 1996. Radiation injury. *Surgical Oncology Clinics of North America*, 5:809.824.

Matzke, M., Kanno, T., Daxinger, L., Huettel, B. and Matzke, A.J. 2009. RNA-mediated chromatin-based silencing in plants. *Current Opinion in Cell Biology*, 21:367.376.

Mondal, S., Badigannavar, A.M., Kale, D.M. and Murty, G.S. 2007. An induced dominant seed coat color mutation in groundnut. *Indian Journal of Genetics and Plant Breeding*, 67:177.179.

Mondal, S., Go, Y.S., Lee, S.S., Chung, B.Y. and Kim, J.H. 2016. Characterization of histone modifications associated with DNA damage repair genes upon exposure to gamma rays in *Arabidopsis* seedlings. *Journal of Radiation Research*, 57:646.654.

Ochando, I., Jover-Gil, S., Ripoll, J.J., Candela, H., Vera, A., Ponce, M.R., Martínez-Laborda, A. and Micol, J.L. 2006. Mutations in the microRNA complementarity site of the INCURVATA4 gene perturb meristem function and adaxialize lateral organs in *Arabidopsis*. *Plant Physiology*, 141:607.619.

Oliver, C., Pradillo, M., Jover-Gil, S., Cuñado, N., Ponce, M.R. and Santos, J.L. 2017. Loss of function of *Arabidopsis* microRNA-machinery genes impairs fertility, and has effects on homologous recombination and meiotic chromatin dynamics. *Scientific Reports*, 7(1):9280. doi: 10.1038/s41598-017-07702-x.

Oliver, C., Santos, J.L. and Pradillo, M. 2014. On the role of some ARGONAUTE proteins in meiosis and DNA repair in *Arabidopsis* thaliana. *Frontiers in Plant Science*, 5:177. doi: 10.3389/fpls.2014.00177.

Onodera, Y., Haag, J.R., Ream, T., Costa Nunes, P., Pontes, O. and Pikaard, C.S. 2005. Plant nuclear RNA polymerase IV mediates siRNA and DNA methylation-dependent heterochromatin formation. *Cell*, 120:613.622.

Reisz, J.A., Bansal, N., Qian, J., Zhao, W. and Furdui, C.M. 2014. Effects of ionizing radiation on biological molecules–mechanisms of damage and emerging methods of detection. *Antioxidants and Redox Signaling*, 21(2):260. doi: 10.1089/ars.2013.5489. doi: 10.1089/ars.2013.5489.

Rinn, J.L. and Chang, H.Y. 2012. Genome regulation by long noncoding RNAs. *Annual Review Biochemistry*, 81: 145–166. https://doi.org/10.1146/annurev-biochem-051410-092902.

Schalk, C., Cognat, V., Graindorge, S., Vincent, T., Voinnet, O. and Molinier, J. 2017. Small RNA-mediated reapir of UV-induced DNA lesions by the DNA damage-binding protein 2 and argonaute 1. *PNAS*, 114(14):E2965.E2974.

Simone, N.L., Soule, B.P., Ly, D., Saleh, A.D., Savage, J.E., Degraff, W., Cook, J., Harris, C.C., Gius, D. and Mitchell, J.B. 2009. Ionizing radiation-induced oxidative stress alters miRNA expression. *PLoS One*, 4(7):e6377. doi: 10.1371/journal.pone.0006377.

Sparrow, A.H. and Miksche, J.P. 1961. Correlation of nuclear volume and DNA content with higher plant tolerance to chronic radiation. *Science*, 134:282–283. https://doi.org/10.1126/science.134.3474.282

Tang, G., Reinhart, B.J., Bartel, D.P. and Zamore, P.D. 2003. A biochemical framework for RNA silencing in plants. *Genes and Development*, 17:49.63

Vanhoudt, N., Horemans, N., Wannijn, J., Nauts, R., Van Hees, M. and Vandenhove, H. 2014. Primary stress responses in *Arabidopsis thaliana* exposed to gamma radiation. *Journal of Environmental Radioactivity*, 129:1–6. https://doi.org/10.1016/j.jenvrad.2013.11.011.

Wang, Z., Schwacke, R. and Kunze, R. 2016. DNA damage-induced transcription of transposable elements and long non-coding RNAs in *Arabidopsis* is rare and ATM-dependent. *Molecular Plant*, 9:1142.1155.

Wang, M., Wang, Q. and Zhang, B. 2013. Response of miRNAs and their targets to salt and drought stresses in cotton (*gossypium hirsutum* l.). *Gene*, 530:26.32.

Wan, G., Mathur, R., Hu, X., Liu, Y., Zhang, X., Peng, G. and Lu, X. 2013a. Long non-coding RNA ANRIL (CDKN2B-AS) is induced by the ATM-E2F1 signalling pathway. *Cell Signal*, 25:1086.1095.

Wan, G., Hu, X., Liu, Y., Han, C., Sood, A.K., Calin, G.A., Zhang, X. and Lu, X. 2013b. A novel non-coding RNA lncRNA-JADE connects DNA damage signalling to histone H4 acetylation. *The EMBO Journal*, 32:2833.2847.

Wei, W., Ba, Z., Gao, M., Wu, Y., Ma, Y., Amiard, S., White, C.I., Rendtlew Danielsen, J.M., Yang, Y.G. and Qi, Y. 2012. A role for small RNAs in DNA double-strand break repair. *Cell*, 149(1):101. doi: 10.1016/j.cell.2012.03.002. doi: 10.1016/j.cell.2012.03.002.

Willers, H., Dahm-Daphi, J. and Powell, S.N. 2004. Repair of radiation damage to DNA. *British Journal of Cancer*, 90:1297.301

Xiao, C., He, N., Liu, Y., Wang, Y. and Liu, Q. 2020. Research progress on biodosimeters of ionizing radiation damage. *Radiation Medicine and Protection*, 1:127.132.

Xie, Z., Johansen, L.K., Gustafson, A.M., Kasschau, K.D., Lellis, A.D., Zilberman, D., Jacobsen, S.E. and Carrington, J.C. 2004. Genetic and functional diversification of small RNA pathways in plants. *PLoS Biology*, 2:E104.

Yao, Y., Bilichak, A., Golubov, A., Blevins, T. and Kovalchuk, I. 2010. Differential sensitivity of *Arabidopsis* siRNA biogenesis mutants to genotoxic stress. *Plant Cell Reports*, 29:1401. doi: 10.1007/s00299-010-0930-9. doi:10.1007/s00299-010-0930-9.

Zhang, M., Liang, S., Hang, X., Xiang, Y., Cheng, Z., Li, W., Shi, J., Huang, L. and Sun, Y. 2011. Identification of heavy-ion radiation-induced microRNAs in rice. *Advances in Space Research*, 47:1054.1061.

Zhou, J., Deng, K., Cheng, Y., Zhong, Z., Tian, L., Tang, X., Tang, A., Zheng, X., Zhang, T., Qi, Y. and Zhang, Y. 2017. CRISPR-Cas9 based genome editing reveals new insights into microRNA function and regulation in rice. *Frontiers in Plant Science*, 8:1598. doi: 10.3389/fpls.2017.01598.

Zhou, X., Wang, G. and Zhang, W. 2007. UV-b responsive microRNA genes in *Arabidopsis thaliana*. *Molecular Systems Biology*, 3:103. doi: 10.1038/msb4100143.

9 Role of Non-Coding RNAs in Abiotic Stress Response and Adaptation in Chickpea

Neeraj Kumar[1], Anjali Soni[1], C. Bharadwaj[1], T. Vinutha[1], Manish Roorkiwal[2], Rajeev K. Varshney[3], and S. K. Chaturvedi[4]

[1]ICAR – Indian Agricultural Research Institute, New Delhi, India
[2]Khalifa Center for Genetic Engineering and Biotechnology (KCGEB), UAE University, UAE
Hyderabad, Telangana, India
[3]State Agricultural Biotechnology Centre, Murdoch University, Murdoch, Australia
[4]Rani Laxmibhai Central Agricultural University, Jhansi, Uttar Pradesh, India

9.1 INTRODUCTION

Recent climatic aberrations leading to severe abiotic stress conditions such as drought, salinity, heat and low temperature, are seriously affecting growth and productivity of agricultural crops. Chickpea (*Cicer arietinum* L.), being the most widely cultivated legume, is considered as an important source of dietary protein and fibre for the human population habituating in the arid and semi-arid regions of the world (Jukanti et al., 2012). Chickpea-producing regions are largely affected by extreme environments such as heat, drought and salt at different time points in the growing season (Kaloki et al., 2019). It belongs to the legume family *Fabaceae* with $2n = 16$ chromosomes, a genome size of 738.09 Mb and an estimated 28,269 genes (Singh et al., 2008; Varshney et al., 2013). It was earlier considered an orphan crop, but is now enriched with the availability of modern genomic resources, which provides an opportunity to perform advanced genomic research (Garg et al., 2011; Jain et al., 2013; Kumar et al., 2021). The crop is cultivated over a 17.8 million-hectare (ha) area with a production of 17.2 million metric tonnes worldwide, and India is the leading producer with a total production of 11.4 million metric tonnes, approximately 66% of the total global production; however, yield has only increased from 0.6 t/ha in 1960 to 0.9 t/ha in 2014, representing a 0.006% increase per annum (FAO, 2018). This increase is most likely not enough to feed the increasing world population, though adoption of modern technologies and breeding approaches can play an important role in increasing the rate of genetic gain.

Globally, abiotic stresses, individually or in combination, cause annual chickpea yield losses that translate into severe financial penalties. These stresses affect different traits, metabolic and physiological processes and elicit different responses from plants. Cellular homeostasis mechanisms were utilized using multi-gene and multiple gene regulatory mechanisms for adaptations in response to survivability under harsh conditions (Kale et al., 2015; Kumar et al., 2021). The role of ncRNAs in these multi-gene regulations has been indicated as one of the critical components in such stress situations (Di et al., 2014; Lv et al., 2016; Xu et al., 2016). Advancements in next-generation sequencing (NGS) techniques and bioinformatics analysis has provided opportunity to investigate the dark matter of the genome, i.e., the regions that do not code protein known as ncRNAs, which regulates diverse biological functions in plants (Song et al., 2019). Today, it is a routine activity to identify

DOI: 10.1201/9781003369288-9

intergenic non-coding regions to study their role in plant growth and regulation with the potential to manipulate gene functions during stress responses.

These ncRNAs include small ncRNAs, comprising miRNAs (21–24 nucleotide [nt] long) (Kim, 2005), small interfering RNAs (siRNAs) (Chen, 2009) and lncRNAs (>200 nt long) (Wang and Chang, 2011), which are the main classes of ncRNA that has been documented for various abiotic responses in several plant species. These contribute to various plant development pathways, leading to acclimatization to the abiotic stress condition. We have identified the role of ncRNAs and tried to collate the available information and literature in chickpea to their path in abiotic stress tolerance. The chapter also identifies the roles played by ncRNAs in both biotic and abiotic stress and the present research status of the mechanisms played by miRNA and lncRNA.

9.2 ORIGIN, BIOGENESIS AND MODE OF ACTION OF ncRNAs IN PLANTS

NcRNAs comprise various classes of RNAs that do not encode a protein. These ncRNAs broadly include microRNAs (miRNAs), siRNAs and long non-coding RNAs (lncRNAs). They are supposed to be actively involved in the transcriptional and post-transcriptional modulation of gene expression affecting RNA stability and translation (Lima et al., 2012; Khraiwesh et al., 2012; Kumar et al., 2021). This section summarizes the origin, biogenesis and mode of regulation of miRNAs, siRNAs and lncRNAs involved in abiotic stress responses.

9.2.1 miRNA

miRNAs are endogenous ncRNAs, about 20–24 nt long, originated from the transcription of miRNA genes aided by RNA pol II and catalyzed by DICER-like (DCL) proteins (Muñoz-Mérida et al., 2012; Kim, 2005). Most of the miRNAs are coded from intergenic regions. Biogenesis is complex and regulated via the coordination of several proteins and enzymes (Wang et al., 2017). After transcription of miRNA genes, the primary transcripts (pri-miRNA) with 5′cap and polyadenylation are produced, which is then sequentially processed by the DCL1 containing microprocessor complex into precursor miRNAs (pre-miRNA) and then into mature miRNAs. The conversion is facilitated by RNA-binding protein HYPONASTIC LEAVES1 (HYL1) and a C2H2 zinc finger protein SERRATE in the nucleus (Mallory and Vaucheret, 2006; Fei et al., 2013). Eventually, after dual processing miRNA by DCL-1, mature miRNA duplexes are stabilized by a methyl transferase, Hua Enhancer 1 (HEN1) and methyl groups are added to the 3′ terminal nucleotides of each strand. This prevents 3′ uridylation and degradation (Chae et al., 2015; Rhee et al., 2015). Ultimately, the mature miRNA is assembled with the ARGONAUTE protein to form the miRNA-induced silencing complex (miRISC) (Ha and Kim, 2014). The steps in miRNA biogenesis are depicted in Figure 9.1.

Studies shows that miRNAs are evolutionarily highly conserved among plant species (Addo-Quaye et al., 2009; Kozomara and Griffiths-Jones, 2014). Conserved miRNAs have major roles in regulating common plant developmental processes. Species-specific and genotype-condition-specific miRNAs may regulate special traits such as modulation of cellular mechanisms, nodulation in legumes and promotion of adaptation and tolerance of environmental conditions via controlling and coordinating the network of crucial genes, TFs and phytohormones (Sunkar et al., 2012; Shriram et al., 2016; Jatan and Lata, 2019). Essentially, miRNAs not only play a critical role at the post-transcription level by base-pairing with the target mRNA, leading to degrading or inhibiting mRNA translation (Bartel, 2004; Voinnet, 2009), but they can also interact with other regulatory networks, affecting plant development, responses to biotic and abiotic stresses and plant environment interactions (Kohli et al., 2014; Song et al., 2019).

9.2.2 siRNA

SiRNAs are generally 21–24 nt in length ncRNAs which are produced by the processing of long double-stranded RNA (dsRNA) in a non-overlapping or phased manner from DCL-catalyzed processing of dsRNA precursors (Chen, 2009; Dai and Zhao, 2011). Depending on site of origin,

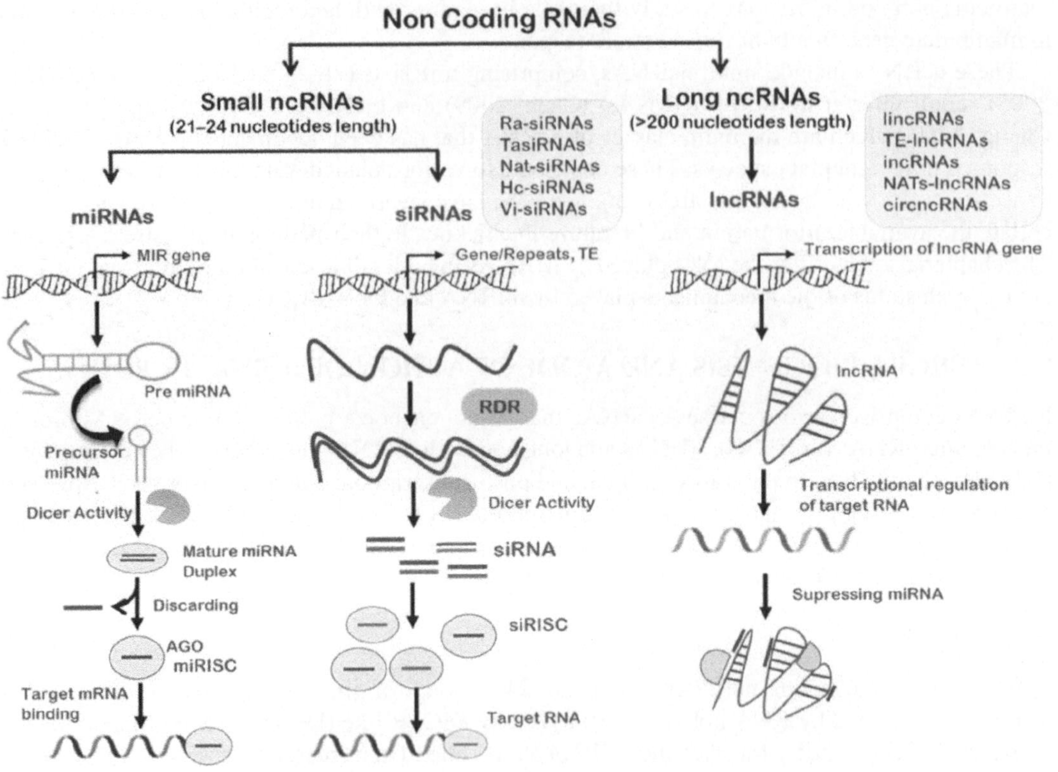

FIGURE 9.1 Generalized scheme to illustrate the various steps involved in plant miRNA, siRNA and lncRNA biogenesis and activity.

siRNAs are classified as: (a) ra-siRNAs, which are derived from transposable elements (TEs) and repetitive DNAs (Guo et al., 2016) and have a critical role in silencing retrotransposons and various stress factors in plants (Mallory and Vaucheret, 2009; Barber et al., 2012); (b) transacting siRNAs (ta-siRNAs), derived from long non-coding single-stranded RNAs, which have crucial regulatory roles post-transcriptional silencing (Katiyar et al., 2015) and have been studied for drought and salinity stress in plants (Hunter et al., 2006; Matsui et al., 2014); (c) natural antisense transcript (nat-siRNAs) derived from the annealed region's natural antisense transcript (NAT) pairs (Yu et al., 2016) and have roles in regulating various biological processes such as phosphate homeostasis and stress response (Wang et al., 2005; Lapidot and Pilpel, 2006; Jabnoune et al., 2013); (d) heterochromatic (hc-siRNAs) are derived from heterochromatic intergenic regions including repeats and transposons (Castel and Martienssen, 2013) and have a critical role in modulating DNA methylation and histone modification machinery, leading to responses in several biotic and abiotic stress factors (Ku et al., 2015; Medina et al., 2018); and (e) Vi-siRNAs originated from dsRNA replicative intermediates of viruses induce antiviral immunity (Ding and Lu, 2011) and their role has been reported in response to viral pathogens (Yadav and Chattopadhyay, 2011).

SiRNAs have critical roles in DNA methylation, chromatin modification and repression of distinct mRNA targets (Fei et al., 2013). The siRNAs can target endogenous as well as exogenous sequences, serving as the first line of defence. The biogenesis process is assisted by DCLs leading the long dsRNA conversion to mature siRNA, which is further associated with the AGO protein to form the catalytic core of the RNA-induced silencing complex (RISC) to facilitate gene silencing (Sanan-Mishra et al., 2021). The guide strand directs the RISC complex to its target transcript for cleavage, while the other strand, known as the passenger strand, escapes and undergoes degradation. The siRNAs also mediate transcriptional gene silencing through the RNA-induced transcriptional silencing (RITS) complex (**Figure 9.1**) (Macrae et al., 2006). Overall, the siRNAs are involved in regulating gene expression, maintaining genome stability and aiding plant defence.

9.2.3 lncRNA

Long ncRNAs (lncRNAs), usually >200 nt in size, are located in the cytoplasm and nucleus and have very crucial roles in genome and epigenetic-aided regulation (Nejat and Mantri, 2018; Yu et al., 2019). These ncRNAs lack protein coding potential but can interact with coding regions to affect the specific functions (Jha et al., 2020). Based on their biogenesis and location as described in **Figure 9.1**, the lncRNAs can be classified into several sub-groups: (a) Long intergenic (lincRNAs) are 200 to 2000 nt in length and derived from the intergenic region. Very long intervening ncRNAs also called macroRNAs (Quek et al., 2015). They are localized at least 5 kb away from protein-coding regions and are, therefore, said to lie in gene deserts. LincRNAs are further sub-classified as enhancer RNAs or eRNAs (Szcześniak et al., 2016). They have tissue-specific expression and trans-regulatory roles (Jin et al., 2013). (b) Transposable element (TE-lncRNAs) are derived from the TEs and can serve as precursors to miRNAs and siRNAs (Li et al., 2014; Burgess et al., 2020). They have been reposted to be involved in abiotic stresses in various plant species (Wang et al., 2017). (c) Intron-derived lncRNAs (incRNAs) originated from the introns of protein coding genes (Ma et al., 2013) and include totally intronic RNA (TIN) or partially intronic RNA (PIN). IncRNAs have been reported to be regulated by various transcription activation pathways. (d) Natural antisense transcripts (NATs-lncRNA) are derived from coding regions (both exon and intron) of genomes. They are equipped with both cis and trans-actions to regulate gene expression by silencing (Lindsay et al., 2013; Yu et al., 2013). Their binding triggers the production of specific siRNAs, which also exhibits a predominantly trans-mode of action. (e) Circular lncRNAs (circncRNAs) are more stable, low abundance and highly conserved than the linear lncRNAs (Li et al., 2018). They originate in the nucleus from the back-splicing of exons in pre-mRNAs (Zhang et al., 2013; Chu et al., 2020). Studies show that these classes of ncRNAs can affect the splicing of their linear counterparts. They have potential to regulate cell development via acting as endogenous target mimics (eTM) and miRNA sponges (Qin et al., 2020).

The major class of lncRNAs are long intergenic RNAs, transcribed by RNA polymerase II or III and polymerase IV/V (Herr et al., 2005). These lncRNAs can serve as precursors of miRNAs and siRNAs and act as eTM, competing for various miRNAs (Franco-Zorrilla et al., 2007). Moreover, they participate in, alternative spicing (Heo and Sung, 2011), post-translational regulation (Jabnoune et al., 2013), protein re-localization and chromatin topology modification (Sousa et al., 2001).

9.3 TOOLS AND DATABASE OF ncRNAs IN PLANTS

The application of computational tools to identify and predict ncRNAs have taken a new dimension in computational biology. The establishment of high-throughput sequencing and better bioinformatics pathways have clearly helped in the prediction mechanisms of ncRNAs. Similarly, comparative analysis of these ncRNAs has also led to the formation of various databases to improve our understanding of the conservation of ncRNA genes with their precise mode of function across various species in the plant kingdom (Table 9.1).

MiRNA prediction tools such as miRPlant (An et al., 2014), miRanalyzer (Hackenberg et al., 2009), miRA (Evers et al., 2015) and miRDeep-P (Yang and Li, 2011) can be effectively used to predict new miRNAs from sequencing data. Tools such as miRU (Zhang, 2005) and psRNA-Target (Dai and Zhao, 2011) are used to predict plant miRNA targets. MiRNAs target prediction has become more possible these today due to advances in bioinformatics search modules and tools (**Table 9.1**). For example, MTide (Zhang et al., 2015) is the combination of a modified

TABLE 9.1
A Comprehensive List of Key Tools and Databases for Identifying and Annotating ncRNAs in Plants

Sr. No.	Tools	Reference	Sr. No.	Database	Reference
1	BioVLAB-MMIA-NGS	Chae et al., 2015	1	ASRP	Backman et al., 2007
2	CNCI	Sun et al., 2013	2	CANTATAdb	Szcześniak et al., 2016
3	CPAT	Wang et al., 2013	3	DMD	Chiang et al., 2015
4	CPC	Kong et al., 2007	4	GREENC	Gallart et al., 2016
5	DeepLNC	Tripathi et al., 2016	5	lncRNAdb	Quek et al., 2015
6	iSeeRNA	Sun et al., 2013	6	miRBase	Kozomara and Griffiths-Jones, 2014.
7	lncRNATargets	Hu and Sun, 2016	7	miRFANs	Liu et al., 2012
8	MicroPC	Mhuantong and Wichadakul, 2009	8	miRTarBase	Hsu et al., 2011
9	miRA	Evers et al., 2015	9	NONCODE	Zhao et al., 2016
10	miRanalyzer	Hackenberg et al., 2009	10	PASmiR	Zhang et al., 2013
11	miRDeep-P	Yang and Li, 2011	11	PlanTE-MIR DB	Lorenzetti et al., 2016
12	miRPlant	An et al., 2014	12	PLncDB	Jin et al., 2013
13	miRU	Zhang, 2005	13	PLNlncRbase	Xuan et al., 2015
14	MTide	Zhang et al., 2015	14	PmiRKB	Meng et al., 2010
15	PlantMirnaT	Rhee et al., 2015	15	PMRD	Zhang et al., 2010
16	PLncPRO	Singh et al., 2017	16	PNRD	Yi et al., 2015
17	psRNATarget	Dai and Zhao, 2011	17	Rfam	Nawrocki et al., 2015
18	Semirna	Muñoz-Mérida et al., 2012	18	WMP	Remita et al., 2016

miRDeep2 (Huang et al., 2006), a modified CleaveLand4 (Addo-Quaye et al., 2009), and other useful scripts, which are used to calculate expression profiles of known MIRs, predict novel miRNAs and identify target mRNAs. Similarly, several robust databases are available for sequence and homology identification of miRNAs from experimental or computational identification of various species, such as miRBase (Kozomara and Griffiths-Jones, 2014) and Rfam (Nawrocki et al., 2015). Annotated databases and miRNA–target interaction databases such as the plant microRNA database (PMRD) (Zhang et al., 2010), Plant microRNA Knowledge Base (PmiRKB) (Meng et al., 2010) and miRTarBase (Hsu et al., 2011) are routinely used annotation databases in miRNA study. For specific stress-related miRNA, databases such as PASmiR (Zhang et al., 2013) and WMP (Remita et al., 2016) are used to identify and search for miRNAs involved in stress regulatory functions.

For functional characterization and identification of plant lncRNAs, several novel tools such as PhyloCSF (Lin et al., 2011), CPC (Kong et al., 2007), CPAT (Wang et al., 2013), CNCI (Sun et al., 2013), iSeeRNA (Sun et al., 2013) and DeepLNC (Tripathi et al., 2016) are various regression model-based regression algorithms that facilitate sequence analysis of transcriptome data. For rapid functional characterization and action mechanisms of lncRNAs, tools such as lncRNA targets (Hu and Sun, 2016) are used. LncRNA databases such as NONCODE (Zhao et al., 2016) and lncRNAdb (Quek et al., 2015) hold complete collection and annotation databases of lncRNAs along with the information of their experimentally verified biological functions. Numerous plant-specific databases such as NONCODE and lncRNAdb are available in the public domain. In addition, some of the software also maintain all of the plant experimental lncRNAs such as PlncDB, GREENC (Gallart et al., 2016) and PLNlncRbase (Xuan et al., 2015). In addition, PNRD (Yi et al., 2015) and CANTATAdb (Szcześniak et al., 2016) are the two largest plant lncRNA databases used in routine lncRNA analysis.

9.4 ncRNAs IN CHICKPEA ADAPTATION TO ABIOTIC STRESS

Abiotic stress impacts chickpea production globally, severely affecting all plant growth and development and ultimately reducing crop yields (Flowers et al., 2010; Kale et al., 2015; Kumar et al., 2020). Chickpea adapts to harsh environments by evoking various physiological, biochemical, metabolic and molecular mechanisms (Kumar et al., 2021). Many QTL/genes contributing towards major abiotic stresses like drought, salt and heat have been studied in chickpea (Soren et al., 2020; Jha et al., 2021). However, very few studies are available to intercept the regulatory roles of ncRNAs and their corresponding target gene(s) controlling the abiotic stress response in chickpea (Table 9.2). MiRNA and lncRNA have been characterized to a limited extent mainly for drought and salt stress (Figure 9.2). In this section, we will discuss the brief advancement and developments of deciphering ncRNA roles exclusively for the abiotic stress response in chickpea.

9.4.1 miRNA-Aided Abiotic Stress Response

High-throughput transcriptome sequencing has led to the identification of a large number of conserved salt-responsive miRNAs, including miR390, miR172, miR171, miR169, miR408, miR159, miR396, miR2111, miR5213, miR397, miR393, miR162, miR168, miR166, miR167, miR156, miR530, miR399, miR160, miR319, miR164, miR398, miR2118 and miR394. These miRNAs have been extensively characterized and studied in the chickpea ICC 4598 genotype under elevated salinity. The upregulation of miR156, miR396 and miR319 were recorded in response to salt stress and have clearly shown involvement with several TFs families such as CZF1 and protein kinases associated with transcriptional regulation, suggesting their significant role under stress. Studies also indicated that miR172, miR319, miR171, miR390 and miR396 have serine/threonine protein kinases and MAPK protein kinases as their target genes, which involve signalling pathways crucial for initiating defence mechanisms under stress conditions via transcriptional regulators (Kohli et al., 2014).

TABLE 9.2
Role of ncRNAs in Chickpea Plants Regulating Growth, Development and Adaptation to Abiotic Stress Responses

Sr. No.	Name/Number of ncRNA	Targeted Traits/Genes	References
1	59 novel miRNAs miR156_1, miR156_10, car-miR008, car-miR011, car-miR015	Salinity stress targeting, Squamosa promoter-binding protein.	Kohli et al., 2014
2	157 miRNA loci miR156, miR159, miR160, miR162, miR164, miR168, miR172, miR393, miR408	Stress response and development processes, targeting SPB factor, MYB transcription factor, ARFs, DCL1, HD-Zip, Argonaute 1, AP2, F-box protein.	Srivastava et al., 2015
3	3457 high-confidence lncRNAs	Drought and salinity.	Singh et al., 2017
4	284 unique miRNAs	Drought and salinity. LACCASE4, COPPER. Increase lateral root formation and improves uptake of K+ under salinity stress.	Khandal et al., 2017
5	3452 novel lncRNAs	Salinity tolerance.	Kumar et al., 2021
6	miR408	Drought overexpression represses plantacyanin encoding targeting DREB regulation under water stress.	Hajyzadeh et al., 2015
7	miR167, miR168, miR171	Auxin response factors, WD-repeat and scarecrow-like transcription factors.	Singh et al., 2021

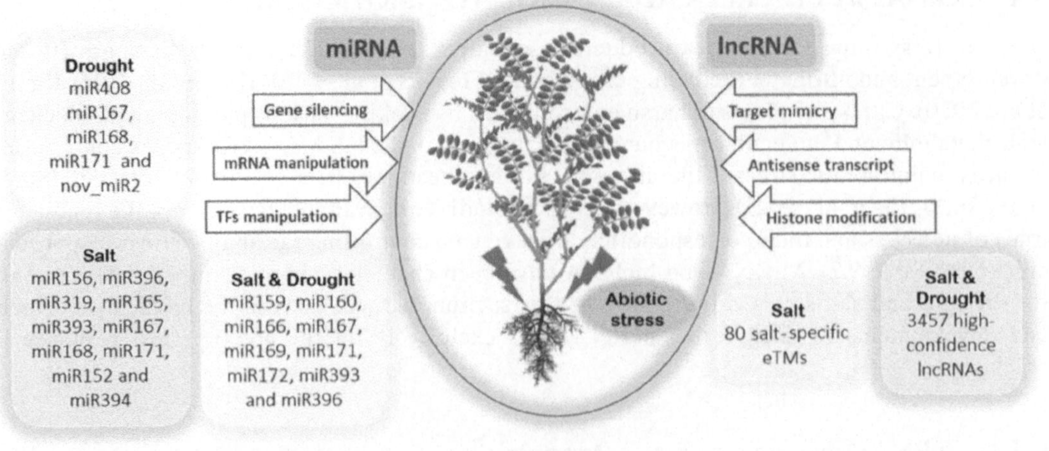

FIGURE 9.2 NcRNA module (miRNA and lncRNA) controlling various abiotic response (drought and salinity) genes, TFs and pathways in chickpea.

Results of MicroRNA profiling have provided insight into post-transcriptional regulation of gene expression in the chickpea root apex under salinity and water deficiency (Khandal et al., 2017), where, 609 mRNA targets related to diverse cellular processes were predicted for 244 miRNAs. Studies have suggested the crucial roles of miRNA-mediated post-transcriptional regulation of gene expression influencing the re-patterning of root hair cells, lateral root formation and K+ uptake. These studies revealed a complex regulatory mechanism of gene regulation influencing stress signalling and root development.

Predicted targets of miRNAs like Car-miR-156d, Car-miR-7031, Car-miR-361.1, Car-miR615, Car-miR6562.1, Car-miR1532, Car-miR172e, Car-miR7907b.1, Car-miRK12, Car-novmiR2 and Car-miR5507 were critically regulating the genes involved in root development and abiotic stress response. The targeted genes included homeobox protein GLABRA2 (GL2), proline-rich receptor-like protein kinase (PERK13), LATERAL ORGAN BOUNDARIES (LOB), ABC transporter genes, ARGINASE 1, ARGINASE 2, HAK5 and CIPK23 genes, which are associated with the regulation of root development and abiotic stress responsiveness in various plant species (Masucci et al., 1996; Noh et al., 2001; Kant et al., 2007; Flores et al., 2008; Won et al., 2009).

MiRNAs associated with drought tolerance also have been exclusively characterized and studied in chickpea. MiR408 conserved in plants tends to regulate copper-related genes to empower plants to withstand severe drought. Overexpression studies of miR408 in chickpea transgenic lines induced drought tolerance via influencing plantacyanin transcript repression caused the regulation of DREB and other drought responsive genes (Hajyzadeh et al., 2015). Similarly, under severe drought conditions, Singh et al. (2021) reported significant upregulation (>3-fold) in the expression of miR167, miR168 and miR171 in a tolerant genotype. These miRNAs target auxin response factors, WD-repeat and scarecrow-like transcription factors, respectively known to play crucial roles in drought stress and provided an understanding of post-transcriptional gene regulation under drought stress conditions in chickpea at the molecular level.

Additionally, sequencing of small RNA components of leaves and inflorescence of chickpea also aided in the discovery of a new class of small ncRNA, i.e., phased siRNAs (phasiRNAs), which are abundantly produced in legumes along with conserved and novel miRNAs. The predicted targets of these novel small RNAs were mostly transcription factors including superoxide dismutases, plantacyanin, laccases and F-box proteins that play vital roles in abiotic stress response (Srivastava et al., 2015).

9.4.2 lncRNA-Aided Abiotic Stress Response

Involvement of lncRNA in imparting abiotic stress tolerance in chickpea has been demonstrated. But to date, there are only two studies highlighting the role and mechanism of abiotic stress tolerance in chickpea (Singh et al., 2017; Kumar et al., 2021).

The first study utilizes high-throughput transcriptome data using the novel tool PLncPRO to predict lncRNAs in chickpea. A total of 3457 high-confidence lncRNAs responding to drought and salinity stress were identified. The drought-sensitive genotype ICC1882 showed the least differentially expressed lncRNAs as compared to the ICC4958 drought-tolerant chickpea genotype (Singh et al., 2017). In this study, a novel tool, PLncPRO, was reported to be effective in aiding the discovery of lncRNAs in plants. This study also provides a comprehensive list of lncRNAs expressed under drought and/or salinity stress, which serves as a useful resource to analyze their exact function in abiotic stress responses and other biological process.

The second study was specifically focused on understanding the regulatory role of lncRNA under the salt stress condition in chickpea (Kumar et al., 2021). A total of 3452 novel lncRNAs were identified to be distributed across all 8 chickpea chromosomes when comparing salt tolerant (ICCV 10, JG 11) and salt-sensitive cultivars (DCP 92–3, Pusa 256). Additionally, 80 unique lncRNAs were identified as interacting with 136 different miRNAs as eTM targets of miRNAs and implicated them in the regulatory network of salt stress response. Functional analysis of these lncRNA revealed their association in targeting salt stress response-related genes such as potassium transporter, transporter family genes, serine/threonine-protein kinase, aquaporins such as TIP1-2 and PIP2-5 and transcription factors such as AP2, NAC, bZIP, ERF, MYB and WRKY.

Additionally, multiple sets of interaction were detected and several eTMs were predicted as a precursor of numerous miRNA that were likely to be involved in stress tolerance, such as CatmiR159g-3p, Cat-miR156j.2 and Cat-NovmiR100a, via competitively inhibiting the degradation of mRNAs (Jin et al., 2020; Sun et al., 2020). The miRNA–eTM interaction (car-miRNA015, Cat-miR159g-3p and Cat-miR172c.2) also reflected the role of these TFs under stress. More than 16 target eTMs were annotated with SQUAMOSA promoter-binding (SBP) protein-like proteins (SPLs). These protein families are plant-specific TFs aided with a conserved DNA binding domain SBP-box consisting of two zinc finger structures (Moreno et al., 1997). This protein family has essential roles in plant development and morphogenesis (Chen et al., 2010). Recently, their role has been investigated and validated in abiotic stress tolerance. These SPL families mainly regulate the genes involved in signal transduction, anthocyanin metabolism, reactive oxygen species (ROS) scavenging and proline synthesis under elevated salt/drought stress conditions (Wang et al., 2019). Several other eTMs, such as the MATE efflux family protein, also played a significant role under salt stress. The MATE gene family mostly encodes transporter genes involved in various physiological process in plants (Nimmy et al., 2015). The eTM targets (Cat-NovmiR105, CatmiR171h.4) annotated with pentatricopeptide repeat proteins (PPR) were enriched. Expression of this protein family enhances ABA regulation, which induces stomatal closure and restricts stomatal opening, and plants withstand multiple abiotic stresses including salinity and drought. It also provides high tolerance to salt stress in the germination and post-germination stages with no negative effect on plant growth (Chen et al., 2018; Xing et al., 2018).

9.5 CONCLUSION AND FUTURE PERSPECTIVE

The gradually increasing global population and climate change incidences have challenged the agriculture sector. Intensity of variable weather aberrations and environmental stresses are projected to increase in the near future, adding more limitations. To tackle these problems, it is important to develop crops that can withstand these abiotic stress conditions. Chickpea, being the most important global legume crop, provides a protein source to the population inhabiting arid and semi-arid regions of the world. The impact of these abiotic stresses has been rampant on its productivity and production. To develop these tolerant cultivars, it is essential to understand the regulatory and molecular mechanism of plants.

With the advent of NGS and development of advanced bioinformatics tools, the study of ncRNAs and their functional annotation has received considerable interest for understanding the molecular mechanisms controlling response of plants to various abiotic stress such as salt and drought (Kumar et al., 2021; Singh et al., 2021). As ncRNAs are dynamic, they are rapidly being discovered and functionally characterized in various plant species, including chickpea. Understanding the key roles of ncRNAs can help to learn the molecular mechanics underlying metabolites and phytohormonal and gene-regulatory pathways to withstand stress. Eventually, this would pave the way to harness molecular breeding tactics to elevate chickpea breeding programs aimed at achieving abiotic stress resilience by employing molecular breeding techniques.

REFERENCES

Addo-Quaye, C., Miller, W. and Axtell, M.J., 2009. CleaveLand: A pipeline for using degradome data to find cleaved small RNA targets. *Bioinformatics*, *25*(1), pp.130–131.

An, J., Lai, J., Sajjanhar, A., Lehman, M.L. and Nelson, C.C., 2014. miRPlant: An integrated tool for identification of plant miRNA from RNA sequencing data. *BMC Bioinformatics*, *15*(1), pp.1–4.

Backman, T.W., Sullivan, C.M., Cumbie, J.S., Miller, Z.A., Chapman, E.J., Fahlgren, N., Givan, S.A., Carrington, J.C. and Kasschau, K.D., 2007. Update of ASRP: The Arabidopsis small RNA project database. *Nucleic Acids Research*, *36*(suppl_1), pp.D982–D985.

Barber, W.T., Zhang, W., Win, H., Varala, K.K., Dorweiler, J.E., Hudson, M.E. and Moose, S.P., 2012. Repeat associated small RNAs vary among parents and following hybridization in maize. *Proceedings of the National Academy of Sciences*, *109*(26), pp.10444–10449.

Bartel, D.P., 2004. MicroRNAs: Genomics, biogenesis, mechanism, and function. *Cell*, *116*(2), pp.281–297.

Burgess, D., Li, H., Zhao, M., Kim, S.Y. and Lisch, D., 2020. Silencing of mutator elements in maize involves distinct populations of small RNAs and distinct patterns of DNA methylation. *Genetics*, *215*(2), pp.379–391.

Castel, S.E. and Martienssen, R.A., 2013. RNA interference in the nucleus: Roles for small RNAs in transcription, epigenetics and beyond. *Nature Reviews Genetics*, *14*(2), pp.100–112.

Chae, H., Rhee, S., Nephew, K.P. and Kim, S., 2015. BioVLAB-MMIA-NGS: microRNA–mRNA integrated analysis using high-throughput sequencing data. *Bioinformatics*, *31*(2), pp.265–267.

Chen, X., 2009. Small RNAs and their roles in plant development. *Annual Review of Cell and Developmental*, *25*, pp.21–44.

Chen, X., Zhang, Z., Liu, D., Zhang, K., Li, A. and Mao, L., 2010. SQUAMOSA promoter-binding protein-like transcription factors: Star players for plant growth and development. *Journal of Integrative Plant Biology*, *52*(11), pp.946–951.

Chen, G., Zou, Y., Hu, J. and Ding, Y., 2018. Genome-wide analysis of the rice PPR gene family and their expression profiles under different stress treatments. *BMC Genomics*, *19*(1), pp.1–14.

Chiang, K., Shu, J., Zempleni, J. and Cui, J., 2015. Dietary MicroRNA database (DMD): An archive database and analytic tool for food-borne microRNAs. *PLoS One*, *10*(6), p.e0128089.

Chu, Q., Bai, P., Zhu, X., Zhang, X., Mao, L., Zhu, Q.H., Fan, L. and Ye, C.Y., 2020. Characteristics of plant circular RNAs. *Briefings in Bioinformatics*, *21*(1), pp.135–143.

Dai, X. and Zhao, P.X., 2011. psRNATarget: A plant small RNA target analysis server. *Nucleic Acids Research*, *39*(suppl_2), pp.W155–W159.

Ding, S.W. and Lu, R., 2011. Virus-derived siRNAs and piRNAs in immunity and pathogenesis. *Current Opinion in Virology*, *1*(6), pp.533–544.

Di, C., Yuan, J., Wu, Y., Li, J., Lin, H., Hu, L., Zhang, T., Qi, Y., Gerstein, M.B., Guo, Y. and Lu, Z.J., 2014. Characterization of stress-responsive lnc RNA s in Arabidopsis thaliana by integrating expression, epigenetic and structural features. *The Plant Journal*, *80*(5), pp.848–861.

Evers, M., Huttner, M., Dueck, A., Meister, G. and Engelmann, J.C., 2015. miRA: Adaptable novel miRNA identification in plants using small RNA sequencing data. *BMC Bioinformatics*, *16*(1), pp.1–10.

Fei, Q., Xia, R. and Meyers, B.C., 2013. Phased, secondary, small interfering RNAs in posttranscriptional regulatory networks. *The Plant Cell*, *25*(7), pp.2400–2415.

Flores, T., Todd, C.D., Tovar-Mendez, A., Dhanoa, P.K., Correa-Aragunde, N., Hoyos, M.E., Brownfield, D.M., Mullen, R.T., Lamattina, L. and Polacco, J.C., 2008. Arginase-negative mutants of Arabidopsis exhibit increased nitric oxide signaling in root development. *Plant Physiology*, *147*(4), pp.1936–1946.

Flowers, T.J., Gaur, P.M., Gowda, C.L., Krishnamurthy, L., Samineni, S., Siddique, K.H., Turner, N.C., Vadez, V., Varshney, R.K. and Colmer, T.D., 2010. Salt sensitivity in chickpea. *Plant, Cell & Environment*, *33*(4), pp.490–509.

Food and Agriculture Organization of the United Nations. FAOSTAT, 2018. Available at: http://www.fao.org/faostat/en/#data/QC. Accessed 20 Jan 2022.

Franco-Zorrilla, J.M., Valli, A., Todesco, M., Mateos, I., Puga, M.I., Rubio-Somoza, I., Leyva, A., Weigel, D., García, J.A. and Paz-Ares, J., 2007. Target mimicry provides a new mechanism for regulation of microRNA activity. *Nature Genetics*, *39*(8), pp.1033–1037.

Gallart, A.P., Pulido, A.H., De Lagrán, I., Sanseverino, A.M. and Cigliano, W., 2016. GREENC: A wiki-based database of plant lncRNAs. *Nucleic Acids Research*, *44*(Database issue), p.D1161.

Garg, R., Patel, R.K., Jhanwar, S., Priya, P., Bhattacharjee, A., Yadav, G., Bhatia, S., Chattopadhyay, D., Tyagi, A.K. and Jain, M., 2011. Gene discovery and tissue-specific transcriptome analysis in chickpea with massively parallel pyrosequencing and web resource development. *Plant Physiology*, *156*(4), pp.1661–1678.

Guo, Q., Liu, Q., Smith, N.A., Liang, G. and Wang, M.B., 2016. RNA silencing in plants: Mechanisms, technologies and applications in horticultural crops. *Current Genomics*, *17*(6), pp.476–489.

Hackenberg, M., Sturm, M., Langenberger, D., Falcon-Perez, J.M. and Aransay, A.M., 2009. miRanalyzer: A microRNA detection and analysis tool for next-generation sequencing experiments. *Nucleic Acids Research*, *37*(suppl_2), pp.W68–W76.

Hajyzadeh, M., Turktas, M., Khawar, K.M. and Unver, T., 2015. miR408 overexpression causes increased drought tolerance in chickpea. *Gene*, *555*(2), pp.186–193.

Ha, M. and Kim, V.N., 2014. Regulation of microRNA biogenesis. *Nature Reviews Molecular Cell Biology*, *15*(8), pp.509–524.

Heo, J.B. and Sung, S., 2011. Vernalization-mediated epigenetic silencing by a long intronic noncoding RNA. *Science*, *331*(6013), pp.76–79.

Herr, A.J., Jensen, M.B., Dalmay, T. and Baulcombe, D.C., 2005. RNA polymerase IV directs silencing of endogenous DNA. *Science*, *308*(5718), pp.118–120.

Hsu, S.D., Lin, F.M., Wu, W.Y., Liang, C., Huang, W.C., Chan, W.L., Tsai, W.T., Chen, G.Z., Lee, C.J., Chiu, C.M. and Chien, C.H., 2011. miRTarBase: A database curates experimentally validated microRNA–target interactions. *Nucleic Acids Research*, *39*(suppl_1), pp.D163–D169.

Huang, H.Y., Chien, C.H., Jen, K.H. and Huang, H.D., 2006. RegRNA: An integrated web server for identifying regulatory RNA motifs and elements. *Nucleic Acids Research*, *34*(suppl_2), pp.W429–W434.

Hunter, C., Willmann, M.R., Wu, G., Yoshikawa, M., de la, L., Gutiérrez-Nava, M. and Poethig, S.R.., 2006. Trans-acting siRNA-mediated repression of ETTIN and ARF4 regulates heteroblasty in Arabidopsis. *Development*, 133, pp.2973–2981.

Hu, R. and Sun, X., 2016. lncRNATargets: A platform for lncRNA target prediction based on nucleic acid thermodynamics. *Journal of Bioinformatics and Computational Biology*, *14*(04), p.1650016.

Jabnoune, M., Secco, D., Lecampion, C., Robaglia, C., Shu, Q. and Poirier, Y., 2013. A rice cis-natural antisense RNA acts as a translational enhancer for its cognate mRNA and contributes to phosphate homeostasis and plant fitness. *The Plant Cell*, *25*(10), pp.4166–4182.

Jain, M., Misra, G., Patel, R.K., Priya, P., Jhanwar, S., Khan, A.W., Shah, N., Singh, V.K., Garg, R., Jeena, G. and Yadav, M., 2013. A draft genome sequence of the pulse crop chickpea (*Cicer arietinum* L.). *The Plant Journal*, *74*(5), pp.715–729.

Jatan, R. and Lata, C., 2019. Role of MicroRNAs in abiotic and biotic stress resistance in plants. *Proceedings of the Indian National Science Academy*, *85*, pp.553–567.

Jha, U.C., Nayyar, H., Jha, R., Khurshid, M., Zhou, M., Mantri, N. and Siddique, K.H., 2020. Long non-coding RNAs: Emerging players regulating plant abiotic stress response and adaptation. *BMC Plant Biology*, *20*(1), pp.1–20.

Jha, U.C., Nayyar, H., Palakurthi, R., Jha, R., Valluri, V., Bajaj, P., Chitikineni, A., Singh, N.P., Varshney, R.K. and Thudi, M., 2021. Major QTLs and potential candidate genes for heat stress tolerance identified in chickpea (*Cicer arietinum* L.). *Frontiers in Plant Science*, *12*, 655103.

Jin, Z., Gao, S., Ma, W., Lyu, X., Cao, X. and Yao, Y., 2020. Identification and functional prediction of salt stress-related long noncoding RNAs in grapevine roots. *Environmental and Experimental Botany*, *179*, p.104215.

Jin, J., Liu, J., Wang, H., Wong, L. and Chua, N.H., 2013. PLncDB: Plant long non-coding RNA database. *Bioinformatics*, *29*(8), pp.1068–1071.

Jukanti, A.K., Gaur, P.M., Gowda, C.L.L. and Chibbar, R.N., 2012. Nutritional quality and health benefits of chickpea (*Cicer arietinum* L.): A review. *British Journal of Nutrition*, *108*(S1), pp.S11–S26.

Kale, S.M., Jaganathan, D., Ruperao, P., Chen, C., Punna, R., Kudapa, H., Thudi, M., Roorkiwal, M., Katta, M.A., Doddamani, D. and Garg, V., 2015. Prioritization of candidate genes in "QTL-hotspot" region for drought tolerance in chickpea (*Cicer arietinum* L.). *Scientific Reports*, *5*(1), pp.1–14.

Kaloki, P., Devasirvatham, V. and Tan, D.K., 2019. Chickpea abiotic stresses: Combating drought, heat and cold. *Abiotic and Biotic Stress in Plants*. 139–162. doi: 10.5772/intechopen.83404

Kant, P., Kant, S., Gordon, M., Shaked, R. and Barak, S., 2007. STRESS RESPONSE SUPPRESSOR1 and STRESS RESPONSE SUPPRESSOR2, two DEAD-box RNA helicases that attenuate Arabidopsis responses to multiple abiotic stresses. *Plant Physiology*, *145*(3), pp.814–830.

Katiyar, A., Smita, S., Muthusamy, S.K., Chinnusamy, V., Pandey, D.M. and Bansal, K.C., 2015. Identification of novel drought-responsive microRNAs and trans-acting siRNAs from sorghum bicolor (L.) Moench by high-throughput sequencing analysis. *Frontiers in Plant Science*, *6*, p.506.

Khandal, H., Parween, S., Roy, R., Meena, M.K. and Chattopadhyay, D., 2017. MicroRNA profiling provides insights into post-transcriptional regulation of gene expression in chickpea root apex under salinity and water deficiency. *Scientific Reports*, *7*(1), pp.1–14.

Khraiwesh, B., Zhu, J.K. and Zhu, J., 2012. Role of miRNAs and siRNAs in biotic and abiotic stress responses of plants. *Biochimica et Biophysica Acta (BBA)-Gene Regulatory Mechanisms*, *1819*(2), pp.137–148.

Kim, V.N., 2005. MicroRNA biogenesis: Coordinated cropping and dicing. *Nature Reviews Molecular Cell Biology*, *6*(5), pp.376–385.

Kohli, D., Joshi, G., Deokar, A.A., Bhardwaj, A.R., Agarwal, M., Katiyar-Agarwal, S., Srinivasan, R. and Jain, P.K., 2014. Identification and characterization of wilt and salt stress-responsive microRNAs in chickpea through high-throughput sequencing. *PloS One*, *9*(10), p.e108851.

Kong, L., Zhang, Y., Ye, Z.Q., Liu, X.Q., Zhao, S.Q., Wei, L. and Gao, G., 2007. CPC: Assess the protein-coding potential of transcripts using sequence features and support vector machine. *Nucleic Acids Research*, *35*(suppl_2), pp.W345–W349.

Kozomara, A. and Griffiths-Jones, S., 2014. miRBase: Annotating high confidence microRNAs using deep sequencing data. *Nucleic Acids Research*, *42*(D1), pp.D68–D73.

Kumar, N., Bharadwaj, C., Sahu, S., Shiv, A., Shrivastava, A.K., Reddy, S.P.P., Soren, K.R., Patil, B.S., Pal, M., Soni, A. and Roorkiwal, M., 2021. Genome-wide identification and functional prediction of salt-stress related long non-coding RNAs (lncRNAs) in chickpea (*Cicer arietinum* L.). *Physiology and Molecular Biology of Plants*, *27*(11), pp.2605–2619.

Kumar, N., Bharadwaj, C., Soni, A., Sachdeva, S., Yadav, M.C., Pal, M., Soren, K.R., Meena, M.C., Roorkiwal, M., Varshney, R.K. and Rana, M., 2020. Physio-morphological and molecular analysis for salt tolerance in chickpea (*Cicer arietinum*). *Indian Journal of Agricultural Sciences*, *90*(4), pp.132–136.

Kumar, N., Bharadwaj, C., Soren, K.R., Meena, M.C., Priya, P.S., Joshi, N., Soni, A., Patil, B.S., Pal, M., Roorkiwal, M. and Varshney, R.K., 2021. Response of shoot ionic (Na+/K+) distribution on yield and agro-physiological traits in chickpea (*Cicer arietinum* L.) Recombinant inbred lines population under salinity. *Indian J. Genet*, *81*(4), pp.590–593.

Kumar, N., Soren, K.R., Bharadwaj, C., Shrivastava, A.K., Pal, M., Roorkiwal, M., Kumar, K., Patil, B.S., Soni, A. and Nimmy, M.S., 2021. Genome-wide transcriptome analysis and physiological variation modulates gene regulatory networks acclimating salinity tolerance in chickpea. *Environmental and Experimental Botany*, *187*, p.104478.

Ku, Y.S., Wong, J.W.H., Mui, Z., Liu, X., Hui, J.H.L., Chan, T.F. and Lam, H.M., 2015. Small RNAs in plant responses to abiotic stresses: Regulatory roles and study methods. *International Journal of Molecular Sciences*, *16*(10), pp.24532–24554.

Lapidot, M. and Pilpel, Y., 2006. Genome-wide natural antisense transcription: Coupling its regulation to its different regulatory mechanisms. *EMBO Reports*, *7*(12), pp.1216–1222.

Li, L., Eichten, S.R., Shimizu, R., Petsch, K., Yeh, C.T., Wu, W., Chettoor, A.M., Givan, S.A., Cole, R.A., Fowler, J.E. and Evans, M.M., 2014. Genome-wide discovery and characterization of maize long non-coding RNAs. *Genome Biology*, *15*(2), pp.1–15.

Lima, J.C.D., Loss-Morais, G. and Margis, R., 2012. MicroRNAs play critical roles during plant development and in response to abiotic stresses. *Genetics and Molecular Biology*, *35*, pp.1069–1077.

Lindsay, M.A., Griffiths-Jones, S., Wight, M. and Werner, A., 2013. The functions of natural antisense transcripts. *Essays in Biochemistry*, *54*, pp.91–101.

Lin, M.F., Jungreis, I. and Kellis, M., 2011. PhyloCSF: A comparative genomics method to distinguish protein coding and non-coding regions. *Bioinformatics*, *27*(13), pp.i275–i282.

Liu, H., Jin, T., Liao, R., Wan, L., Xu, B., Zhou, S. and Guan, J., 2012. miRFANs: An integrated database for Arabidopsis thaliana microRNA function annotations. *BMC Plant Biology*, *12*(1), pp.1–8.

Li, X., Yang, L. and Chen, L.L., 2018. The biogenesis, functions, and challenges of circular RNAs. *Molecular Cell*, *71*(3), pp.428–442.

Lorenzetti, A.P.R., de Antonio, G.Y.A., Paschoal, A.R. and Domingues, D.S., 2016. PlanTE-MIR DB: A database for transposable element-related microRNAs in plant genomes. *Functional and Integrative Genomics*, *16*, pp.235–242.

Lv, Y., Liang, Z., Ge, M., Qi, W., Zhang, T., Lin, F., Peng, Z. and Zhao, H., 2016. Genome-wide identification and functional prediction of nitrogen-responsive intergenic and intronic long non-coding RNAs in maize (*Zea mays* L.). *BMC Genomics*, *17*(1), pp.1–15.

Ma, L., Bajic, V.B. and Zhang, Z., 2013. On the classification of long non-coding RNAs. *RNA Biology*, *10*(6), pp.924–933.

Macrae, I.J., Li, F., Zhou, K., Cande, W.Z. and Doudna, J.A., 2006. Structure of Dicer and mechanistic implications for RNAi. In *Cold Spring Harbor symposia on quantitative biology*, *71*, pp.73–80. Cold Spring Harbor Laboratory Press.

Mallory, A.C. and Vaucheret, H., 2006. Functions of microRNAs and related small RNAs in plants. *Nature Genetics*, *38*(6), pp.S31–S36.

Mallory, A.C. and Vaucheret, H., 2009. ARGONAUTE 1 homeostasis invokes the coordinate action of the microRNA and siRNA pathways. *EMBO Reports*, *10*(5), pp.521–526.

Masucci, J.D., Rerie, W.G., Foreman, D.R., Zhang, M., Galway, M.E., Marks, M.D. and Schiefelbein, J.W., 1996. The homeobox gene GLABRA2 is required for position-dependent cell differentiation in the root epidermis of Arabidopsis thaliana. *Development*, *122*(4), pp.1253–1260.

Matsui, A., Mizunashi, K., Tanaka, M., Kaminuma, E., Nguyen, A.H., Nakajima, M., Kim, J.M., Nguyen, D.V., Toyoda, T. and Seki, M., 2014. tasiRNA-ARF pathway moderates floral architecture in Arabidopsis plants subjected to drought stress. *BioMed Research International*, 2014. https://doi.org/10.1155/2014/303451

Medina, C., Da Rocha, M., Magliano, M., Raptopoulo, A., Marteu, N., Lebrigand, K., Abad, P., Favery, B. and Jaubert-Possamai, S., 2018. Characterization of siRNAs clusters in Arabidopsis thaliana galls induced by the root-knot nematode meloidogyne incognita. *BMC Genomics*, *19*(1), pp.1–16.

Meng, Y., Gou, L., Chen, D., Mao, C., Jin, Y., Wu, P. and Chen, M., 2010. PmiRKB: A plant microRNA knowledge base. *Nucleic Acids Research*, *39*(suppl_1), pp.D181–D187.

Mhuantong, W. and Wichadakul, D., 2009. MicroPC (μPC): A comprehensive resource for predicting and comparing plant microRNAs. *BMC Genomics*, *10*(1), pp.1–8.

Moreno, M.A., Harper, L.C., Krueger, R.W., Dellaporta, S.L. and Freeling, M., 1997. liguleless1 encodes a nuclear-localized protein required for induction of ligules and auricles during maize leaf organogenesis. *Genes & Development*, *11*(5), pp.616–628.

Muñoz-Mérida, A., Perkins, J.R., Viguera, E., Thode, G., Bejarano, E.R. and Pérez-Pulido, A.J., 2012. Semirna: Searching for plant miRNAs using target sequences. *Omics: A Journal of Integrative Biology*, *16*(4), pp.168–177.

Nawrocki, E.P., Burge, S.W., Bateman, A., Daub, J., Eberhardt, R.Y., Eddy, S.R., Floden, E.W., Gardner, P.P., Jones, T.A., Tate, J. and Finn, R.D., 2015. Rfam 12.0: Updates to the RNA families database. *Nucleic Acids Research*, *43*(D1), pp.D130–D137.

Nejat, N. and Mantri, N., 2018. Emerging roles of long non-coding RNAs in plant response to biotic and abiotic stresses. *Critical Reviews in Biotechnology*, *38*(1), pp.93–105.

Nimmy, M.S., Kumar, V., Singh, A.K., Jain, P.K. and Srinivasan, R., 2015. Expression analysis of a MATE-type transporter gene of Arabidopsis and its orthologues in rice and chickpea under salt stress. *Indian J Genet Plant Breed*, *75*, pp.478–485.

Noh, B., Murphy, A.S. and Spalding, E.P., 2001. Multidrug resistance–like genes of Arabidopsis required for auxin transport and auxin-mediated development. *The Plant Cell*, *13*(11), pp.2441–2454.

Qin, T., Li, J. and Zhang, K.Q., 2020. Structure, regulation, and function of linear and circular long non-coding RNAs. *Frontiers in Genetics*, *11*, p.150.

Quek, X.C., Thomson, D.W., Maag, J.L., Bartonicek, N., Signal, B., Clark, M.B., Gloss, B.S. and Dinger, M.E., 2015. lncRNAdb v2. 0: Expanding the reference database for functional long noncoding RNAs. *Nucleic Acids Research*, *43*(D1), pp.D168–D173.

Remita, M.A., Lord, E., Agharbaoui, Z., Leclercq, M., Badawi, M.A., Sarhan, F. and Diallo, A.B., 2016. A novel comprehensive wheat miRNA database, including related bioinformatics software. *Current Plant Biology*, *7*, pp.31–33.

Rhee, S., Chae, H. and Kim, S., 2015. PlantMirnaT: miRNA and mRNA integrated analysis fully utilizing characteristics of plant sequencing data. *Methods*, *83*, pp.80–87.

Sanan-Mishra, N., Jailani, A.A.K., Mandal, B. and Mukherjee, S.K., 2021. Secondary siRNAs in plants: Biosynthesis, various functions, and applications in virology. *Frontiers in Plant Science*, *12*, 610283

Shriram, V., Kumar, V., Devarumath, R.M., Khare, T.S. and Wani, S.H., 2016. MicroRNAs as potential targets for abiotic stress tolerance in plants. *Frontiers in Plant Science*, *7*, p.817.

Singh, U., Khemka, N., Rajkumar, M.S., Garg, R. and Jain, M., 2017. PLncPRO for prediction of long non-coding RNAs (lncRNAs) in plants and its application for discovery of abiotic stress-responsive lncRNAs in rice and chickpea. *Nucleic Acids Research*, *45*(22), pp.e183–e183.

Singh, L., Kohli, D., Gaikwad, K., Kansal, R., Dahuja, A., Paul, V., Bharadwaj, C. and Jain, P.K., 2021. Effect of drought stress on morphological, biochemical, physiological traits and expression analysis of microRNAs in drought-tolerant and sensitive genotypes of chickpea. *Indian J. Genet*, *81*(2), pp.266–276.

Singh, R., Sharma, P., Varshney, R.K., Sharma, S.K. and Singh, N.K., 2008. Chickpea improvement: Role of wild species and genetic markers. *Biotechnology and Genetic Engineering Reviews*, *25*(1), pp.267–314.

Song, X., Li, Y., Cao, X. and Qi, Y., 2019. MicroRNAs and their regulatory roles in plant–environment interactions. *Annual Review of Plant Biology*, *70*, pp.489–525.

Soren, K.R., Madugula, P., Kumar, N., Barmukh, R., Sengar, M.S., Bharadwaj, C., Sharma, P.C., Singh, S., Bhandari, A., Singh, J. and Sanwal, S.K., 2020. Genetic dissection and identification of candidate genes for salinity tolerance using Axiom® CicerSNP array in chickpea. *International Journal of Molecular Sciences*, *21*(14), p.5058.

Sousa, C., Johansson, C., Charon, C., Manyani, H., Sautter, C., Kondorosi, A. and Crespi, M., 2001. Translational and structural requirements of the early nodulin gene enod40, a short-open reading frame-containing RNA, for elicitation of a cell-specific growth response in the alfalfa root cortex. *Molecular and Cellular Biology*, *21*(1), pp.354–366.

Srivastava, S., Zheng, Y., Kudapa, H., Jagadeeswaran, G., Hivrale, V., Varshney, R.K. and Sunkar, R., 2015. High throughput sequencing of small RNA component of leaves and inflorescence revealed conserved and novel miRNAs as well as phasiRNA loci in chickpea. *Plant Science*, *235*, pp.46–57.

Sun, K., Chen, X., Jiang, P., Song, X., Wang, H. and Sun, H., 2013. iSeeRNA: Identification of long intergenic non-coding RNA transcripts from transcriptome sequencing data. *BMC Genomics*, *14*(2), pp.1–10.

Sunkar, R., Li, Y.F. and Jagadeeswaran, G., 2012. Functions of microRNAs in plant stress responses. *Trends in Plant Science*, *17*(4), pp.196–203.

Sun, L., Luo, H., Bu, D., Zhao, G., Yu, K., Zhang, C., Liu, Y., Chen, R. and Zhao, Y., 2013. Utilizing sequence intrinsic composition to classify protein-coding and long non-coding transcripts. *Nucleic Acids Research*, *41*(17), pp.e166–e166.

Sun, X., Zheng, H., Li, J., Liu, L., Zhang, X. and Sui, N., 2020. Comparative transcriptome analysis reveals new lncRNAs responding to salt stress in sweet sorghum. *Frontiers in Bioengineering and Biotechnology*, *8*, p.331.

Szcześniak, M.W., Rosikiewicz, W. and Makałowska, I., 2016. CANTATAdb: A collection of plant long non-coding RNAs. *Plant and Cell Physiology*, *57*(1), pp.e8–e8.

Tripathi, R., Patel, S., Kumari, V., Chakraborty, P. and Varadwaj, P.K., 2016. DeepLNC, a long non-coding RNA prediction tool using deep neural network. *Network Modeling Analysis in Health Informatics and Bioinformatics*, *5*(1), pp.1–14.

Varshney, R.K., Song, C., Saxena, R.K., Azam, S., Yu, S., Sharpe, A.G., Cannon, S., Baek, J., Rosen, B.D., Tar'an, B. and Millan, T., 2013. Draft genome sequence of chickpea (*Cicer arietinum*) provides a Resource for trait improvement. *Nature Biotechnology*, *31*(3), pp.240–246.

Voinnet, O., 2009. Origin, biogenesis, and activity of plant microRNAs. *Cell*, *136*(4), pp.669–687.

Wang, K.C. and Chang, H.Y., 2011. Molecular mechanisms of long noncoding RNAs. *Molecular Cell*, *43*(6), pp.904–914.

Wang, X.J., Gaasterland, T. and Chua, N.H., 2005. Genome-wide prediction and identification of cis-natural antisense transcripts in Arabidopsis thaliana. *Genome Biology*, *6*(4), pp.1–11.

Wang, L., Park, H.J., Dasari, S., Wang, S., Kocher, J.P. and Li, W., 2013. CPAT: Coding-potential assessment tool using an alignment-free logistic regression model. *Nucleic Acids Research*, *41*(6), pp.e74–e74.

Wang, D., Qu, Z., Yang, L., Zhang, Q., Liu, Z.H., Do, T., Adelson, D.L., Wang, Z.Y., Searle, I. and Zhu, J.K., 2017. Transposable elements (TE s) contribute to stress-related long intergenic noncoding RNA s in plants. *The Plant Journal*, *90*(1), pp.133–146.

Wang, H., Wang, H., Duan, X., Liu, C. and Li, Z., 2017. Digital quantitative analysis of microRNA in single cell based on ligation-depended polymerase colony (Polony). *Biosensors and Bioelectronics*, *95*, pp.146–151.

Wang, J., Ye, Y., Xu, M., Feng, L. and Xu, L.A., 2019. Roles of the SPL gene family and miR156 in the salt stress responses of tamarisk (*Tamarix chinensis*). *BMC Plant Biology*, *19*(1), pp.1–11.

Won, S.K., Lee, Y.J., Lee, H.Y., Heo, Y.K., Cho, M. and Cho, H.T., 2009. Cis-element-and transcriptome-based screening of root hair-specific genes and their functional characterization in Arabidopsis. *Plant Physiology*, *150*(3), pp.1459–1473.

Xing, H., Fu, X., Yang, C., Tang, X., Guo, L., Li, C., Xu, C. and Luo, K., 2018. Genome-wide investigation of pentatricopeptide repeat gene family in poplar and their expression analysis in response to biotic and abiotic stresses. *Scientific Reports*, *8*(1), pp.1–9.

Xuan, H., Zhang, L., Liu, X., Han, G., Li, J., Li, X., Liu, A., Liao, M. and Zhang, S., 2015. PLNlncRbase: A Resource for experimentally identified lncRNAs in plants. *Gene*, *573*(2), pp.328–332.

Xu, X.W., Zhou, X.H., Wang, R.R., Peng, W.L., An, Y. and Chen, L.L., 2016. Functional analysis of long intergenic non-coding RNAs in phosphate-starved rice using competing endogenous RNA network. *Scientific Reports*, *6*(1), pp.1–12.

Yadav, R.K. and Chattopadhyay, D., 2011. Enhanced viral intergenic region–specific short interfering RNA accumulation and DNA methylation correlates with resistance against a geminivirus. *Molecular Plant-Microbe Interactions*, *24*(10), pp.1189–1197.

Yang, X. and Li, L., 2011. miRDeep-P: A computational tool for analyzing the microRNA transcriptome in plants. *Bioinformatics*, *27*(18), pp.2614–2615.

Yi, X., Zhang, Z., Ling, Y., Xu, W. and Su, Z., 2015. PNRD: A plant non-coding RNA database. *Nucleic Acids Research*, *43*(D1), pp.D982–D989.

Yu, D., Meng, Y., Zuo, Z., Xue, J. and Wang, H., 2016. NATpipe: An integrative pipeline for systematical discovery of natural Antisense transcripts (NATs) And phase-distributed nat-siRNAs from de novo assembled transcriptomes. *Scientific Reports*, *6*(1), pp.1–6.

Yu, X., Yang, J., Li, X., Liu, X., Sun, C., Wu, F. and He, Y., 2013. Global analysis of cis-natural antisense transcripts and their heat-responsive nat-siRNAs in Brassica rapa. *BMC Plant Biology*, *13*(1), pp.1–13.

Yu, Y., Zhang, Y., Chen, X. and Chen, Y., 2019. Plant noncoding RNAs: Hidden players in development and stress responses. *Annual Review of Cell and Developmental Biology*, *35*, pp.407–431.

Zhang, Y., 2005. miRU: An automated plant miRNA target prediction server. *Nucleic Acids Research*, *33*(suppl_2), pp.W701–W704.

Zhang, Z., Jiang, L., Wang, J., Gu, P. and Chen, M., 2015. MTide: An integrated tool for the identification of miRNA–target interaction in plants. *Bioinformatics*, *31*(2), pp.290–291.

Zhang, S., Yue, Y., Sheng, L., Wu, Y., Fan, G., Li, A., Hu, X., ShangGuan, M. and Wei, C., 2013. PASmiR: A literature-curated database for miRNA molecular regulation in plant response to abiotic stress. *BMC Plant Biology*, *13*(1), pp.1–8.

Zhang, Z., Yu, J., Li, D., Zhang, Z., Liu, F., Zhou, X., Wang, T., Ling, Y. and Su, Z., 2010. PMRD: Plant microRNA database. *Nucleic Acids Research*, *38*(suppl_1), pp.D806–D813.

Zhang, Y., Zhang, X.O., Chen, T., Xiang, J.F., Yin, Q.F., Xing, Y.H., Zhu, S., Yang, L. and Chen, L.L., 2013. Circular intronic long noncoding RNAs. *Molecular Cell*, *51*(6), pp.792–806.

Zhao, Y., Li, H., Fang, S., Kang, Y., Wu, W., Hao, Y., Li, Z., Bu, D., Sun, N., Zhang, M.Q. and Chen, R., 2016. NONCODE 2016: An informative And valuable data source of long non-coding RNAs. *Nucleic Acids Research*, *44*(D1), pp.D203–D208.

10 Plant Functional Long Non-Coding RNAs

Yesterday's Dark Matter, Today's Regulatory Component for Abiotic Stress Response

Sayanti Mandal,[1] Deepak Khairnar,[2] and Swati Hazra[3]

[1]Department of Chemistry and Biochemistry, Sharda School of Basic Sciences & Research, Sharda University, Greater Noida, Uttar Pradesh, India;
Institute of Bioinformatics and Biotechnology (IBB), Savitribai Phule Pune University (SPPU), Pune, Maharashtra, India

[2]Department of Biotechnology, Dr. D. Y. Patil Arts, Commerce & Science College (affiliated to Savitribai Phule Pune University), Sant Tukaram Nagar, Pimpri, Pune, Maharashtra, India;
National Centre for Cell Science, Savitribai Phule Pune University Campus, Ganeshkhind, Pune, Maharashtra, India

[3]Sharda School of Agricultural Sciences, Sharda University, Greater Noida, Uttar Pradesh, India

10.1 INTRODUCTION

The discovery of regulatory information inside the non-protein-coding regions of DNA has spurred attempts to find non-coding RNA (ncRNA) molecules. This has been accelerated by advancement in RNA sequencing technology, which has led to the identification of ncRNAs, and elucidation of their structures and functions. Noncoding RNAs have recently gained prominence in recent decades, especially as our knowledge of their regulatory roles has greatly expanded. The technical advances of genome sequencing, especially RNA sequencing (RNA-seq), next-generation sequencing (NGS), and sophisticated computational tools have led to the discovery of new regulatory ncRNAs in many biological processes in both animals and plants [Yu et al., 2019; Budak et al., 2020]. Recent breakthroughs in large-scale transcriptomic analysis and high-throughput gene sequencing analyses have revealed that a substantial part of the eukaryotic genome is transcribed into RNAs that do not code for proteins.

These transcripts, ncRNAs, can be distinguished with respect to the length of their mature product. These ncRNAs are majorly classified into three types: short ncRNAs (sRNAs; 18–30 nt), medium ncRNAs (31–200 nt), and long ncRNAs (lncRNAs; >200 nt) [Wang et al., 2017a]. MicroRNAs (miRNAs) and small interfering RNAs (siRNAs) are the two primary types of sRNAs found in plants, and they differ in their biogenesis and mechanisms of action. RNAs with improperly base-paired hairpin structures are used to make miRNAs, which are typically 21–24 nt long. RNA-dependent RNA polymerases (RDRs) may be required to generate siRNAs from perfect complementary long double-stranded RNAs (dsRNAs) [Borges and Martienssen, 2015]. Considering the previously mentioned classification, long non-coding RNAs (lncRNAs) are generally defined as

 DOI: 10.1201/9781003369288-10

transcripts longer than 200 nt which lack the protein-coding ability. LncRNAs are the most abundant and functionally versatile subclass of these ncRNAs. Certainly, the description of lncRNA that is only based on length is arbitrary. Amaral et al., 2011 suggest a stronger biological rationale for distinguishing lncRNAs from small ncRNAs, defining lncRNAs as those ncRNAs that operate either as primary or spliced transcripts, irrespective of existing recognized types of small ncRNAs. As a result, certain lncRNAs are shorter than the arbitrary length criterion (*BC1* and *snaR*, for example, are less than or near to 200 nt, yet are included in lncRNAdb database.) [Amaral et al., 2011; Shin et al., 2018; Sun et al., 2020; Corona-Gomez et al., 2020].

The emergence of plant lncRNAs as potential "biological regulators" played a key role in modulating the basic cellular machinery have received considerable attention over the past few years. But only a small percentage of lncRNAs have been extensively investigated, although the association of lncRNAs with abiotic stress responses is now well established. According to research, lncRNAs have been implicated in transcriptional, post-transcriptional, and translation regulation, as well as post-translational modification [Dykes and Emanueli, 2017]. Despite that the mechanisms of plant lncRNAs remain unknown, they have been found to govern essential agronomic traits, making them significantly interesting for plant biologist and breeders. These new lncRNAs may function at the epigenome, transcription, and post-transcription levels as a regulatory hub for monitoring multiple abiotic responsive hormonals signaling pathways. Therefore, altering the lncRNAs expression in various crops make these ncRNAs an appealing target to engineer crops for improved stress resilience.

The molecular mechanism by which lncRNAs function and mediate gene regulation is not well known in plants. Numerous lncRNAs have also been found to be tissue-specific, a characteristic that makes their identification more challenging [Cabili et al., 2011; Kang and Liu, 2015; Khemka et al., 2016; Liu et al., 2016; Song et al., 2016]. Therefore, ascribing functions to a particular lncRNA has proven to be far more difficult than expected. In comparison to protein-coding genes, our knowledge of the lncRNA mechanism and function is extremely limited, which may be correlated with various lncRNA features [Quinn and Chang, 2016; Esposito et al., 2019]. First, the majority of lncRNAs in relation to mRNAs and miRNAs have low levels of expression, therefore, capturing and annotating lncRNA from transcriptomic data is a challenge [Kopp and Mendell, 2018]. Second, lncRNAs lack significant open reading frames (ORFs), in contrast to protein-coding genes, which have easily identifiable ORFs that allow for the prediction of 2D or 3D structure, structural domains, and protein-protein interactions [Kopp and Mendell, 2018]. Third, unlike miRNAs and mRNAs, it is difficult to find functional information through evolutionary studies, because most of the lncRNAs have poor sequence conservation among different species [Derrien et al., 2012]. Furthermore, several lncRNAs have distinct temporal and spatial expression patterns.

Overall, lncRNAs display evidence of rapid evolution for which the mechanisms are still unclear and under investigation. Few studies, however, have looked at lncRNAs throughout the genome-wide association study, their complex regulatory molecular processes, and their functional annotation [Szcześniak et al., 2016]. LncRNAs are delineated briefly in this chapter, which covers basic features of lncRNAs, including classification, annotation difficulties, and underlying processes of lncRNA mediated plant abiotic stress tolerance. We want to understand the lncRNA studies that may be critical for the development of agronomically important crops, which will eventually lead to the creation of stable climate-smart crops that are more resilience to climate change. This may potentially open the door for crop breeders to explore lncRNAs.

10.2 FEATURES OF lncRNA

10.2.1 Length

Non-coding transcripts with a length of more than 200 nt that do not translate into proteins are known as lncRNAs. LncRNAs are typically engaged in critical processes such as gene translation, inactivation, and splicing [Ben Amore et al., 2009; Gomes et al., 2013; Liu et al., 2013]. An lncRNA can be more than 2 kb in length, yet it can only code for about 100 amino acids [Cao, 2014]. Plant lncRNAs are transcribed by two polymerases, i.e., RNA polymerase IV or RNA polymerase V (RNA Pol IV or RNA Pol V)[Dinger et al., 2008, 2009; Wierzbicki et al., 2008]. They are processed by splicing or non-splicing, polyadenylation or non-polyadenylation, and may be present in either the nucleus or the cytoplasm.

10.2.2 Location in the Genome

The majority of lncRNAs are found in poorly conserved genomic regions, such as intronic regions of genes [Lapidot and Pilpel, 2006]. In addition, certain lncRNAs are known to be transcribed inside protein-coding region from one of the strands of the DNA sequence [Louro et al., 2008]. The evolutionary conserved lncRNAs are directly related to their genomic locations [Lapidot and Pilpel, 2006, Quinn and Chang, 2016]. Research findings indicate that large number of non-coding RNAs (lncRNAs) are evolutionarily conserved, although to a lesser extent when compared to the protein-coding genes [Ponjavic et al., 2007; Guttman et al., 2009]. Interestingly, the promoter-regions of lncRNAs are more conserved than the lncRNA sequence [Pang et al., 2006]. The existence of ORF in some lncRNAs makes it difficult to distinguish protein-coding RNAs [Dinger et al., 2011].

10.2.3 Action

Various lncRNA families exert different methods of action to regulate gene expression and protein synthesis. These ncRNAs may have secondary structures for interacting with protein, DNA and RNA, or may behave as scaffolding in subnuclear domains. LncRNAs are expressed in a cell-specific manner. In addition, it has been revealed that transcription of particular lncRNAs takes place at a specified moment; as a result, they may act as molecular signals to react to a variety of stimuli [Wang and Chang, 2011].

10.2.4 *Cis*- and *Trans*-Regulating Action

Natural antisense transcripts (NATs) are a specific category of RNAs that have base pair or sequence complementarity with other RNA transcripts. Like miRNAs, *trans*-NATs and their targets are physically located in different loci on the genome. The *cis*-NATs and their targets reside on the same locus, but opposite the DNA strands. These *cis*-NATs were initially discovered in viruses, followed by prokaryotes, and subsequently eukaryotes. It is estimated that about 5–29% of the transcription sequences in eukaryotes, except nematodes, are engaged in the overlap [Lapidot and Pilpel, 2006]. It has been shown that RNA polymerase II is involved in mRNA processing via transcription of the *cis*-NAT sequences. The interplay between sense and antisense transcripts indicates that NATs have a function in gene expression control. Furthermore, it has been shown that in the event of transcription of gene locus and RNA hybrid formation in both directions, it is possible to induce gene silence or an induced immunological response [Wight and Werner, 2015].

10.2.5 Abundance of lncRNA Transcripts

In a variety of plant species including model plant *Arabidopsis thaliana* and agronomically important crops, lncRNAs have also been discovered (Table 10.1). The number of lncRNAs detected varies depending on the technology used for identification in each species, and large-scale studies have reported between 6480 and 6510 lncRNAs in *Arabidopsis* [Liu et al., 2012; Zhao et al., 2018]. The most distinguishing feature of lncRNAs is that they are much shorter than protein-coding mRNAs and have fewer exons, but still include ORFs capable of producing smaller peptides [Lin et al., 2020].

10.2.6 Expression Specificity

In contrast to mRNAs that translate into protein, lncRNAs are usually produced in a more tissue-specific manner. Typically, tissue-specific lncRNAs are expressed higher than protein-coding genes.

TABLE 10.1
List of Abundant Plant-Specific lncRNAs in Various Species

S. No.	Plant Species	No. of lncRNA	Tissue	References
1.	*Arabidopsis thaliana* (thale cress)	6480	Inflorescence, seedling	Liu et al., (2012)
2.	*Setaria italic* (foxtail millet)	584	Shoots	Qi et al., (2013)
3.	*Oryza sativa* (rice)	2224	Shoot, pistil, anther, seed,	Zhang et al., (2014)
4.	*Zea mays* (maize)	20,163	Shoot, leaf, and root	Li et al., (2014)
5.	*Populus tomentosa* (poplar)	1377	Anthers, pistils, seeds	Chen et al., (2015)
6.	*Solanum lycopersicum* cv. Ailsa Craig (tomato)	3679	Fruits	Zhu et al., (2015)
7.	*S. lycopersicum* (tomato)	1565	Leaves	Wang et al., (2015)
8.	*Medicago truncatula* (barrelclover)	23,324	Leaves and roots	Wang et al., (2015a)
9.	*Fragaria vesca* (strawberry)	5884	Fruit and flower	Kang and Liu (2015)
10.	*Gossypium hirsutum* (cotton)	35,268	Hypocotyl, leaf, root, and flowers	Wang et al., (2015b)
11.	*Gossypium arboretum* (cotton)	5996	Fiber	Zou et al., (2016)
12.	*P. tomentosa* (poplar)	7655	Leaves	Tian et al., (2016)
13.	*Morus notabilis* (mulberry)	1133	Leaf, flower, root, bark, and bud	Song et al., (2016)
14.	*Z. mays* L. (maize)	7245	Leaf	Lv et al., (2016)
15.	*Helianthus annus* (sunflower)	25,327	Floral bud	Flórez-Zapata et al., (2016)
16.	*Brassica napus* (canola)	3181	Leaves	Joshi et al., (2016)
17.	*Brassica rapa* (Bcajh97-01A/B' GMSA/B)	12,051	Pollen	Huang et al., (2018)
18.	*Arachis hypogaea* L. (peanut)	1442	Leaves, root, and seed	Tian et al., (2020)
19.	*B. rapa* L. ssp. *Pekinensis* (Chinese cabbage)	3711	Seedlings	Zhang et al., (2021)
20.	*B. rapa* (var. *pekinensis*)	2088	Leaves	Mehraj et al., (2021)

Experiments using real-time reverse transcription-PCR (qRT-PCR) in *Arabidopsis* revealed that about 32% of lncRNAs had organ-specific expression that is well documented [Liu et al., 2012]. LncRNAs are highly expression specific and are appropriate markers for tissue markers and the developmental stage. The low expression level of lncRNAs as well as difficulties in identification by conventional mRNA-sequencing methods may account for some of the apparent selectivity.

10.2.6.1 Tissue-Specificity

LncRNA expression levels in animals have been extensively found to be tissue-specific [Dinger et al., 2008; Tsoi et al., 2015]. This suggests that these transcripts are involved in numerous processes such as differentiation, development, repair, and maintenance as well as in a number of other activities. Recently, there have been many studies in plants that have shown tissue-specific lncRNAs that are only expressed during certain developmental phases. The findings indicate that lncRNAs are thought to be involved in fiber growth in *Gossypium arboretum*, fruit and flower development in *Fragaria vesca*, flower development in *Cicer arietinum*, and root and floral organ development in *Morus notabilis* [Kang and Liu, 2015; Khemka et al., 2016; Song et al., 2016; Zou et al., 2016].

10.2.6.2 Cell-Type-Specificity

In studies showing the tissue specificity of lncRNAs, a general tendency of lower expression levels for lncRNAs relative to protein-coding genes has been found. The low amounts of lncRNAs in all the cells may be responsible for this, or their expression is found only in certain cell populations or cell subpopulations [Cabili et al., 2011; Liu et al., 2016]. In both animals and plants, cell-type lncRNAs were frequently discovered. Cell-type-specific long intergenic ncRNAs (lincRNAs) were discovered in a recent study on the *Arabidopsis* root expression map; nevertheless, even in specialized cell types, lincRNA expression levels were lower than mRNAs [Li et al., 2016].

10.2.7 Similarity to mRNAs

Although lncRNAs do not have the capacity to code for proteins, they are comparable to mRNAs in many aspects, such as the RNA polymerases that are responsible for lncRNA polyadenylation, 50 capping, transcription, and alternative splicing patterns [Derrien et al., 2012; Yotsukura et al., 2016]. LncRNAs are spliced with exon/intron lengths that resemble those of mRNA-coding genes in the majority of cases [Gerstein et al., 2007; Guttman et al., 2009]. However, outliers found in plants, such as *G. arboretum* (cotton) and *C. arietinum* (chickpea), have shown exceptions to this rule, with the average exon length of lincRNAs being significantly greater than that of mRNAs [Khemka et al., 2016; Zou et al., 2016].

10.2.8 Comparison with miRNA

NcRNAs include both miRNAs and lncRNAs. MiRNAs are just 22 nt long, but lncRNAs are 8–10 times larger. Although the precise function of lncRNA is not fully understood, both miRNA and lncRNA have been demonstrated to function as regulators for regulating biological processes at the level of post-transcriptional repression of protein-coding genes [Wang et al., 2017a, 2017b]. In addition, lncRNAs can function as miRNA sponges and reduce the regulatory impact they have on mRNA [Paraskevopoulou and Hatzigeorgiou, 2016]. In addition, lncRNAs may serve as a precursor for siRNAs and miRNAs that regulate alternative splicing of pre-mRNAs and as endogenous target mimic (eTM) that compete with other miRNAs [Jiang et al., 2020; Zhou et al., 2020a]. The interaction between miRNAs and lncRNAs has been discovered in recent research. In addition to acting as targets or sources of miRNAs, certain lncRNAs may influence the synthesis and function of miRNAs. LncRNAs and miRNAs interact significantly in regulating numerous cellular processes like development, nutrients, biotic, and abiotic stress [Zhou et al., 2020b].

10.2.9 LncRNA Discovery, Biogenesis and Classification

An early marker for nodule organogenesis, early nodulin 40 (*Enod40*), was the first discovered plant lncRNA in *Medicago truncatula* [Crespi et al., 1994]. MtRBP1, a nuclear RNA binding protein, was shown to change its subcellular localization in response to *Enod40* [Crespi et al., 1994]. It has now been established that plant lncRNAs regulate miRNA activity, chromatin structure, and epigenetic control, among other things [Franco-Zorrilla et al., 2007; Swiezewski et al., 2009; Kim and Sung 2017; Wu et al., 2020]. Furthermore, the antisense lncRNAs *MAF4* antisense RNA (*MAS*) and *LRK* Antisense Intergenic RNA (*LAIR*) were discovered to interact with WDR5a in *Oryza sativa* and *Arabidopsis*, respectively, controlling flowering timing [Wang et al., 2018c; Zhao et al., 2018].

These lncRNAs are separate transcription units that exist in sequence space and do not overlap with coding sequences or protein-coding genes. LncRNAs are largely differentiated from mRNAs due to the absence of discernible coding capability. Most RNA polymerase II transcripts are similar to mRNAs and are capped with 50 ends, alternative splicing, and 30 Poly-A tails [Chekanova, 2015]. Two more polymerases, RNA Pol IV or RNA Pol V, may transcribe lncRNAs from plants [Dinger et al., 2008; Wierzbicki et al., 2008; Dinger et al., 2009]. In contrast to Pol II transcripts, these lncRNAs are less well-characterized and exhibit a lack of Poly-A tails [Zhou and Law, 2015]. Due to their low expression and high instability, it is especially challenging to detect RNA Pol IV- or PolV-transcribed lncRNAs [Ariel et al., 2015]. On the other hand, these transcripts are the primary drivers of RNA-mediated DNA methylation (RdDM). To ensure genomic integrity, plants have evolved a highly complex RNA interference-dependent RdDM mechanism [Matzke and Mosher, 2014]. In this mechanism, RNA Pol IV transcribes lncRNA, which is subsequently processed into siRNAs of 24 nt. Simultaneously with chromatin modifying enzymes, the siRNA-AGO complex identifies lncRNA generated by RNA Pol V and guides it to the chromatin target site. Additionally, methyltransferases and proteins are used by cytosine residues in the target area after contact with the AGO complex, resulting in gene silence [Wierzbicki et al., 2008]. The transcripts of RNA Pol IV are primarily precursors of siRNA, whereas RNA Pol V and some RNA Pol II are the targets for sRNA.

Based on their location and surrounding protein-coding genes, lncRNAs are widely categorized as (i) lincRNAs, (ii) intronic ncRNAs (incRNAs), (iii) NATs, and (iv) circular long non-coding RNAs (circRNAs) [Cech and Steitz, 2014; Laurent et al., 2015; Sun et al., 2018]. LincRNAs are transcribed from intronic regions and perform *trans* (remote gene) regulatory functions. IncRNAs are produced from intergenic regions and feature poorly spliced, polyadenylated tissue-specific expression [Ulisky et al., 2013; Wang et al., 2014; Wu et al., 2017]. LincRNAs come from areas that exhibit poorly spliced, tissue-specific polyadenylated expression, and are regulated by *trans* (distant gene) whereas incRNAs from intronic regions are transcribed. NATs are *cis*-and *trans*-regulatory proteins that arise from complementary DNA strands of sense coding areas. NATs are derived from additional sensor coding regions DNA strands and include both *cis* and *trans*-regulatory measures [Cech and Steitz, 2014; Yang et al., 2014]. CircRNAs, however, are low in number, arise from the "back splicing process in premRNA" [Zhang et al., 2014a; Wu et al., 2017], have a structure that is covalently closed, and show a greater sequence conservation than linear lncRNAs [Salzman et al., 2012; Zhang et al., 2013].

In recent years, growing evidence has shown that lncRNAs are playing more prominent roles in ribonucleoprotein regulation, chromatin remodeling, transcriptional regulation, RNA processing, and molecular imprinting throughout plant life. We outline the classification of lncRNAs based on their stated functions as decoys, scaffolds, guides, precursors, and signals for siRNAs [Ma et al., 2013]. LncRNAs may act as molecular signals in a variety of biological processes and integrate developmental cues due to their selectivity and variable spatiotemporal expression in response to different biological or environmental stimuli.

10.2.10 Structure of lncRNAs

Secondary structures exist in lncRNAs, which may be required for their function. In most cases, lncRNAs contain two types of functional sites: interaction sites, which are required for sequence-based interactions with structural sites, and RNA-binding proteins, which are necessary for the identification of secondary and/or tertiary structures that direct the interaction of interacting partners [Fabbri et al., 2019]. Among the *COOLAIR* motif sequences involved in vernalization are two right-hand turn motifs and a multi-way junction motif [Hawkes et al., 2016]. These two motifs are among the most well-conserved secondary structures in the *Brassicaceae* family.

10.2.11 Localization of lncRNAs in the Subcellular Space

Translation occurs when mRNAs are transported from the nucleus to the cytoplasm. It has been demonstrated that after processing, lncRNAs may stay in the nucleus or transported to subcellular locations and organelles, such as mitochondria, via ribosome profiling or RNA FISH [Carlevaro-Fita and Johnson, 2019]. In contrast to mRNAs, animal model studies indicate that lncRNAs are more likely to be localized in the nucleus than in the cytoplasm [Derrien et al., 2012]. RNA binding proteins and sequence elements inside lncRNAs impact their cytosolic or nuclear localization, which influences their biological roles and activity [Carlevaro-Fita and Johnson, 2019]. Human lncRNAs with *Arthrobacter luteus* (ALU) elements repeats, for example, are more likely to be maintained in the nucleus due to the binding of particular splicing factors such as heterogeneous nuclear ribonucleoprotein K (HNRNPK) [Lubelsky and Ulitsky, 2018]. In addition to being connected to both mono- and polyribosomal complexes, certain cytosolic lncRNAs may also have a role in the production of short peptides, according to the findings [Hsu et al., 2016]. In mammalian cells, chromatin binds a collection of nuclear lncRNAs, and this localization can be maintained by U1 small nuclear ribonucleoprotein particle (U1 snRNP) [Yin et al., 2020]. Transcription factor (TF) binding or enhancer functioning may be influenced by chromatin-associated lncRNAs. LncRNAs with association may influence TF binding or enhancer functioning [Shlyueva et al., 2014]. Although these animal model systems data show complicated mechanisms behind the distribution of lncRNAs in the cell, there is less knowledge about plant lncRNAs. Many of the lncRNAs that have been discovered in plants (e.g., *DRIR*, *COOLAIR*) are found in the nucleus and act there. For instance, cold-induced *COOLAIR* covers the *FLOWERING LOCUS C* (*FLC*) locus in the nucleus and works to inhibit FLC expression by altering the dynamics of histone modification (H3K36me3) status [Rosa et al., 2016; Wu et al., 2020]. In contrast, *cis*-NAT is found in the cytoplasm of cells that overlap with protein-coding genes, and some of these transcripts have the potential to interfere with the translation of mRNAs [Deforges et al., 2019]. In short, distinct types of subcellular locations indicate diverse molecular methods of action of lncRNAs in the transcriptional and post-transcription regulation of gene expression.

10.2.12 lncRNA Degradation

It has been shown that the half-lives of lncRNAs are usually shorter than those of mRNAs, which indicates that the lncRNA metabolism in plants is complexly regulated [Szabo et al., 2020]. LncRNAs are inefficiently produced and decay rapidly [Mukherjee et al., 2017]. In plants, lncRNAs may be degraded via 3′–5′ exonucleolytic degradation via the nuclear exosome, as well as 5′–3′ exonucleolytic degradation via exonucleases such XRN3 and XRN2 [Kurihara et al., 2012]. A collection of "promoter upstream transcripts" (PROMPT) and "cryptic unstable transcripts" (CUT)-like specialized lncRNAs were found in exosome subunits mutated from the TSSs of mRNAs [Chekanova, 2015; Thieffry et al., 2020]. Exosome-regulated lncRNAs appear to control enhancer activity in humans, with the exosome resolving detrimental R-loop topologies [Pefanis et al., 2015; Nair et al., 2020]. In plants, the nonsense-mediated mRNA decay (NMD) pathway monitors the quality of plant lncRNAs, much as it does for mRNAs [Kurihara et al., 2009; Drechsel et al., 2013]. Contrary to expectations, mutants of the up-frameshift (upf) gene, which are deficient in a component of the NMD process, accumulate excessive amounts of transcripts generated from intergenic regions and antisense transcription [Kurihara et al., 2009]. This indicates that lncRNA stability is tightly regulated through a variety of molecular regulatory pathways.

10.3 UNDERSTAND THE ROLES OF ABIOTIC STRESS-RESPONSIVE lncRNAs

Abiotic stress including salinity, drought, heat, and low temperature has a significant effect on plant growth and productivity. Due to their widespread nature and devastating effects on plant development and yield, abiotic stress has become a serious problem. However, plants have evolved complex mechanisms for sensing and responding to environmental changes. In recent years, several protein-coding genes have been discovered to monitor the reactions of the plants to abiotic stress; nevertheless, our understanding of the regulative processes involved in this response remains restricted [Borah et al., 2017]. Plants rely on these post-transcriptional regulators to restore and re-establish cellular homeostasis during stress and recovery stages. Plants assign lncRNAs as essential post-transcriptional gene-expression regulators to suppress plant growth and development under stress conditions, according to recent research, albeit how this is accomplished at the molecular level remains unknown. LncRNAs influence gene expression via directing translational repression, chromatin remodeling, mRNA cleavage, and/or DNA methylation [Kim and Sung, 2012]. Several studies have verified that the aberrant expressions of lncRNAs in plants are induced by abiotic stress. In recent years, whole genome sequencing and computational methods have been utilized to uncover a substantial number of stress-related lncRNAs. Several lncRNAs, as well as their potential target gene(s) and functioning mechanisms (summarized in Table 10.2), have been discovered as

TABLE 10.2
Summary of Plant lncRNA Involved in Response to Different Abiotic Stress

S. No.	Crops/Species	LncRNA	Target Gene	Reference
		Heat Stress		
1.	*Brassica juncea*	TCONS_00051908	–	Bhatia et al., 2019
		TCONS_00088973	–	
2.	*Brassica rapa* (Chinese cabbage, GHA: heat sensitive variety and XK: heat tolerant variety)	TCONS_00048391 and NAC1 (Bra030820)	–	Wang et al., 2019
		TCONS_00010856 and Bra021232	–	
		TCONS_00004594	–	
3.	*Cucumis sativus* (cucumber)	TCONS_00031790	–	He et al., 2020
		TCONS_00014332	–	
4.	*Triticum aestivum* (TAM107, Chinese spring)	TahlnRNA27	–	Xin et al., 2011
		TalnRNA5	–	
		TahlnRNA12	–	
		TalnRNA21	–	
		TahlnRNA23 and TahlnRNA29	–	
		Cold Stress		
5.	*Arabidopsis thaliana* (Col-0)	*SVALKA*	*CBF1*	Kim and Sung, 2017; Kindgren et al., 2018; Zhao et al., 2018 and Calixto et al., 2019
		COLDWRAP	*FLC*	
		TAS1a	–	
		MAS	*MAF4* gene	
6.	*Brachypodium distachyon*	BdCOOLAIR1	BdODDSOC1,	Jiao et al., 2019
		BdCOOLAIR2	BdODDSOC2	
7.	*B. rapa* (Chinese cabbage, RJKB-T24)	*MSTRG.259 and MSTRG.491*	–	Shea et al., 2019
		MSTRG.17153	–	
		MSTRG21908	related to vernalization	
		MSTRG.4795 and MSTRG.18513	*BrFLC* and *BrMAF* genes	
8.	*Manihot esculenta* (Cassava; TMS60444)	lincRNA159	NAM, ATAF1/2 and CUC2	Li et al., 2017
9.	*Medicago truncatula* (Jemalong A17)	lncRNA MtCIR1	MtCBF genes	Zhao et al., 2020a
10.	*Vitis vinifera* (Cabernet Sauvignon)	*VIT_ 203s0017n00360*	–	Wang et al., 2019
		VIT_ 201s0011n00530	*VIT_ 208s0058g00960*	
		VIT_200s0225n00020	*VIT_ 202s0154g00610*	
		VIT_ 207s0031n00070	*VIT_ 216s0100g00380*	
		VIT_ 209s0002n00340	*VIT_ 215s0046g02110*	
		VIT_ 213s0067n00110	*VIT_ 200s0246g00150*	
		VIT_ 213s0158n00020	*VIT_ 202s0025g01280*	
		Drought Stress		
11.	*Brassica napus* (Qinyou 8 [drought-sensitive] and Q2 [drought-tolerant])	*XLOC_044363*, *XLOC_052298* and *XLOC_076449*	–	Tan et al., 2020
		XLOC_042431, *XLOC_074677* and *XLOC_ 093758*	*BnaC06g05090D* and *BnaC02g25200D*	
		XLOC_019521 and *XLOC_ 042894*	*BnaC07g44670D*	
		XLOC_075476 and *XLOC_074677,*	*BnaC02g25020D, BnaC02g25150D,*	
		XLOC_ 095305 and XLOC_100682	*BnaA01g17750D*	

(Continued)

TABLE 10.2 (*Continued*)
Summary of Plant lncRNA Involved in Response to Different Abiotic Stress

S. No.	Crops/Species	LncRNA	Target Gene	Reference
12.	*Cleistogenes songorica*	*MSTRG.43964.1* and *MSTRG.4400.2*	Genes related to abscisic acid (ABA)	Yan et al., 2019
13.	*M. esculenta* (Cassava; TMS60444 and Ku50)	TCONS_00149293, TCONS_00003360 and TCONS_00015102	–	Li et al., 2017; Ding et al., 2019; Wu et al., 2019
		lincRNA340	Nuclear factor Y (NF-Y)	
		TCONS_00060863 and TCONS_00068353	*CYP707A1*	
		TCONS_00069665, TCONS_00097416 and *TCONS_00040721*	*CSLD5, DOX1, ERL1, GRF1, HDG11, HB51, LAX2, SPCH* and *SCR*	
14.	*Oryza sativa* (Ilmi and DXWR)	*NAT Os02g0180800–01*	cinnamoyl-CoA reductase, *Os02g0180700–01*	Chung et al., 2016; Weidong and Hongping, 2020
		NAT Os02g0250700–01	late embryogenesis abundant protein, *Os02g0250600–01*	
		lncRNAs MSTRG69391 lncRNAs MSTRG65848	Transcription factor, calmodulin protein gene etc.	
		MSTRG41712 and MSTRG68635	HSP genes and mitochondrial carrier	
15.	*Panicum virgatum* L (Alamo)	*XLOC_053020*	Pavir.Ia01153	Zhang et al., 2018
		XLOC_014465	Pavir.Bb00347	
		XLOC_033252	Pavir.Eb01847	
		XLOC_01692, XLOC_090250, and *XLOC_ 067866*	Pavir.J23169 and Pavir. Ca01179	
16.	*Populus trichocarpa* (Nisqually 1)	lincRNA20 and lincRNA2752	–	Shuai et al., 2014
		lincRNA2962 and lincRNA1039	–	
17.	*T. aestivum* L. (Kiziltan, TR39477and TTD-22)	*c70772_g2_i1 and c90557_g1_i1*	*c69036_g1_i1* and c9653_g1_i2	Cagirici et al., 2017
18.	*Zea mays* (B73)	lncRNA MSTRG6838.1	V-ATPase encoding gene, vpp4	Pang et al., 2019
		Salinity Stress		
19.	*A. thaliana*	*DRIR*	ANNAT7, NAC3 and *WRKY8*	Qin et al., 2017
20.	*Gossypium hirsutum* (SN91–11)	lnc_388, lnc_883, lnc_973 and lnc_ 253	*Gh_A09G1182,* and *Gh_ D03G0339 genes*	Deng et al., 2018; Zhang et al., 2019
		lncRNA973	*SOD, CAT, POD and P5CS,* *RBOHB, RBOHD, NHX7, MYB5, WRKY46, ERF62, NAC29*	

(*Continued*)

TABLE 10.2 (*Continued*)
Summary of Plant lncRNA Involved in Response to Different Abiotic Stress

S. No.	Crops/Species	LncRNA	Target Gene	Reference
21.	*M. truncatula* (Jemalong A17)	TCONS_ 00097188 TCONS_00046739 TCONS_00100258 and TCONS_ 00118328 TCONS_00047650 TCONS_ 00020253 and TCONS_00116877	*Medtr6g006990, cytochrome P450* *Transmembrane proteins gene* *Medtr3g069280, Medtr1g081900* and *Medtr7g094600*	Wang et al., 2015a
22.	*Pistacia vera* L. (Sarakhs and Ghazvini)	5 NAT-lncRNAs	*CERK1, LEA, Laccase genes*	Jannesar et al., 2020
		Nutrient Deficiency		
	Nitrogen starvation			
23.	*A. thaliana*	*trans-acting siRNA3* (*TAS3*)	NRT2.4 SEC14p-like protein/Patellin-4, Nitrate transporter 2	Fukuda et al., 2019
24.	*Hordeum vulgare* (landrace B968, Liuzhutouzidamai)	*lnc00090* and *lnc000248*	–	Chen et al., 2020
25.	*O. sativa* cv. nipponbare	$cis\text{-}NAT_{AMT1.1}$, *cis- NATAMT1.2*	*AMT1.1* and *AMT1.2*	Shin et al., 2018
		Phosphate Starvation		
26.	*A. thaliana*	*XLOC_001691, XLOC_013661* and *XLOC_020833*	PHOTOLYASE 1 (PHR1)	Yuan et al., 2016
27.	*M. truncatula* (Jemalong A17)	*PDIL1*, *PDIL2* and *PDIL3*	$MtPHO_2$ and Medtr1g074930	Wang et al., 2017c

contributing to various abiotic stress responses in diverse agronomically important crops. These results show that lncRNAs might serve as possible targets for genetic manipulations to engineering abiotic stress tolerance in various crop plants.

10.3.1 Heat-Stress-Responsive lncRNAs

Heat stress is one of the crucial abiotic stresses that greatly influences plant growth at the morphological, physiological, and biochemical levels. Phenological changes in plants and adaptation mechanisms severely influence agricultural output as global temperatures rise, posing a threat to global food security. Plants have evolved dynamic responses at morphological, physiological, and biochemical levels that enable them to withstand adverse environmental circumstances. Phytohormones, metabolites, antioxidant enzymes, and heat shock proteins (HSPs)/chaperones, are all produced by plants and play a role in adapting to heat stress [Jha et al., 2017]. Research findings show that lncRNAs may serve as a crucial molecular regulatory layer, and they offer an understanding of how plants respond to heat stress. Using the Solexa sequencing and Affymetrix wheat GeneChip-based microarray, 54 potential heat-stress-responsive lncRNAs were discovered in wheat to investigate their role in conferring response against the abiotic stress [Xin et al., 2011]. Four miRNA precursors (viz., *miR2004*, *miR2066,* and *Ta-miR2010,*) and 26 siRNA precursors were found among the lncRNA transcripts identified. The qRT-PCR study verified the up-regulation of lncRNAs *TalnRNA5* and *TahlnRNA27*, as well as their target miR2004 and miR2010, demonstrating their importance in the heat stress effect on wheat. However, regulation of *TalnRNA5* shows higher tissue-specific expressions in seeds when exposed to the heat-stress response [Xin et al., 2011]. The majority of the lncRNA sequences are poorly conserved and their expression is tissue-specific, varies at different developmental stages, and is even specific to a particular plant species [Swiezewski et al., 2009]. Under drought stress, tissue/development-specific expression of lncRNAs has been observed in *Zea mays* L. [Huanca-Mamani et al., 2018; Pang et al., 2019] and *M. esculenta* [Ding et al., 2019]. In diverse crops, a multitude of differentially expressed lncRNAs, as well as their corresponding heat-stress-regulated miRNAs and target protein-coding genes, have been reported [Sun et al., 2013; Wang et al., 2019a]. In *Brassica rapa* under heat stress, Wang et al., 2019b analyzed the regulation of lncRNAs and differentially expressed genes (DEG)s involved in jasmonic acid (JA), Ethylene (ET), salicylic acid (SA), ABA, auxin and brassinosteroid (BRs) phytohormone signaling pathways, as well as DEGs involved in the regulation of various HSPs across the genome, using strand-specific RNA-seq. Further, qRT-PCR validated the *cis*-regulatory expression of heat-responsive, DElncRNA*TCONS_00004594* downstream to the protein-coding gene *Bra021232* [Wang et al., 2019b]. However, under heat stress, lncRNAs *TCONS_00048391* and *TCONS_00010856* also served as eTMs for bra-miR164a. Thus, heat tolerance in "XK" variety displayed down-regulation of lncRNA*TCONS_00048391* and the target *Bra030820* gene and up-regulation of bra-miR164a. Similarly in *Cleistogenes songorica*, drought-specific lncRNA *MSTRG.42613.1* was discovered to be the target mimic of conserved miRNA164, which is regulated under drought stress [Yan et al., 2019].

10.3.2 Cold-Stress-Responsive lncRNAs

Plant development and yield are challenged by low-temperature stress, which is a major abiotic stress [Jha et al., 2017; Zhang et al., 2019]. Plants are orchestrated into a variety of sophisticated regulatory networks, including the cold-regulated genes (CORs) and C-repeated binding factors (CBFs), which are regulated by miRNA. Several Quantitative trait loci (QTLs) and possible candidate genes have been identified responsive to cold tolerance in plants due to advancement in genetic and genomic techniques [Jha et al., 2017]. *A. thaliana* [Swiezewski et al., 2009; Shea et al., 2019], *Manihot esculenta* [Li et al., 2017], *Musa acuminata*, *Vitis vinifera* [Wang et al., 2019], *B. rapa* [Shea et al., 2019; Corona-Gomez et al., 2020], and *Brachypodium distachyon* [Jiao et al., 2019] have all demonstrated emerging evidence of lncRNAs regulating cold stress modulating cold adaptation. Plant species that are well-adapted to cold regions practice vernalization, a naturally occurring process that inhibits flowering throughout the period of vegetative growth during the winter and enables flowering in the reproductive phase under optimal circumstances [Kim et al., 2009]. In *A. thalania, FLC* epigenetically regulates the locus that controls flowering time and suppresses flowering during cold stress [Whittaker and Dean, 2017, Bastow et al., 2004]. During vernalization, there is chromatin modification (reduced active histone mark H3K36me3 and improved repressive histone mark H3K27me3) through Polycomb-mediated epigenetic regulation in the *FLC* locus, which is regulated by lncRNAs, thus, inhibiting expression of the FLC under cold stress [Swiezewski et al., 2009; Sun et al., 2013; Csorba et al., 2014]. In *Arabis lyrata*, *A. alpina*, and *A. thaliana*, class I antisense COOLAIR (COLD ASSISTED INTRONIC NON-CODING RNA) that control *FLC* repression through vernalization are evolutionarily conserved [Castaings et al., 2014]. However, *COLDAIR*, transcribed from intron1 of the *FLC* gene recruits the Polycomb Repressive Complex 2 (PRC2) that leads to chromatin modification (increase H3K27me3) of the FLC locus thereby repressing its expression [Heo and Sung, 2011. *COLD-WRAP*, a Polycomb-binding lncRNA, further acts in a co-operative manner to stabilize the repression of the *FLC* locus in *Arabidopsis* during vernalization [Kim and

Sung, 2017]. Furthermore Kindergren et al. recognized the unique function of cryptic antisense CBF1 (*asCBF1*) and *SVALKA* lncRNAs in *Arabidopsis* induced under cold stress. RNA polymerase II (RNAPII) collision, generated from lncRNAs *SVALKA* and *asCBF1*, suppresses transcription of the *CBF1* gene, regulating cold acclimatization. Similarly, to investigate the role of lncRNAs in regulating the cold-stress response, strand-specific RNA-sequencing (ssRNA-seq) revealed 2460 lincRNAs and 4050 NAT lncRNAs identified in *Arabidopsis* as cold-responsive lncRNAs [Zhao et al., 2018]. Zhao et al. used the complex WDR5a, which deposits H3K4me3 at the MAF4 gene for activation, to understand the role of a cis-acting NAT lncRNA, MAS (NAT lncRNA_2962), transcribing matching cold-responsive MADS AFFECTING FLOWERING 4 (MAF4, an FLC family member). Under cold conditions, the active gene eventually inhibits flowering. Additionally, in *B. rapa*, three *FLC* paralogs have been identified as floral repressors during vernalization [Itabashi et al., 2018; Shea et al., 2018]. Under cold stress, NATs have been found to interact with the *FLC2* locus of *B. rapa* [Li et al., 2016]. A 2088 *B. rapa* lncRnA was identified through RNA-seq-driven transcriptome analysis, with three *BrFLC* loci contributing to the control of cold-stress-only *BrFLC2*, which harbored NAT *BrFLC2as* (*MSTRG.2765*) and had the homology to *A. thalian* COOLAIR transcript, which has been up-regulated under cold stress [Shea et al., 2019]. COOLAIR functions as a "*cis*-NAT with regard to the *AtFLC* locus" [Csorba et al., 2014], but the role of *BrFLC2as* a *cis*- or *trans*-acting mode requires more research. In *B. rapa*, at various developmental stages under cold stress, differentially expressed lincRNAs (NATs) identified as lncRNAs *MSTRG.21908*, *MSTRG.18513*, and *MSTRG.4795* were up-regulated and *MSTRG.17153*, *MSTRG.491*, and *MSTRG.259* were down-regulated [Shea et al., 2019]. RNA-seq analysis revealed 182 novel down-regulated lncRNAs, 284 novel up-regulated lncRNAs, 242 differentially expressed lncRNA (DElncRNAs) targeting 326 protein-coding genes, and different stress-responsive genes such as late embryogenesis-abundant protein genes, WRKY TF genes, and CBF4 TF genes in a genome-wide search for cold-responsive lncRNAs in *Vitis vinifera* [Wang et al., 2019]. Under cold-stress tolerance in *V. vinifera,* qRT-PCR confirmed down-regulation of *VIT_209s0002n00020*, *VIT_208s0007n00270*, and *VIT_201s0010n00070* and up-regulation of *VIT_207s0005n0048*, *VIT_207s0141n00070*, and *VIT_200s0179n00030*, which confirms the role of the selected lncRNAs [Wang et al., 2019]. In the cassava, genome wide 318 lncRNAs were identified under cold and drought stress [Li et al., 2017]. *LincRNA207, 234*, and *419* were up-regulated under cold stress, which was further validated through q-RT-PCR. Under cold stress, *lincRNA159* serving as a target mimic for miR164, reduced the expression of NAC genes, which was used to unravel the regulatory network of stress-responsive gene, miRNAs, and lncRNAs that govern cold tolerance [Li et al., 2017]. Alternative splicing of primary microRNA (pri-miRNAs) and lncRNAs could participate in regulating the cold-stress response in *Arabidopsis.* The induction of TAS1a lncRNA, controlled by alternative splicing under cool stress, was discovered in *Arabidopsis* among the 135 lncRNAs identified with differential alternative splicing and cold-dependent differential expression. The unspliced AT2G27400.1 intron-containing transcript from *TAS1a* had "miR173binding site and the generating site for *tasiRNAs*," but the AT2G27400 ID1-spliced transcript remained intronless. As the temperature drops, the transcript of AT2G27400 ID1 decreased during the first 6 hours following cold stress, whereas unspliced AT2G27400.1 increased quickly during the first 3 hours. It then began to decrease over the next 12 hours [Calixto et al., 2019]. Alternative splicing of lncRNAs, therefore, plays a major role in controlling the tolerance to cold stress.

10.3.3 Drought-Stress-Responsive lncRNAs

Drought stress-related events are becoming more prominent throughout the world as a result of unpredictable rainfall patterns, which impairs plant development and represent a major threat to global food security. To respond to drought stress, plants have numerous biochemical, physiological, and complex molecular networks, including signal transduction cascades [Lamaoui et al., 2018]. The underlying genes/QTL, numerous complex gene regulatory networks, and molecular signaling cascades that govern the plant drought-stress response were already identified by advancements in molecular biology [Gupta et al., 2017]. Following that, the identification of drought-stress-responsive miRNAs and their potential target genes in a variety of crop plants unraveled the molecular processes underlying drought stress adaption. Based on a gene co-expression network analysis that comprises miRNAs, lncRNAs, TFs, and protein-coding genes, recent evidence suggests that lncRNAs play an important role in plants under drought stress [Zhang et al., 2018; Ding et al., 2019; Pang et al., 2019]. A significant number of drought-responsive lncRNAs have been identified in different crop species as a result of the widespread application of high-throughput sequencing technologies in agriculture. There are 6 in *Arabidopsis*, 1597 in *Panicum virgatum*, 664 in *Z. mays*, 504 in Populus spp., 98 in *O. sativa*, 185 in *M. esculenta*, and 19 in *Setaria italica*, among other plant species [Ben Amor et al., 2009; Qi et al., 2013; Shuai et al., 2014; Zhang et al., 2014b; Chung et al., 2016; Li et al., 2017; Zhang et al., 2018]. LncRNAs may have an impact on the drought-stress response via enlisting multiple processes based on antisense transcription-mediated regulation, eTM, chromatin modification, or direct influence on the transcription of genes responsive to drought stress [Li et al., 2017; Wang et al., 2017; Yan et al., 2019; Weidong and Hongping, 2020]. A comprehensive *Panicum italicum* L. (foxtail millet) sequencing revealed 584 lncRNAs, 17 of which displayed differentiation during drought stress, together with 2 NAT-lncRNAs. Among the drought-reactive lncRNAs in foxtail millet, only one exhibited significant similarity to the drought-reactive lncRNA in sorghum, demonstrating the poor conservation of lncRNAs in the plant kingdom [Qi et al., 2013]. RNA-seq study of *Populus trichocarpa* revealed a complete landscape of >2500 lncRNAs, of which 504 were drought-responsive, as part of a systematic approach. Functional validation of 8 drought-responsive lncRNAs from a pool of 504 drought-response lncRNAs using RT-qPCR showed that 6 lincRNAs were up-regulated and 2 lincRNAs were down-regulated under water stress, confirming previous findings [Shuai et al., 2014]. Drought-responsive lncRNA genes were identified in the *M. esculenta* genome using ssRNA-seq data. A total of 153 NAT lncRNAs and 318 lncRNAs were detected in response to both cold and drought stress [Li et al., 2017]. Of the 51 lncRNAs expressed differently (DElncRNAs), 40 were shown to be excessively regulated in the event of drought stress. Under drought stress, qRT-PCR was performed for functional validation of the selected lincRNAs and revealed that *lincRNA101*, *lincRNA356,* and *lincRNA391* were up-regulated while *lincRNA64*, *lincRNA182*, *lincRNA350*, and *lincRNA392*, were down-regulated. In addition, the use of the target mimic mechanism has enhanced *lincRNA340* expression under drought, reducing target miR169 activity and eventually increasing the expression of the gene *nuclear transcription factor Y* (*NF-Y*) [Li et al., 2017]. Under drought stress, Ding et al., found 124 DElncRNAs in *M. esculenta*, 11 of which acted as target mimics for miR164, miR169, miR156, and miR172 [Ding et al., 2019]. *TCONS 00068353* lncRNA was functionally validated as a target mimic for miR172c and miR156k, which regulated several abiotic stress-responsive genes, whereas *TCONS 00097416* and *TCONS 00060863* lncRNAs were involved in regulating the ET and ABA signaling pathways, respectively, during drought stress [Ding et al., 2019]. According to a study, 1535 lncRNA were discovered in genome-wide research in *Z. mays* L. at different developmental stages [Pang et al., 2019]. At the R1 phase (silk phase), the lncRNAs had a key function in resistance to drought stress. Additionally, the V-ATPase encoding gene (vpp4) was discovered as a potential target gene for *lncRNAMSTRG.6838.1*; *vpp4* and the detected lncRNA may work as *cis*-acting pairs. LncRNAs might control the transcription of numerous drought-stress-inducible genes in addition to functioning as eTM or NAT. Using RNA-seq analysis, 98 drought-responsive lncRNAs were found in *O. sativa* L., including two significant drought-responsive lncRNAs, NAT *Os02g0180800-01* (targeting *cinnamoyl-CoA reductase* gene) and NAT *Os02g0250700-01* (targeting *late embryogenesis abundant protein* gene)

[Chung et al., 2016]. The expression of these two lncRNAs remained negatively associated with their respective target genes. *DROUGHT INDUCED lncRNA* (*DRIR*), a novel ncRNA localized in the nucleus responsible for regulating drought-stress response, has been extensively studied in *A. thaliana*. This *DRIR* consists of a 755 nt long nuclear localized lincRNA that regulates various drought-induced genes, including *NIP1* and *TIP4* (aquaporin), *RD29A*, *RD29B*, *ABI5*, and *P5CS1*(ABA-signaling), *ANNAT7* (*annexin*), *FUT4* (*FUCOSYL TRANSFERASE4*), and *WARKY8* and *NAC3*, and TF at the transcription level [Qin et al., 2017]. The *Arabidopsis* line with *drir*^{D} mutation (T-DNA insertion) and *DRIR* overexpression displayed higher drought tolerance than wild-type, as evident by the higher-fold expression of these genes. As a result of its role as a positive regulator, the lncRNA *DRIR* imparted water-deficit stress resistance. LncRNAs involved in regulating drought responsive genes involved starch and sucrose synthesis, ET and ABA synthesis, and calcium signaling, have been found in *C. songorica*, *M. esculenta*, *O. sativa*, *P. virgatum* and *P. betulifolia* [Zhang et al., 2018; Wang et al., 2018b; Ding et al., 2019; Wu et al., 2019; Weidong and Hongping,, 2020]. *P. virgatum* L. under drought 441 DElncRNAs are expressed at different developmental stages. These few lncRNAs, such as *XLOC_074836* regulating ET signaling, *XLOC_053020*, *XLOC_033252*, and *XLOC_014465* controlling ABA synthesis, and *XLOC_005809* involved in trehalose phosphate synthase metabolism, are significantly up-regulated in drought stress, showing that they have a key influence in tolerance [Zhang et al., 2018].

10.3.4 Salinity Stress-Responsive lncRNAs

The uncontrolled practice of overirrigating agricultural land and the fast depletion of groundwater are two key causes linked to the global rise in salinity-related issues. Worldwide salinity stress threatens 45 million hectares of agricultural lands and 32 million hectares of hardy land worldwide [FAO, 2015]. As a result, soil salinization continues to be a major limitation to world food production. Plants are subjected to an excessive load of harmful ions when subjected to salinity stress, which decreases plant growth and development as well as grain yield [Ismail and Horie, 2017]. To respond to salt stress, plants have developed a number of cellular and physiological processes. On the molecular level, the controls on salinity stress in different plants include an abundance of ion transporting proteins encoded by gene(s)/QTLs and other regulatory genes [Ismail and Horie, 2017]. Also, evidence of lncRNAs regulating functions that allow plants to survive salt stress has helped us to better understand the molecular processes for controlling the salinity stress response [Qin et al., 2017; Ben Amor et al., 2009]. LncRNAs, including *TCONS_00020253* (up-regulating expression of Na+/H+ exchanger gene *Medtr1g081900* in roots), *TCONS_00097188* (regulating photosynthesis by up-regulating *Medtr6g006990* gene), *TCONS_00116877* (up-regulating the *Medtr7g094600* gene encoding glutathione peroxidase in roots), and *TCONS_00046739* (regulating cytochrome P450 in roots), were identified in regulating osmotic stress and salinity stress in *M. truncatula* under [Wang et al., 2015]. In *A. thaliana*, the nuclear localized lncRNA *DRIR* regulates the expression of the *Atrboh*B, *ANNAT7*, *FUT4*, *NAC3*, *P5CS1*, *RD29A*, and *RD29B* genes that regulate tolerance to salt stress. Salt and drought tolerance were increased in *A. thaliana* overexpressing the *DRIR* gene, as well as higher survival rates when exposed to salt- and drought-stress conditions [Qin et al., 2017]. Deep transcriptome sequencing of *Gossypium hirsutum* salt-treated leaf tissue revealed 44 differentially expressed lincRNAs from 1117 unique lncRNAs. Functional validation of selected lincRNAs under salinity stress revealed the up-regulation of lnc_883lncRNA targeting on *Gh_D03G0339MS* (channel protein-coding gene) and *lnc_388* on (*cis*-acting) target on *Gh_A09G1182* (leucine-rich repeat 8 gene) [Deng et al., 2018]. Under salt stress, Deng et al., 2018 further verified the involvement of lncRNAs *lnc_253* and *lnc_973* in acting as a target mimic for ghr-156e and ghr-miR399. In addition, *de novo* transcriptome analysis of leaf and root tissue-under combined salinity and boron stress identified 1710 lncRNAs in *Z. mays* L. cv. Amylacea (Lluteño landrace). Under combined salinity and boron stress, 848 differentially expressed *trans*-NAT lncRNAs were expressed and 12 *trans*-NAT lncRNAs were validated through q-PCR that regulate numerous stress regulatory gene expressions [Huanca-Mamani et al., 2018].

10.3.5 Nutrient Deficiency Stress-Responsive lncRNAs

The absorption of soil nutrients continues to be an essential physiological mechanism for controlling plant growth and development [Islam, 2019]. Plant nutrient homeostasis is mediated by a number of molecular pathways, including numerous nutrient transporters [Wang et al., 2018]. Many plants have evidence of both lncRNAs and miRNAs that are capable of regulating nutrient absorption [Franco-Zorrilla et al., 2007; Shin et al., 2018; Wang et al., 2018]. Phosphorus (P) is a key component of plant development and growth among the main nutrients, and is also its one of the major sources for ATP production [Adnan et al., 2018]. Several factors limit the availability of inorganic phosphate (Pi) from soil, which limits the overall plant growth and development [Puga et al., 2017]. Furthermore, the intricate regulatory network of P homeostasis in plants is poorly understood [Wang et al., 2017, 2018]. Plants are activated to enhance the availability of soil-inorganic phosphates and to raise the efficiency of phosphorus usage (PUE) through several molecular and biochemical processes [Puga et al., 2017; Wang et al., 2017; Wang et al., 2018]. The involvement of miRNAs in regulating the availability of phosphates has been described in several plants in details [Song et al., 2019]. Similarly, in *A. thaliana* [Franco-Zorrilla et al., 2007], *O. sativa* [Jabnoune et al., 2013; Shin et al., 2018], and the legume *M. truncatula* [Wang et al., 2017], the increasing function of lncRNAs in controlling phosphate concentration is well documented. The mechanism by which miR399 and its target gene *PHOSPHATE2* (*PHO2*) regulate phosphate concentration in *Arabidopsis* is well understood [Chiou et al., 2006; Huang et al., 2013]. In *Arabidopsis, lncRNA INDUCED BY PHOSPHATE STARVATION1* (*IPS1*) acts as an eTM or decoy for miR399 and regulates the *PHO2* gene expression and phosphate homeostasis [Franco-Zorrilla et al., 2007]. Furthermore, in the presence of high phosphate, lncRNA *IPS1* interacts with miR399 and inhibits it from targeting the *PHO2* gene, likely eliminating the phosphate transporters' functional role by ubiquitination pathway, thereby preventing root uptake of excessive phosphate [Bari et al., 2006]. The *PHO2* gene is repressed under phosphate-deficient circumstances because miR399 destroys the *PHO2* transcript, allowing phosphate transporters to collect phosphate [Chiou et al., 2006]. Additionally, in *M. truncatula*, the phosphate regulation mechanism has been demonstrated, which is based on the "*PHR1*–miR399–*PHO2*" pathway in conjunction with the phosphate deficiency-responsive lncRNA*PDIL1*, a paralog of Mt4 [Wang et al., 2017]. The lncRNAs*PDIL2* and *PDIL3* negatively regulats the expression of *Medtr1g074930*, the phosphate transporter gene. In *O. sativa*, lncRNA *cis-NATPHO1;2* acts as a translational enhancer of the gene *PHO1;2* to maintain phosphate homeostasis [Jabnoune et al., 2013]. Nitrogen (N), like phosphate P, acts as an important nutrient for regulation of plant growth and development as well as a N source for ATP and amino acids [Hawkesford and Griffiths, 2019]. Several QTLs allegedly increase efficiency of N utilization (NUE) in different agricultural crops [Hawkesford and Griffiths, 2019]. Along with QTL, there are several regulatory genes and transporter genes influencing NUE in agricultural plants have been discovered with advances in functional genomics techniques [Curci et al., 2017]. In plants, however, the full molecular process of N absorption is unknown [Simons et al., 2014]. We now have a better knowledge of how N-responsive lncRNAs contribute to plant N homeostasis thanks to cutting-edge deep transcriptome sequencing using RNA-seq. Diverse N-responsive lncRNAs have been discovered in a variety of plant species, including rice, maize, and poplar [Chen et al., 2016; Lv et al., 2016; Shin et al., 2018]. In *O. sativa*, the *Ammonium transporter 1* (*AMT1*) gene for N homeostasis is regulated by lncRNAs *cis-NAT*$_{AMT1.1}$ and *cis-NAT*$_{AMT1.2}$ is a well-known mechanism [Shin et al., 2018]. In *A. thaliana* under N-starved conditions, a key lincRNA was identified as *trans-siRNA3* (*TAS3*) that targets the *nitrate transporter 2* (NRT2) gene, thereby regulating transport of N under nutrient-deficient conditions [Fukuda et al., 2019]. One of the most significant micronutrients in plants is boron (B), which is essential for plant growth and development as well as cell-wall synthesis and membrane integrity. In addition, 729 up-regulated and 721 down-regulated lncRNAs with stress from B-deficient conditions have been discovered using expression profiling. Under B-deficient stress in *Poncirus trifoliata*, functional validation of chosen lncRNAs revealed target genes implicated in calcium signaling and phytohormone signal transduction pathways [Zhou et al., 2019].

10.4 LNCRNAS: CHALLENGES IN DECODING THEIR FUNCTIONS

We are currently in the early stages of discovering plant lncRNAs and their functional processes. Despite that high-throughput sequencing has identified several plant lncRNAs with unknown activities, not all of them are functional. Until now, the low-throughput molecular biology and biochemical functional assays have only been used to investigate a limited number of lncRNAs. In fact, the most recent update of the EVLncRNAs database (https://www.sdklab-biophysics-dzu.net/EVLncRNAs2/) found only 506 functional lncRNAs and 41 interactions of lncRNA macromolecules in 56 plant species (Zhou et al., 2020b). These data represent just a small portion of what may be discovered. For the vast amount of lncRNAs with unknown structures and functions that exist, a significant amount of further research is required. Thus, deciphering the structures, functions, and mechanisms of plant ncRNAs, particularly lncRNAs, remains a major challenge.

Although RNA-seq technology is one of the principal tools for the unbiased profiling of transcriptomes in complex organisms, it is still a long way from reaching the transcriptome's limit, since the majority of lncRNAs are expressed at extremely low levels [Mercer et al., 2011]. Because of the development of novel techniques, such as RNA capture sequencing (CaptureSeq) [Mercer et al., 2011] and single-cell transcriptome profiling [Islam et al., 2011], the likelihood of identifying rare transcriptional events occurring in particular genomic areas or cell types of interest has risen significantly. The continuous improvement of the direct RNA-seq technique where a single RNA molecule is directly sequenced without a previous conversion to cDNA promises a decrease in the number of artifacts/noises associated with existing RNA-seq technology during cDNA synthesis [Ozsolak and Milos, 2011]. Further technological innovation and revolution are necessary nevertheless to regularly detect uncommon transcript in animals, and especially plants, on a genome-wide scale. The prediction of ORFs is currently the only way to distinguish between mRNAs and lncRNAs. However, there are a number of things to be considered. The primary factor determining the overall accuracy of ORF prediction is the full-length status of the input sequences. Second, there are certain transcripts, such as *Enod40*, that do not include lengthy ORFs, but do contain short ORF(s) that have the ability to produce small peptides [Rohrig et al., 2002]. When working with such transcripts, it is not easy to ascribe a function to the short ORF(s) or to an individual RNA molecule. Though the functionality of short ORFs can be assessed by using evolutionary conservation of amino acid sequences, and the intracellular localization of an RNA transcript offers information on its functionality, the most effective solution is to conduct biochemical experiments on the molecule or peptide to determine their activity [Kageyama et al., 2011]. Third, some protein-coding RNAs may have additional structural functions unrelated to their translation product, as demonstrated by the identification of a class of coding transcripts that contribute to the maintenance of a decondensed and biologically active interphase chromatin conformation in human and mouse cells via their RNA molecules [Caudron-Herger et al., 2011]. Thus, when it comes to defining the functional properties of single transcripts, the presence of an ORF does not exclude the possibility of additional regulatory activities at the RNA level, and vice versa [Solda et al., 2009]. While it has been shown that a variety of lncRNAs guide chromatin modification complexes to particular genomic loci, the precise nature of the chromatin binding sites has not yet been identified. Chromatin Isolation by RNA purification (ChiRP-Seq) is a novel method for genome-wide analysis of chromatin-binding sites [Chu et al., 2011]. Also using this approach on three lncRNAs showed that lncRNA binding sites are focal, abundant, and sequence-specific, similar to TF binding sites

[Chu et al., 2011]. However, with the advent of genome editing technology, the task of identification will become less challenging.

Most protein-coding genes can be effectively knocked down using siRNA, RNA interference (RNAi), or short hairpin RNAs (shRNA), which is less successful for lncRNAs, owing to their location and expression [Gutschner et al., 2011; Maamar et al., 2013]. The Clustered regularly interspaced palindromic repeats (CRISPR)/ CRISPR-associated (Cas) proteins 9 system has emerged as a powerful and flexible tool for manipulating a range of genomic components, including lncRNAs; nevertheless, some limitations exist [Aparicio-Prat et al., 2015]. Because insertion-deletion (indel) mutations may induce frameshifts and consequent depletion of the encoded protein, it is generally adequate to utilize a single guide RNA (sgRNA) to target the ORF for protein-coding genes [Shalem et al., 2014]. On the other hand, since lncRNAs don't contain ORFs, there is little chance they can be inactivated in this approach. However, lncRNAs have been targeted using modified versions of the standard CRISPR/Cas9 method. This includes the use of double guide RNA (gRNAs) for transcriptional regulation (activation or inhibition) and deletion [Aparicio-Prat et al., 2015; Joung et al., 2017; Liu et al., 2017] since lncRNAs not translated into a protein provide a greater challenge from a technical perspective. CRISPR-mediated genome editing has been helpful in elucidating the biological roles of lncRNAs in cancer and normal physiology. Thus, modification in the CRISPR system can be effectively used to ascribe functions to plant lncRNA.

10.5 CONCLUSIONS AND FUTURE DIRECTIONS

There has been notable growth in studies on plant lncRNAs since the first report of plant lncRNAs in the early 2000s. These small ncRNA molecules regulate important agronomic traits and response to environmental abiotic stresses as its one of the major threats to agriculture yields. In recent years, miRNAs have long been known to be important components of cellular regulatory networks, but lncRNAs and siRNAs are now being recognized as important regulatory molecules. However, studies of plant phased secondary small interfering RNAs (phasiRNAs), lncRNAs, and circRNAs, on the other hand, are still in their infancy, and there are many unanswered issues that need to be addressed. Thus, understanding the complexity, biogenesis, and function of plant lncRNAs continues to be a difficult task. In addition, new types of ncRNAs in plants are expected to be discovered. Novel lncRNA species, such as short nucleolar RNA (snoRNA)-ended and transfer RNA (tRNA)-ended lncRNAs, have recently been identified in mammals, which are processed by noncanonical pathways, but are yet to be discovered in plants [Wu et al., 2017]. LncRNAs are one of the most significant kinds of ncRNAs, and they appear to be important regulators of genes, gene expression, and genome stability in both animals and plants. Identification of lncRNAs has been aided by the emergence of new technologies such as high-throughput RNA-seq and other novel methods. LncRNAs have been reported to play an important part in regulating gene expression during abiotic stress. Several studies have shown that precise control of lncRNAs is important for functional stress responses, and such regulation is reported to be connected to a wide array of responses. As it turns out, new research has shown their regulatory functions in photomorphogenesis, resistance to biotic stress, fertility, and so on. Novel roles, in addition to those mentioned here, are expected to be discovered as research into lncRNAs progresses, with the goal of uncovering the regulatory mechanisms by which these lncRNAs function. Interactions between RNA-binding proteins are critical throughout the lifecycle of plants, and they are essential for the maintenance of cellular functions as well as pathogenesis. LncRNA-protein interactions may be beneficial, and finding proteins involved in lncRNA-mediated biological processes is essential. As a result, expanding sophisticated methods such as RNA-binding will certainly advance our knowledge of lncRNA biology and its systematic characterization.

Genome editing through CRISPR/Cas can target both miRNAs and lncRNAs for developing climate resilient crops [Zhang et al., 2020]. Recently, a few significant reports on CRISPR/Cas9-mediated gene editing for miRNA were reported [Li et al., 2016; Zhou et al., 2017]. It is challenging to find a protospacer adjacent motif (PAM) sequence that is essential for the CRISPR/Cas genome editing due to the short length of miRNAs. As new Cas nuclease proteins are discovered and existing Cas proteins are being constantly improved, the PAM requirement will be reduced and more genomic loci will become accessible by CRISPR/Cas system, including lncRNAs as they exceed the length of miRNA [Zhang and Zhang, 2020]. As a result, lncRNAs offer prospective targets for imparting abiotic stress tolerance in plants by altering their expression; however, there is still a long way to go before this approach can be used effectively to generate abiotic stress-resistant crops.

REFERENCES

Adnan M, Zahir S, Fahad S, et al. Phosphate-solubilizing bacteria nullify the antagonistic effect of soil calcification on bioavailability of phosphorus in alkaline soils. *Sci Rep.* 2018, *8*: 4339.

Amaral PP, Clark MB, Gascoigne DK, Dinger ME, Mattick JS. lncRNAdb: a reference database for long noncoding RNAs. *Nucleic Acids Res.* 2011, *39* (Database issue): D146–51; http://dx.doi.org/10.1093/nar/gkq1138. PMID:21112873

Aparicio-Prat E, Arnan C, Sala I, Bosch N, Guigo R, Johnson R. DECKO: single-oligo, dual-CRISPR deletion of genomic elements including long non-coding RNAs. *BMC Genom.* 2015, *16*: 846.

Ariel F, Romero-Barrios N, Jégu T, Benhamed M, Crespi M. Battles and hijacks: noncoding transcription in plants. *Trends Plant Sci.* 2015, *20* (6): 362–71; http://dx.doi.org/10.1016/j.tplants.2015.03.003

Bari R, Datt Pant B, Stitt M, Scheible WR. PHO2, MicroRNA399, and PHR1 define a phosphate-signaling pathway in plants. *Plant Physiol.* 2006, *141*: 988–99.

Bastow R, Mylne JS, Lister C, Lippman Z, Martienssen RA, Dean C. Vernalization requires epigenetic silencing of FLC by histone methylation. *Nature.* 2004, *427*: 164–67.

Ben Amor B, Wirth S, Merchan F, Laporte P, d'Aubenton-Carafa Y, Hirsch J, Maizel A, Mallory A, Lucas A, Deragon JM, Vaucheret H, Thermes C, Crespi M. Novel long non-protein coding RNAs involved in Arabidopsis differentiation and stress responses. *Genome Res.* 2009, *19* (1): 57–69; http://dx.doi.org/10.1101/gr.080275.108

Bhatia G, Singh A, Verma D, Sharma S, Singh K. Genome-wide investigation of regulatory roles of lncRNAs in response to heat and drought stress in *Brassica juncea* (Indian mustard). *Environ Exp Bot.* 2019, *171*: 103922.

Borah P, Sharma E, Kaur A, Chandel G, Mohapatra T, Kapoor S, Khurana JP. Analysis of drought-responsive signalling network in two contrasting rice cultivars using transcriptome-based approach. *Sci Rep.* 2017, *9* (7): 42131; http://dx.doi.org/10.1038/srep42131.

Borges F, Martienssen RA. The expanding world of small RNAs in plants. *Nat Rev Mol Cell Biol.* 2015, *6* (12): 727–41; http://dx.doi.org/10.1038/nrm4085. Epub 2015 Nov 4. PMID: 26530390; PMCID: PMC4948178

Budak H, Kaya SB, Cagirici HB. Long non-coding RNA in plants in the era of reference sequences. *Front Plant Sci.* 2020, *11*: 276; http://dx.doi.org/10.3389/fpls.2020.00276

Cabili MN, Trapnell C, Goff L, Koziol M, Tazon-Vega B, Regev A, Rinn JL. Integrative annotation of human large intergenic noncoding RNAs reveals global properties and specific subclasses. *Genes Dev.* 2011, *25*: 1915–27.

Cagirici HB, Alptekin B, Budak H. RNA sequencing and co-expressed long non-coding RNA in modern and wild wheats. *Sci Rep.* 2017, *7*: 10670.

Calixto CPG, Tzioutziou NA, James AB, et al. Cold-dependent expression and alternative splicing of Arabidopsis long non-coding RNAs. *Front Plant Sci.* 2019, *10*: 235.

Cao J. The functional role of long non-coding RNAs and epigenetics. *Biol Proced Online.* 2014, *16*: 11; http://dx.doi.org/10.1186/1480-9222-16-11

Carlevaro-Fita J, Johnson R. Global positioning system: understanding long noncoding RNAs through subcellular localization. *Mol Cell.* 2019, *73*: 869–83; http://dx.doi.org/10.1016/j.molcel.2019.02.008

Castaings L, Bergonzi S, Albani MC, Kemi U, Savolainen O, Coupland G. Evolutionary conservation of cold-induced antisense RNAs of *FLOWERING LOCUS C* in *Arabidopsis thaliana* perennial relatives. *Nat Commun.* 2014, *5*: 4457

Caudron-Herger M, Muller-Ott K, Mallm JP, Marth C, Schmidt U, Fejes-Toth K, Rippe K. Coding RNAs with a non-coding function: maintenance of open chromatin structure. *Nucleus* 2011, *2*: 410–24.

Cech TR, Steitz JA. The noncoding RNA revolution-trashing old rules to forge new ones. *Cell.* 2014, *157*: 77–94.

Chekanova JA. Long non-coding RNAs and their functions in plants. *Curr Opin Plant Biol.* 2015, *27*: 207–16; http://dx.doi.org/10.1016/j.pbi.2015.08.003.

Chen Z, Jiang Q, Jiang P, et al. Novel low-nitrogen stress-responsive long non-coding RNAs (lncRNA) in barley landrace B968 (Liuzhutouzidamai) at seedling stage. *BMC Plant Biol.* 2020, *20*: 142.

Chen J, Quan M, Zhang D. Genome-wide identification of novel long non-coding RNAs in Populus tomentosa tension wood, opposite wood and normal wood xylem by RNA-seq. *Planta* 2015, *241*: 125–43.

Chen M, Wang C, Bao H, Chen H, Wang Y. Genome-wide identification and characterization of novel lncRNAs in Populus under nitrogen deficiency. *Mol Genet Genomics.* 2016, *291*: 1663–80.

Chiou TJ, Aung K, Lin SI, Wu CC, Chiang SF, Su CL. Regulation of phosphate homeostasis by microRNA in Arabidopsis. *Plant Cell.* 2006, *18*: 412–21.

Chu C, Qu K, Zhong FL, Artandi SE, Chang HY. Genomic maps of long noncoding RNA occupancy reveal principles of RNA-chromatin interactions. *Mol Cell* 2011, *44*: 667–78.

Chung PJ, Jung H, Jeong DH, Ha SH, Choi YD, Kim JK. Transcriptome profiling of drought responsive non-coding RNAs and their target genes in rice. *BMC Genomics.* 2016, *17*: 563.

Corona-Gomez JA, Garcia-Lopez IJ, Stadler PF, Fernandez-Valverde SL. Splicing conservation signals in plant long non-coding RNAs. *RNA.* 2020; http://dx.doi.org/10.1261/rna.074393.119

Crespi MD, Jurkevitch E, Poiret M, d'Aubenton-Carafa Y, Petrovics G, Kondorosi E, Kondorosi A. enod40, a gene expressed during nodule organogenesis, codes for a non-translatable RNA involved in plant growth. *EMBO J.* 1994, *13* (21): 5099–112.

Csorba T, Questa JI, Sun Q, Dean C. Antisense COOLAIR mediates the coordinated switching of chromatin states at FLC during vernalization. *Proc Natl Acad Sci U S A.* 2014, *111*: 16160–5.

Curci PL, AieseCigliano R, Zuluaga DL, Janni M, Sanseverino W, Sonnante G. Transcriptomic response of durum wheat to nitrogen starvation. *Sci Rep.* 2017, *7*: 1–14.

Deforges J, Reis RS, Jacquet P, Sheppard S, Gadekar VP, Hart-Smith G, Tanzer A, Hofacker IL, Iseli C, Xenarios I, Poirier Y. Control of cognate sense mRNA translation by cis-natural antisense RNAs. *Plant Physiol.* 2019, *80* (1): 305–22; http://dx.doi.org/10.1104/pp.19.00043. Epub 2019 Feb 13

Deng F, Zhang X, Wang W, Yuan R, Shen F. Identification of *gossypium hirsutum* long non-coding RNAs (lncRNAs) under salt stress. *BMC Plant Biol.* 2018, *18*: 23.

Derrien T, Johnson R, Bussotti G, et al. The GENCODE v7 catalog of human long noncoding RNAs: analysis of their Gene structure, evolution, and expression. *Genome Res.* 2012, *22*: 1775–89.

Dinger ME, Gascoigne DK, Mattick JS. The evolution of RNAs with multiple functions. *Biochimie.* 2011, *93*: 2013–18.

Dinger ME, Pang KC, Mercer TR, Crowe ML, Grimmond SM, Mattick JS. NRED: a database of long noncoding RNA expression. *Nucleic Acids Res.* 2009, *37* (Database issue): D122–26; http://dx.doi.org/10.1093/nar/gkn617

Dinger ME, Pang KC, Mercer TR, Mattick JS. Differentiating protein-coding and noncoding RNA: challenges and ambiguities. *PLoS Comput Biol.* 2008, *4* (11): e1000176; http://dx.doi.org/10.1371/journal.pcbi.1000176

Ding Z, Tie W, Fu L, et al. Strand-specific RNA-seq based identification and functional prediction of drought-responsive lncRNAs in cassava. *BMC Genomics.* 2019, *20*: 214.

Drechsel G, Kahles A, Kesarwani AK, Stauffer E, Behr J, Drewe P, Rätsch G, Wachter A. Nonsense-mediated decay of alternative precursor mRNA splicing variants is a major determinant of the Arabidopsis steady state transcriptome. *Plant Cell.* 2013, *25* (10): 3726–42; http://dx.doi.org/10.1105/tpc.113.115485

Dykes IM, Emanueli C. Transcriptional and post-transcriptional gene regulation by long non-coding RNA. *Genom Proteom Bioinform.* 2017, *15*: 177–86; http://dx.doi.org/10.1016/j.gpb.2016.12.005

Esposito R, Bosch N, Lanzos A, Polidori T, Pulido-Quetglas C, Johnson R. Hacking the cancer genome: profiling therapeutically actionable long non-coding RNAs using CRISPR-Cas9 screening. *Cancer Cell* 2019, *35*: 545–57.

Fabbri M, Girnita L, Varani G, Calin GA. Decrypting noncoding RNA interactions, structures, and functional networks. *Genome Res.* 2019, *29* (9): 1377–88; http://dx.doi.org/10.1101/gr.247239.118

FAO. FAO land and plant nutrition management service. Rome: Food and Agriculture Organization of the United Nations; 2015.

Flórez-Zapata NM, Reyes-Valdés MH, Martínez O. Long non-coding RNAs are major contributors to transcriptome changes in sunflower meiocytes with different recombination rates. *BMC Genomics* 2016, *17*: 490.

Franco-Zorrilla JM, Valli A, Todesco M, Mateos I, Puga MI, Rubio-Somoza I, Leyva A, Weigel D, García JA, Paz-Ares J. Target mimicry provides a new mechanism for regulation of microRNA activity. *Nat Genet.* 2007, *39* (8): 1033–37; http://dx.doi.org/10.1038/ng2079

Fukuda M, Nishida S, Kakei Y, Shimada Y, Fujiwara T. Genome-wide analysis of long intergenic noncoding RNAs responding to low-nutrient conditions in *Arabidopsis thaliana*-possible involvement of trans-acting siRNA3 in response to low nitrogen. *Plant Cell Physiol.* 2019, *60*: 1961–73.

Gerstein MB, Bruce C, Rozowsky JS, Zheng D, Du J, Korbel JO, Emanuelsson O, Zhang ZD, Weissman S, Snyder M. What is a gene, post-ENCODE? History and updated definition. *Genome Res.* 2007, *17*: 669–81.

Gomes AQ, Nolasco S, Soares H. Non-coding RNAs: multi-tasking molecules in the cell. *Int J Mol Sci.* 2013, *14* (8): 16010–39; http://dx.doi.org/10.3390/ijms140816010

Gupta PK, Balyan HS, Gahlaut V. QTL analysis for drought tolerance in wheat: present status and future possibilities. *Agronomy.* 2017, *7*: 5.

Gutschner T, Baas M, Diederichs S. Noncoding RNA gene silencing through genomic integration of RNA destabilizing elements using zinc finger nucleases. *Genome Res.* 2011, *21*: 1944–54.

Guttman M, Amit I, Garber M, et al. Chromatin signature reveals over a thousand highly conserved large non-coding RNAs in mammals. *Nature* 2009, *458*: 223–27; http://dx.doi.org/10.1038/nature07672.

Hawkes EJ, Hennelly SP, Novikova IV, Irwin JA, Dean C, Sanbonmatsu KY. COOLAIR antisense RNAs form evolutionarily conserved elaborate secondary structures. *Cell Rep.* 2016, *16*: 3087–96; http://dx.doi.org/10.1016/j.celrep.2016.08.045

Hawkesford MJ, Griffiths S. Exploiting genetic variation in nitrogen use efficiency for cereal crop improvement. *Curr Opin Plant Biol.* 2019, *49*: 35–42.

He X, Guo S, Wang Y, Wang L, Shu S, Sun J. Systematic identification and analysis of heat-stress-responsive lncRNAs, circRNAs and miRNAs with associated co-expression and ceRNA networks in cucumber (*Cucumis sativus* L.). *Physiol Plant.* 2020, *168*: 736–54.

Heo JB, Sung S. Vernalization-mediated epigenetic silencing by a long intronic noncoding RNA. *Science.* 2011, *331*: 76–79.

Hsu PY, Calviello L, Wu HL, Li FW, Rothfels CJ, Ohler U, Benfey PN. Super-resolution ribosome profiling reveals unannotated translation events in *Arabidopsis. Proc Natl Acad Sci U S A.* 2016, *8*, *113* (45): E7126–35; http://dx.doi.org/10.1073/pnas.1614788113

Huanca-Mamani W, Arias-Carrasco R, Cárdenas-Ninasivincha S, Rojas-Herrera M, Sepúlveda-Hermosilla G, Caris-Maldonado JC, Bastías E, Maracaja-Coutinho V. Long non-coding RNAs responsive to salt and boron stress in the hyper-arid lluteño maize from atacama desert. *Genes (Basel).* 2018, 20, *9* (3): 170; http://dx.doi.org/10.3390/genes9030170

Huang L, Dong H, Zhou D, Li M, Liu Y, Zhang F, Feng Y, Yu D, Lin S, Cao J. Systematic identification of long non-coding RNAs during pollen development and fertilization in *Brassica rapa. Plant J.* 2018, *96* (1): 203–22; http://dx.doi.org/10.1111/tpj.14016

Huang TK, Han CL, Lin SI, et al. Identification of downstream components of ubiquitin-conjugating enzyme PHOSPHATE2 by quantitative membrane proteomics in Arabidopsis roots. *Plant Cell.* 2013, *25*: 4044–60.

Islam MS. Sensing and uptake of nitrogen in rice plant: a molecular view. *Ric Sci.* 2019, *26*: 343–55.

Islam S, Kjallquist U, Moliner A, Zajac P, Fan JB, Lonnerberg P, Linnarsson S. Characterization of the single-cell transcriptional landscape by highly multiplex RNA-seq. *Genome Res.* 2011, *21*: 1160–67.

Ismail AM, Horie T. Genomics, physiology, and molecular breeding approaches for improving salt tolerance. *Annu Rev Plant Biol.* 2017, *68*: 19.1–19.30

Itabashi E, Osabe K, Fujimoto R, Kakizaki T. Epigenetic regulation of agronomical traits in Brassicaceae. *Plant Cell Rep.* 2018, *37*: 87–101.

Jabnoune M, Secco D, Lecampion C, Robaglia C, Shu Q, Poirier Y. A rice *cis*-natural antisense RNA acts as a translational enhancer for its cognate mRNA and contributes to phosphate homeostasis and plant fitness. *Plant Cell.* 2013, *25*: 4166–82.

Jannesar M, Seyedi SM, Moazzam Jazi M, Niknam V, Ebrahimzadeh H, Botanga C. A genome-wide identification, characterization and functional analysis of salt-related long non-coding RNAs in non-model plant *Pistacia vera* l. Using transcriptome high throughput sequencing. *Sci Rep.* 2020, *27*, *10* (1): 5585; http://dx.doi.org/10.1038/s41598-020-62108-6.

Jha UC, Bohra AB, Jha R. Breeding approaches and genomics technologies to increase crop yield under low-temperature stress. *Plant Cell Rep.* 2017, *36*: 1–35.

Jiang N, Cui J, Hou X, Yang G, Xiao Y, Han L. sl-lncRNA15492 interacts with Sl-miR482a and affects *Solanum lycopersicum* immunity against *Phytophthora infestans. Plant J.* 2020, *103*: 1561–74.

Jiao F, Pahwa K, Manning M, Dichy N, Geuten K. Cold induced antisense transcription of FLOWERING LOCUS C in distant grasses. *Front Plant Sci.* 2019, 10: 72.

Joshi RK, Megha S, Basu U, Rahman MH, Kav NN. Genome wide identification and functional prediction of long non-coding RNAs responsive to *Sclerotinia sclerotiorum* infection in *Brassica napus. PLoS ONE* 2016, *11*: e0158784.

Joung J, Engreitz JM, Konermann S, et al. Genome-scale activation screen identifies a lncRNA locus regulating a gene neighborhood. *Nature* 2017, *548*: 343–46.

Kageyama Y, Kondo T, Hashimoto Y. Coding vs. non-coding: translatability of short ORFs found in putative non-coding transcripts. *Biochimie* 2011, *93*, 1981–86.

Kang C, Liu Z. Global identification and analysis of long non-coding RNAs in diploid strawberry *Fragaria vesca* during flower and fruit development. *BMC Genomics*. 2015, *16*: 815.

Khemka N, Singh VK, Garg R, Jain M. Genome-wide analysis of long intergenic non-coding RNAs in chickpea and their potential role in flower development. *Sci Rep*. 2016, *6*: 1-10.

Kim DH, Doyle MR, Sung S, Amasino RM. Vernalization: winter and the timing of flowering in plants. *Annu Rev Cell Dev Biol*. 2009, *25*: 277–99.

Kim ED, Sung S. Long noncoding RNA: unveiling hidden layer of gene regulatory networks. *Trends Plant Sci*. 2012, *17* (1): 16–21; http://dx.doi.org/10.1016/j.tplants.2011.10.008.

Kim DH, Sung S. Vernalization-triggered intragenic chromatin loop formation by long noncoding RNAs. *Dev Cell*. 2017, *40* (3): 302–312.e4; http://dx.doi.org/10.1016/j.devcel.2016.12.021

Kindgren P, Ard R, Ivanov M, Marquardt S. Transcriptional read-through of the long noncoding RNA SVALKA governs plant cold acclimation. *Nat Commun*. 2018, *9*: 4561.

Kopp F, Mendell JT. Functional classification and experimental dissection of long noncoding RNAs. *Cell* 2018, *172*: 393–407.

Kurihara Y, Matsui A, Hanada K, et al. Genome-wide suppression of aberrant mRNA-like noncoding RNAs by NMD in Arabidopsis. *Proc Natl Acad Sci USA*. 2009, *106*: 2453–58.

Kurihara Y, Schmitz RJ, Nery JR, Schultz MD, Okubo-Kurihara E, Morosawa T, Tanaka M, Toyoda T, Seki M, Ecker JR. Surveillance of 3' Noncoding Transcripts Requires FIERY1 and XRN3 in Arabidopsis. *G3 (Bethesda)*. 2012, *2* (4): 487–98; http://dx.doi.org/10.1534/g3.111.001362

Lamaoui M, Jemo M, Datla R, Bekkaoui F. Heat and drought stresses in crops and approaches for their mitigation. *Front Chem*. 2018, *6*: 26.

Lapidot M, Pilpel Y. Genome-wide natural antisense transcription: coupling its regulation to its different regulatory mechanisms. *EMBO Rep*. 2006, *7*: 1216–22.

Laurent GS, Wahlestedt C, Kapranov P. The landscape of long noncoding RNA classification. *Trends Genet*. 2015, *31*: 239–51.

Li L, Eichten SR, Shimizu R, et al. Genome-wide discovery and characterization of maize long non-coding RNAs. *Genome Biol*. 2014, *15*: 1.

Li M, Li X, Zhou Z, et al. Reassessment of the four yield-related genes Gn1a, DEP1, GS3, and IPA1 in rice using a CRISPR/Cas9 system. *Front Plant Sci*. 2016a, *7*: 377; http://dx.doi.org/10.3389/fpls.2016.00377

Li S, Yamada M, Han X, Ohler U, Benfey PN. High-resolution expression map of the Arabidopsis root reveals alternative splicing and lincRNA regulation. *Dev Cell*. 2016b, *39*: 508–22.

Li S, Yu X, Cheng Z, et al. Genome-wide identification and functional prediction of cold and/or drought-responsive lncRNAs in cassava. *Sci Rep*. 2017, *7*: 45981.

Li X, Zhang S, Bai J, He Y. Tuning growth cycles of Brassica crops via natural antisense transcripts of BrFLC. *Plant Biotechnol J*. 2016, *14*: 905–14.

Lin X, Lin W, Ku YS, Wong FL, Li MW, Lam HM, Ngai SM, Chan TF. Analysis of soybean long non-coding RNAs reveals a subset of small peptide-coding transcripts. *Plant Physiol*. 2020, *182* (3): 1359–74; http://dx.doi.org/10.1104/pp.19.01324.

Liu SJ, Horlbeck MA, Cho SW, et al. CRISPRi-based genome-scale identification of functional long noncoding RNA loci in human cells. *Science*. 2017, *355*: eaah7111.

Liu J, Jung C, Xu J, Wang H, Deng S, Bernad L, Arenas-Huertero C, Chua NH. Genome-wide analysis uncovers regulation of long intergenic noncoding RNAs in Arabidopsis. *Plant Cell*. 2012, *24* (11): 4333–45; http://dx.doi.org/10.1105/tpc.112.102855.

Liu SJ, Nowakowski TJ, Pollen AA, et al. Single-cell analysis of long non-coding RNAs in the developing human neocortex. *Genome Biol*. 2016, *17*: 67.

Liu TT, Zhu D, Chen W, Deng W, He H, He G, Bai B, Qi Y, Chen R, Deng XW. A global identification and analysis of small nucleolar RNAs and possible intermediate-sized non-coding RNAs in *oryza sativa*. *Mol Plant*. 2013, *6* (3): 830–46; http://dx.doi.org/10.1093/mp/sss087.

Louro R, Smirnova AS, Verjovski-Almeida S. Long intronic noncoding RNA transcription: expression noise or expression choice? *Genomics*. 2008, *93*: 291–98

Lubelsky Y, Ulitsky I. Sequences enriched in alu repeats drive nuclear localization of long RNAs in human cells. *Nature*. 2018, *555*: 107–11; http://dx.doi.org/10.1038/nature25757

Lv Y, Liang Z, Ge M, et al. Genome-wide identification and functional prediction of nitrogen-responsive intergenic and intronic long non-coding RNAs in maize (*Zea mays* L.). *BMC Genomics*. 2016, *17*: 350.

Ma L, Bajic VB, Zhang Z. On the classification of long non-coding RNAs. *RNA Biol*. 2013, *10* (6): 925–33; http://dx.doi.org/10.4161/rna.24604. Epub 2013 Apr 15

Maamar H, Cabili MN, Rinn J, Raj A. Linc-HOXA1 is a noncoding RNA that represses Hoxa1 transcription in *cis*. *Genes Dev*. 2013, *27*: 1260–71.

Matzke MA, Mosher RA. RNA-directed DNA methylation: an epigenetic pathway of increasing complexity. *Nat Rev Genet*. 2014, *15* (6): 394–408; http://dx.doi.org/10.1038/nrg3683. Epub 2014 May 8. Erratum in: *Nat Rev Genet*. **2014**, *15* (8): 570.

Mehraj H, Shea DJ, Takahashi S, et al. Genome-wide analysis of long noncoding RNAs, 24-nt siRNAs, DNA methylation and H3K27me3 marks in *Brassica rapa*. *PLoS ONE*. 2021, *16* (3): e0242530; http://dx.doi.org/10.1371/journal.pone.0242530

Mercer TR, Gerhardt DJ, Dinger ME, Crawford J, Trapnell C, Jeddeloh JA, Mattick JS, Rinn JL. Targeted RNA sequencing reveals the deep complexity of the human transcriptome. *Nat Biotechnol*. 2011, *30*: 99–104.

Mukherjee N, Calviello L, Hirsekorn A, de Pretis S, Pelizzola M, Ohler U. Integrative classification of human coding and noncoding genes through RNA metabolism profiles. *Nat Struct Mol Biol*. 2017, *24* (1): 86–96; http://dx.doi.org/10.1038/nsmb.3325

Nair L, Chung H, Basu U. Regulation of long non-coding RNAs and genome dynamics by the RNA surveillance machinery. *Nat Rev Mol Cell Biol*. 2020, *21* (3): 123–36; http://dx.doi.org/10.1038/s41580-019-0209-0

Ozsolak F, Milos PM. RNA sequencing: advances, challenges and opportunities. *Nat Rev Genet*. 2011, *12*: 87–98.

Pang KC, Frith MC, Mattick JS. Rapid evolution of noncoding RNAs: lack of conservation does not mean lack of function. *Trends Genet*. 2006, *22*: 1–5.

Pang J, Zhang X, Ma X, Zhao J. Spatio-temporal transcriptional dynamics of maize long non-coding RNAs responsive to drought stress. *Genes (Basel)*. 2019, *10*: 138.

Paraskevopoulou MD, Hatzigeorgiou AG. Analyzing miRNA–the lncRNA interactions- long non-coding RNAs. *Methods Mol Biol*. 2016, *1402*: 271–86; http://dx.doi.org/10.1007/978-1-4939-3378-5_21

Pefanis E, Wang J, Rothschild G, Lim J, Kazadi D, Sun J, Federation A, Chao J, Elliott O, Liu ZP, Economides AN, Bradner JE, Rabadan R, Basu U. RNA exosome-regulated long non-coding RNA transcription controls super-enhancer activity. *Cell*. 2015, *161* (4): 774–89; http://dx.doi.org/10.1016/j.cell.2015.04.034

Ponjavic J, Ponting CP, Lunter G. Functionality or transcriptional noise? Evidence for selection within long noncoding RNAs. *Genome Res*. 2007, *17*: 556–65.

Puga MI, Rojas-Triana M, de Lorenzo L, Leyva A, Rubio V, Paz-Ares J. Novel signals in the regulation of pi starvation responses in plants: facts and promises. *Curr Opin Plant Biol*. 2017, *39*: 40–49.

Qin T, Zhao HY, Cui P, Albesher N, Xiong LM. A nucleus-localized long noncoding RNA enhances drought and salt stress tolerance. *Plant Physiol*. 2017, *175*: 1321–36

Qi X, Xie S, Liu Y, Yi F, Yu J. Genome-wide annotation of genes and noncoding RNAs of foxtail millet in response to simulated drought stress by deep sequencing. *Plant Mol Biol*. 2013, *83*: 459–73.

Quinn JJ, Chang HY. Unique features of long non-coding RNA biogenesis and function. *Nat Rev Genet*. 2016, *17*: 47–62.

Quinn JJ, Zhang QC, Georgiev P, Ilik IA, Akhtar A, Chang HY. Rapid evolutionary turnover underlies conserved lncRNA–genome interactions. *Genes Dev*. 2016, *30* (2): 191–207; http://dx.doi.org/10.1101/gad.272187.115

Rohrig H, Schmidt J, Miklashevichs E, Schell J, John M. Soybean ENOD40 encodes two peptides that bind to sucrose synthase. *Proc Natl Acad Sci USA*. 2002, *99*: 1915–20.

Rosa S, Duncan S, Dean C. Mutually exclusive sense–antisense transcription at FLC facilitates environmentally induced gene repression. *Nat Commun*. 2016, *7*: 1–7; http://dx.doi.org/10.1038/ncomms13031

Salzman J, Gawad C, Wang PL, Lacayo N, Brown P.O. Circular RNAs are the predominant transcript isoform from hundreds of human genes in diverse cell types. *PLoS ONE*. 2012, *7*: e30733; http://dx.doi.org/10.1371/journal.pone.0030733

Shalem O, Sanjana NE, Hartenian E, et al. Genome-scale CRISPR-Cas9 knockout screening in human cells. *Science*. 2014, *343*: 84–87.

Shea DJ, Itabashi E, Takada S, et al. The role of *FLOWERING LOCUS c* in vernalization of brassica: the importance of vernalization research in the face of climate change. *Crop Pasture Sci*. 2018, *69*: 30–39.

Shea DJ, Nishida N, Takada S, et al. Long noncoding RNAs in *Brassica rapa* l. Following vernalization. *Sci Rep*. 2019, *9*: 9302; http://dx.doi.org/10.1038/s41598-019-45650-w

Shin SY, Jeong JS, Lim JY, et al. Transcriptomic analyses of rice (*Oryza sativa*) genes and non-coding RNAs under nitrogen starvation using multiple omics technologies. *BMC Genomics*. 2018, *19*: 532.

Shlyueva D, Stampfel G, Stark A. Transcriptional enhancers: from properties to genome-wide predictions. *Nat Rev Genet*. 2014, *15*: 272–86

Shuai P, Liang D, Tang S, et al. Genome-wide identification and functional prediction of novel and drought-responsive lincRNAs in *Populus trichocarpa*. *J Exp Bot*. 2014, *65*, 4975–83.

Simons M, Saha R, Guillard L, et al. Nitrogen-use efficiency in maize (*Zea mays* L.): from 'omics' studies to metabolic modelling. *J Exp Bot*. 2014, *65*: 5657–71.

Solda G, Makunin IV, Sezerman OU, Corradin A, Corti G, Guffanti A. An Ariadne's thread to the identification and annotation of noncoding RNAs in eukaryotes. *Briefings Bioinfor* 2009, *10*: 475–89.

Song X, Li Y, Cao X, Qi Y. MicroRNAs and their regulatory roles in plant-environment interactions. *Annu Rev Plant Biol*. 2019, *70*: 489–525; http://dx.doi.org/10.1146/annurev-arplant-050718-100334

Song X, Sun L, Luo H, Ma Q, Zhao Y, Pei D. Genome-wide identification and characterization of long non-coding RNAs from mulberry (*Morus notabilis*) RNA-seq data. *Genes* 2016, *7*: 11.

Sun QW, Csorba T, Skourti-Stathaki K, Proudfoot NJ, Dean C. R-loop stabilization represses antisense transcription at the Arabidopsis FLC locus. *Science*. 2013, *340*: 619–21.

Sun Z, Huang K, Han Z, et al. Genome-wide identification of Arabidopsis long noncoding RNAs in response to the blue light. *Sci Rep*. 2020, *10*: 6229.

Sun X, Zheng H, Sui N. Regulation mechanism of long noncoding RNA in plant response to stress. *Biochem Biophys Res Commun*. 2018, *503*: 402–07.

Swiezewski S, Liu F, Magusin A, Dean C. Cold-induced silencing by long antisense transcripts of an Arabidopsis polycomb target. *Nature*. 2009, *462* (7274): 799–802; http://dx.doi.org/10.1038/nature08618. PMID: 20010688

Szabo EX, Reichert P, Lehniger MK, Ohmer M, de Francisco Amorim M, Gowik U, Schmitz-Linneweber C, Laubinger S. Metabolic labeling of RNAs uncovers hidden features and dynamics of the Arabidopsis transcriptome. *Plant Cell*. 2020, *32* (4): 871–87; http://dx.doi.org/10.1105/tpc.19.00214.

Szcześniak MW, Rosikiewicz W, Makałowska I. CANTATAdb: a collection of plant long non-coding RNAs. *Plant Cell Physiol*. 2016, *l57*: e8.

Tan X, Li S, Hu L, Zhang C. Genome-wide analysis of long non-coding RNAs (lncRNAs) in two contrasting rapeseed (*Brassica napus* L.) Genotypes subjected to drought stress and re- watering. *BMC Plant Biol*. 2020, *20*: 81.

Thieffry A, Vigh ML, Bornholdt J, Ivanov M, Brodersen P, Sandelin A. Characterization of *Arabidopsis thaliana* promoter bidirectionality and antisense RNAs by inactivation of nuclear RNA decay pathways. *Plant Cell*. 2020, *32* (6): 1845–1867; http://dx.doi.org/10.1105/tpc.19.00815.

Tian H, Guo F, Zhang Z, et al. Discovery, identification, and functional characterization of long non-coding RNAs in *Arachis hypogaea* l. *BMC Plant Biol*. 2020, *20*: 308; http://dx.doi.org/10.1186/s12870-020-02510-4

Tian J, Song Y, Du Q, Yang X, Ci D, Chen J, Xie J, Li B, Zhang D. Population genomic analysis of gibberellin-responsive long non-coding RNAs in *Populus*. *J Exp Bot*. 2016, *67*: erw057.

Tsoi LC, Iyer MK, Stuart PE, et al. Analysis of long non-coding RNAs highlights tissue-specific expression patterns and epigenetic profiles in normal and psoriatic skin. *Genome Biol*. 2015, *16*: 1.

Wang KC, Chang HY. Molecular mechanisms of long noncoding RNAs. *MolCell*. 2011, *43* (6): 904–14; http://dx.doi.org/10.1016/j.molcel.2011.08.018

Wang H, Chung PJ, Liu J, Jang IC, Kean MJ. Genome-wide identification of long noncoding natural antisense transcripts and their responses to light in Arabidopsis. *Genome Res*. 2014, *24*: 444–53.

Wang P, Dai L, Ai J, Wang Y, Ren F. Identification and functional prediction of cold-related long noncoding RNA (lncRNA) in grapevine. *Sci Rep*. 2019a, *9*: 6638.

Wang F, Deng M, Xu J, Zhu X, Mao C. Molecular mechanism of phosphate transport and signaling in higher plants. *Semin Cell Dev Biol*. 2018a, *74*: 114–22.

Wang J, Du S, Wang J, Fan W, Wang P, Zhang Z, Xu P, Tang S, Deng Q, Yang W, Yu M. The prognostic value of abnormally expressed lncRNAs in colorectal cancer: a meta-analysis. *PLoS ONE*. 2017a, *12* (6): e0179670; http://dx.doi.org/10.1371/journal.pone.0179670

Wang A, Hu J, Gao C, et al. Genome-wide analysis of long noncoding RNAs unveils the regulatory roles in the heat tolerance of Chinese cabbage (*Brassica rapa* ssp. Chinensis). *Sci Rep*. 2019b, *9*: 5002

Wang J, Lin J, Kan J, et al. Genome-wide identification and functional prediction of novel drought-responsive lncRNAs in *Pyrus betulifolia*. *Genes (Basel)*. 2018b, *9*, 311.

Wang TZ, Liu M, Zhao MG, Chen R, Zhang WH. Identification and characterization of long non-coding RNAs involved in osmotic and salt stress in *Medicago truncatula* using genome-wide high-throughput sequencing. *BMC Plant Biol.* 2015a, *15*: 1.

Wang Y, Luo X, Sun F, Hu J, Zha X, Su W, Yang J. Overexpressing lncRNA LAIR increases grain yield and regulates neighbouring gene cluster expression in rice. *Nat Commun.* 2018c, *9* (1): 3516; http://dx.doi.org/10.1038/s41467-018-05829-7

Wang J, Ye C, Xiong H, Shen Y, Lu Y, Zhou J, Wang L. Dysregulation of long noncoding RNA in breast cancer: an overview of mechanism and clinical implication. *Oncotarget.* 2017b, *8* (3): 5508–22.

Wang J, Yu W, Yang Y, Li X, Chen T, Liu T, Ma N, Yang X, Liu R, Zhang B. Genome-wide analysis of tomato long non-coding RNAs and identification as endogenous target mimic for microRNA in response to TYLCV infection. *Sci Rep.* 2015c, *5*: 16946.

Wang M, Yuan D, Tu L, Gao W, He Y, Hu H, Wang P, Liu N, Lindsey K, Zhang X. Long noncoding RNAs and their proposed functions in fibre development of cotton (gossypium spp.). *New Phytol.* 2015b, *207* (4): 1181–97; http://dx.doi.org/10.1111/nph.13429.

Wang T, Zhao M, Zhang X, et al. Novel phosphate deficiency-responsive long non-coding RNAs in the legume model plant *Medicago truncatula. J Exp Bot.* 2017c, *68*: 5937–48.

Wang W, Zhuang Q, Ji K, Wen B, Lin P, Zhao Y, Li L, Yan C. Identification of miRNA, lncRNA and mRNA-associated ceRNA networks and a potential biomarker for MELAS with mitochondrial DNA A3243G mutation. *Sci Rep.* 2017e, *7*: 41639; http://dx.doi.org/10.1038/srep41639

Weidong Q, Hongping C, Zuozhen Y, et al. Systematic characterization of long non-coding RNAs and their responses to drought stress in Dongxiang wild rice. *Ric Sci.* 2020, *27*: 21–31.

Whittaker C, Dean C. The FLC locus: a platform for discoveries in epigenetics and adaptation. *Annu Rev Cell Dev Biol.* 2017, *33*: 555–75.

Wierzbicki AT, Haag JR, Pikaard CS. Noncoding transcription by RNA polymerase pol IVb/pol v mediates transcriptional silencing of overlapping and adjacent genes. *Cell.* 2008, *135* (4): 635–48; http://dx.doi.org/10.1016/j.cell.2008.09.035

Wight M, Werner A. The functions of natural antisense transcripts. *Essays Biochem.* 2015, *54*: 91–101

Wu C, Ding Z, Chen M, et al. Identification and functional prediction of lncRNAs in response to PEG and ABA treatment in cassava. *Environ Exp Bot.* 2019, *166*, 103809.

Wu Z, Fang X, Zhu D, Dean C. Autonomous pathway: *FLOWERING LOCUS C* repression through an antisense-mediated chromatin-silencing mechanism. *Plant Physiol.* 2020, *182* (1): 27–37; http://dx.doi.org/10.1104/pp.19.01009. Epub 2019 Nov 18.

Wu H, Yang L, Chen LL. The diversity of long noncoding RNAs and their generation. *Trends Genet.* 2017, *33* (8), 540–52; http://dx.doi.org/10.1016/j.tig.2017.05.004

Xin M, Wang Y, Yao Y, Song N, Hu Z, Qin D, Xie C, Peng H, Ni Z, Sun Q. Identification and characterization of wheat long non-protein coding RNAs responsive to powdery mildew infection and heat stress by using microarray analysis and SBS sequencing. *BMC Plant Biol.* 2011, *7*, (11): 61; http://dx.doi.org/10.1186/1471-2229-11-61

Yan Q, Wu F, Yan Z, et al. Differential co-expression networks of long non-coding RNAs and mRNAs in *Cleistogenes songorica* under water stress and during recovery. *BMC Plant Biol.* 2019, *19*: 23

Yang L, Froberg JE, Lee JT. Long noncoding RNAs: fresh perspectives into the RNA world. *Trends Biochem Sci.* 2014, *39*: 35–43

Yin Y, Lu JY, Zhang X, Shao W, Xu Y, Li P, Hong Y, Cui L, Shan G, Tian B, Zhang QC, Shen X. U1 snRNP regulates chromatin retention of noncoding RNAs. *Nature.* 2020, *580* (7801): 147–50; http://dx.doi.org/10.1038/s41586-020-2105-3. Epub 2020 Mar 11

Yotsukura S, Hancock T, Natsume-Kitatani Y, Mamitsuka H. Computational recognition for long non-coding RNA (lncRNA): software and databases. *Brief Bioinform.* 2016, *18*: bbv114.

Yu Y, Zhang Y, Chen X, Chen Y. Plant noncoding RNAs: hidden players in development and stress responses. *Annu Rev Cell Dev Biol.* 2019, *35*: 407–31.

Yuan J, Zhang Y, Dong J, Sun Y, Lim BL, Liu D, Lu ZJ. Systematic characterization of novel lncRNAs responding to phosphate starvation in Arabidopsis thaliana. *BMC Genomics.* 2016, *17*: 655; http://dx.doi.org/10.1186/s12864-016-2929-2

Zhang X, Dong J, Deng F, Wang W, Cheng Y, Song L, Hu M, Shen J, Xu Q, Shen F. The long non-coding RNA lncRNA973 is involved in cotton response to salt stress. *BMC Plant Biol.* 2019, *19* (1): 459; http://dx.doi.org/10.1186/s12870-019-2088-0

Zhang W, Han Z, Guo Q, et al. Identification of maize long non-coding RNAs responsive to drought stress. *PLoS ONE.* 2014, *9*: e98958.

Zhang J, Li XM, Li HX, Chong K. Crop improvement through temperature resilience. *Annu Rev Plant Biol.* 2019, *70*: 753–80.

Zhang YC, Liao JY, Li ZY, Yu Y, Zhang JP, Li QF, Qu LH, Shu WS, Chen YQ. Genome-wide screening and functional analysis identify a large number of long noncoding RNAs involved in the sexual reproduction of rice. *Genome Biol.* 2014a, *15*: 1.

Zhang B, Su T, Li P et al. Identification of long noncoding RNAs involved in resistance to downy mildew in Chinese cabbage. *Hortic Res.* 2021, *8*: 44; http://dx.doi.org/10.1038/s41438-021-00479-1

Zhang C, Tang G, Peng X, Sun F, Liu S, Xi Y. Long non-coding RNAs of switchgrass (*Panicum virgatum* L.) In multiple dehydration stresses. *BMC Plant Biol.* 2018, *18*: 79.

Zhang XO, Wang HB, Zhang Y, Lu X, Chen LL, Yang L. Complementary sequence–mediated exon circularization. *Cell.* 2014b, *159*: 134–47.

Zhang D, Zhang B. SpRY: engineered CRISPR/Cas9 harnesses new genome-editing power. *Trends Genet.* 2020, *36*, 546–48: http://dx.doi.org/10.1016/j.tig.2020.05.001

Zhang Y, Zhang XO, Chen T, Xiang JF, Yin QF, Xing YH, Zhu S, Yang L, Chen LL. Circular intronic long noncoding RNAs. *Mol Cell.* 2013, *51*: 792–806; http://dx.doi.org/10.1016/j.molcel.2013.08.017

Zhang D, Zhang Z, Unver T, Zhang B. CRISPR/Cas: a powerful tool for gene function study and crop improvement. *J Adv Res.* 2020; http://dx.doi.org/10.1016/j.jare.2020.10.003

Zhao X, Li J, Lian B, Gu H, Li Y, Qi Y. Global identification of Arabidopsis lncRNAs reveals the regulation of MAF4 by a natural antisense RNA. *Nat Commun.* 2018, *9* (1): 5056; http://dx.doi.org/10.1038/s41467-018-07500-7

Zhao M, Wang T, Sun T, Yu X, Tian R, Zhang WH. Identification of tissue-specific and cold- responsive lncRNAs in *Medicago truncatula* by high-throughput RNA sequencing. *BMC Plant Biol.* 2020a, *20*: 99

Zhou X, Cui J, Meng J, Luan Y. Interactions and links among the noncoding RNAs in plants under stresses. *Theor Appl Genet.* 2020b, *133*: 3235–48.

Zhou J, Deng K, Cheng Y, et al. CRISPR-cas9 based genome editing reveals new insights into microRNA function and regulation in rice. *Front Plant Sci.* 2017, *8*: 1598; http://dx.doi.org/10.3389/fpls.2017.01598

Zhou B, Ji B, Liu K, et al. EVLncRNAs 2.0: an updated database of manually curated functional long non-coding RNAs validated by low-throughput experiments. *Nucleic Acids Res.* 2020c, *46*: D86–91; http://dx.doi.org/10.1093/nar/gkaa1076

Zhou M, Law JA. RNA pol IV and v in Gene silencing: rebel polymerases evolving away from pol II's rules. *Curr Opin Plant Biol.* 2015, *27*: 154–64; http://dx.doi.org/10.1016/j.pbi.2015.07.005

Zhou GF, Zhang LP, Li BX, et al. Genome-wide identification of long non-coding RNA in trifoliate Orange (*Poncirus trifoliata* (L.) Raf) leaves in response to boron deficiency. *Int J Mol Sci.* 2019, *2*: 5419.

Zhu B, Yang Y, Li R, Fu D, Wen L, Luo Y, Zhu H. RNA sequencing and functional analysis implicate the regulatory role of long non-coding RNAs in tomato fruit ripening. *J Exp Bot.* 2015, *66*: 4483–95.

Zou C, Wang Q, Lu C, Yang W, Zhang Y, Cheng H, Feng X, Prosper MA, Song G. Transcriptome analysis reveals long noncoding RNAs involved in fiber development in cotton (*Gossypium arboreum*). *Sci China Life Sci.* 2016, *59*, 164–71.

11 Machine Learning Approaches for Long Non-Coding RNA Identification in Plants

Sarthak Chordia[1], Nikhil Bhamwani[1], Mehul Gupta[1], Niharika Sharma[2], and Rita Sharma[1]

[1]Department of Biological Sciences, Birla Institute of Technology and Science (BITS), Pilani, Pilani Campus, Rajasthan, India

[2]NSW Department of Primary Industries, Orange Agricultural Institute, Orange, NSW, Australia

11.1 INTRODUCTION

The transcripts that do not code for a protein are called non-coding RNAs (ncRNAs). These include regulatory RNAs and housekeeping RNAs such as transfer RNAs (tRNAs), ribosomal RNAs (rRNAs), small nucleolar RNAs (snoRNAs) and small nuclear RNAs (snRNAs). Based on the length, mode of biogenesis and mechanism of action, regulatory ncRNAs are further categorized as small (sRNAs) or long ncRNAs (lncRNAs). Small ncRNAs comprise a heterogenous but conserved group of RNAs usually smaller than 200 nucleotides (nts), such as short interfering RNAs (siRNAs), microRNAs (miRNAs), circular RNAs (circRNAs), piwi-interacting RNAs (piRNAs), etc. In contrast, lncRNAs comprise a group of poorly conserved ncRNAs usually longer than 200 nts (Morris and Mattick 2014). These may be transcribed from intergenic regions, antisense or overlap with introns, exons, promoters or enhancers of the protein-coding genes (Ponting, Oliver, and Reik 2009). Unlike non-coding small RNAs (sRNAs) and housekeeping ncRNAs, lncRNAs are very similar to mRNAs in their length, structure, mode of transcription and processing. Like mRNAs, many lncRNAs are transcribed by RNA Polymerase II and undergo post-transcriptional processing, including capping, poly-adenylation and splicing (Chen and Carmichael 2010). Although their mechanism of action is still unclear, lncRNAs play crucial roles in several biological processes and stress response by regulating the epigenetic modifications and gene expression at the transcriptional, post-transcriptional, translational, or post-translation levels (Wu, Liu, Qi, Cai, and Xu 2020).

With the large amount of transcriptomics data generated from diverse species, many efforts have been made to accurately and quickly identify lncRNAs using computational approaches. A typical workflow for computational identification of lncRNAs involves filtering of the transcriptomic data based on the length of the transcripts (>200 nts), open reading frame (ORF) length (<120 aa), coding potential and sequence homology with the known protein-coding genes or housekeeping ncRNAs (Paytuvi-Gallart, Sanseverino, and Aiese Cigliano 2019; Kandpal, Dhaka, and Sharma 2021). However, these criteria can easily miss lncRNAs with longer open-reading frames or those encoding short functional peptides (Dinger, Pang, Mercer, and Mattick 2008; Kung, Colognori, and Lee 2013). Also, lncRNAs derived from protein-coding regions would share homologies with protein sequences and may be filtered out. Therefore, comprehensive analysis of lncRNAs remains one of the major challenges. Very low and cell/tissue-specific expression and idiosyncratic features of lncRNAs in different species and tissues further add to this challenge (Cabili et al. 2011; Ulitsky, Shkumatava, Jan, Sive, and Bartel 2011). Furthermore, identification of lncRNAs from poorly annotated or incomplete genes or transcripts reconstructed using *de novo* assembly of RNA sequencing

DOI: 10.1201/9781003369288-11

data is even more challenging. Due to sequencing errors associated with next-generation sequencing (NGS) technologies, the data containing large numbers of insertions or deletions is further prone to erroneous predictions.

To deal with these challenges, several machine learning (ML) models have been trained to develop a wide range of tools for lncRNA prediction over the past decade or so. The accuracy of these tools mainly depends on the quality of the training data, feature prioritization and selection and, the choice of algorithm. A typical workflow to develop prediction models for lncRNA identification involves gathering and preprocessing the positive (lncRNAs) and negative (mRNAs) datasets as the first step (Figure 11.1). The annotation errors in these datasets or an imbalance in the negative and positive training datasets significantly reduce the performance of the prediction tools (Meng, Kang, Chang, and Luan 2021). The data need to be preprocessed to minimize such errors and then split into training and test data for training and evaluating the model, respectively.

Most tools developed to date have been trained using human and animal data. Considering the unique aspects of evolutionary pressure, level of conservation and mode of biogenesis of ncRNAs between plants and animal species, lncRNA predictors trained solely on the animal data exhibit suboptimal performance with the plant data. A comparative assessment of eight popular tools trained using either plant, human, or mixed data revealed that the tools trained using plant data performed better for lncRNA prediction in plants and *vice versa* (Costa Negri, Rossi Paschoal, and Luz Alves 2020). The tools trained using mixed data exhibited average performance with both plants and human data. To address this issue, a few species-neutral and plant-specific models have been trained using a mix of animal and plant datasets or exclusively plant data, respectively. This chapter aims to identify such tools available for lncRNA identification in plants and presents a comparative account to assist users, particularly from the non-bioinformatics backgrounds, in making an informed choice.

The second major step of lncRNA identification is to extract the relevant features for training the model. Most of the tools developed to date use sequence features such as mRNA and ORF length, coding potential, ORF quality and coverage, nucleotide composition, codon usage, guanine-cytosine (GC) content and Fickett score for lncRNA prediction (Fickett 1982). Though lncRNAs lack sequence conservation across species, small binding sites required to maintain the functional interactions between lncRNAs and other biomolecules may be conserved at the structural level (Militti, Maenner, Becker, and Gebauer 2014). Therefore, some researchers have also employed structure-based features for lncRNA prediction. In addition, some researchers leveraged experimental features such as expression specificity and differential binding of ribosomes for lncRNA prediction. Since considering a large number of features does not always enhance the model's performance, most researchers use statistical methods to identify optimal features required for training the model. However, as much as it is vital for the performance of the model, feature selection for training an lncRNA predictor is not a trivial task. The influence of the different features on precision accuracy varies considerably. On the other hand, deep learning models do not require inputs from the user and can predict lncRNAs by identifying and self-learning inherent features associated with lncRNAs and mRNAs used for training (Figure 11.1). Therefore, accuracy of these models largely depends on the size and quality of the training data.

Another important consideration while using ML approaches is the choice of algorithm. So far, support vector machines (SVMs) and random forest algorithms have been most frequently used for lncRNA prediction using shallow learning methods. In contrast, deep learning models have been mainly trained using convolutional neural networks (CNNs). However, the performance of the different models varies significantly with the source and the size of the training data.

Once trained, the model is evaluated for its performance using the test data. This is followed by benchmarking against previously available popular tools (Figure 11.1). The accuracy, precision, sensitivity and F1-score (harmonic average of sensitivity and precision) are commonly used as evaluation metrics for assessing the performance of the newly developed model. However, as most of these tools are trained using human and/or mouse data, these are usually re-trained using plant or hybrid

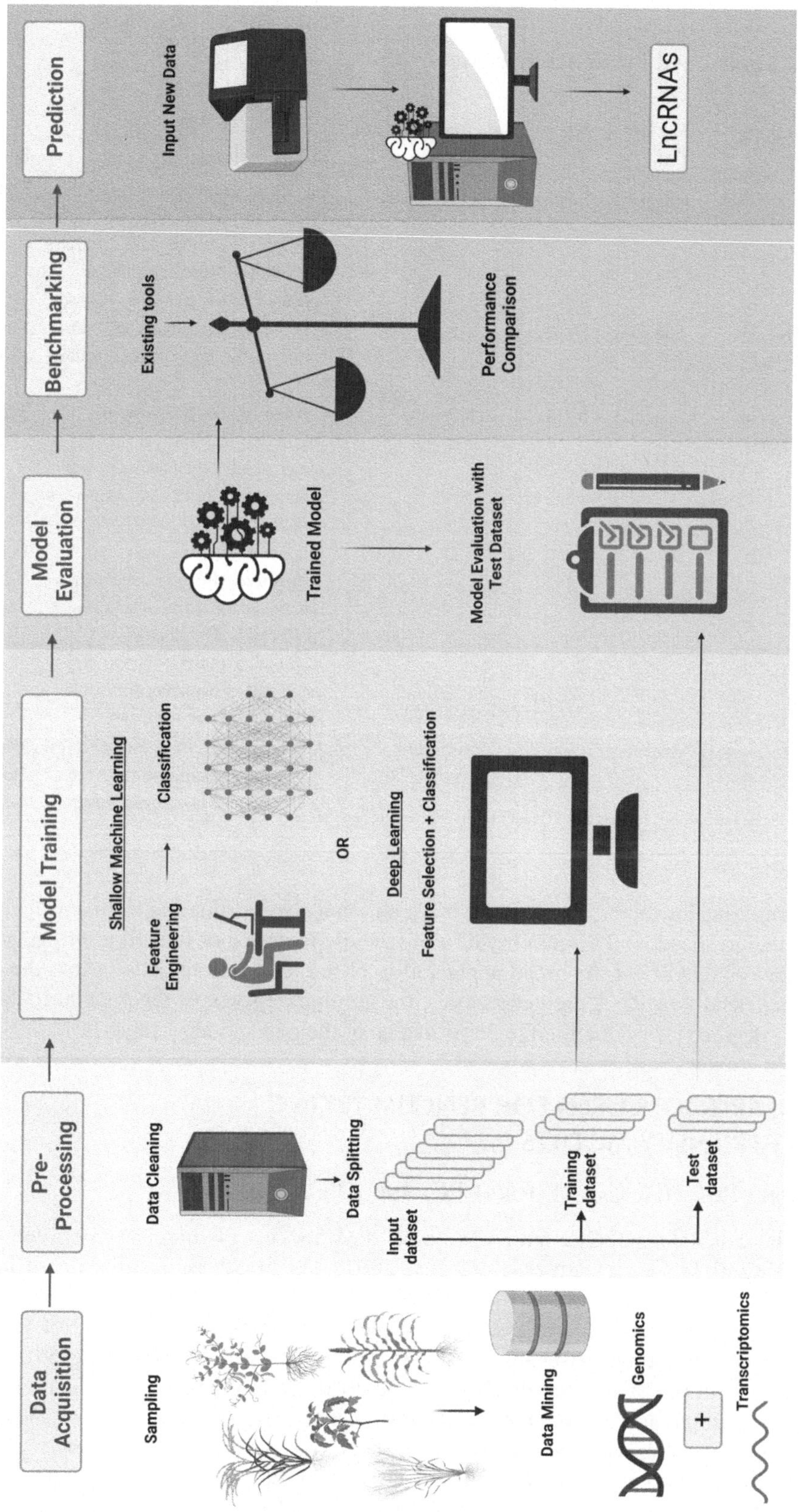

FIGURE 11.1 Schematic illustration of key steps involved in the building and implementation of a machine learning model for lncRNA discovery (created with BioRender.com).

TABLE 11.1
List of Popular Tools Used for Benchmarking

S. No.	Name of the Tool	Training Data	Feature Category	ML Algorithm	Tools used for Benchmarking	Salient Features	Reference
1	CPC	Eukaryotic	Sequence	SVM	CONC	Faster than its predecessor (CONC) with higher accuracy, provides user the option to search UTR databases as well, poor representation of plant sequences in training dataset	Kong et al. 2007
2	CPC2	Human Data + Eukaryotic	Sequence	SVM	CPC	~1000 times faster than CPC with improved accuracy, species neutral	Kang et al. 2017
3	CNCI	Human Data	Sequence	SVM	CPC and phyloCSF	Capable of classifying partial transcripts and sense-antisense pairs, Solely based on nucleotide composition, lacks plant sequences in training dataset	Sun et al. 2013
4	PLEK	Human Data	Sequence	*k*-mer and SVM	CPC, CNCI and phyloCSF	Alignment-free, uses improved *k*-mer usage frequency feature calibrated to size of *k*-mer strings, performs well with *de novo* assembled transcriptomes with high indel sequencing errors, No plant data in training dataset	Li et al. 2014
5	CPAT	Human data	Sequence	Logistic Regression	CPC and phyloCSF	Alignment-free, uses hexamer score to improve discriminatory power, No plant data in training dataset	Wang et al. 2013

data for benchmarking. Further, the ability to work with sequences originating from a wide range of species, computation speed, and availability of user-friendly interface or tunability of the code are some critical factors that affect the broad applicability of the tool. Before delving into the details of plant-specific/friendly lncRNA prediction tools, we summarize some of the popular tools, commonly used as reference for evaluating the performance of the new models (Table 11.1).

11.2 POPULAR TOOLS USED FOR BENCHMARKING PLANT-FRIENDLY MODELS

11.2.1 Coding Potential Calculator (CPC) and CPC2

Coding potential calculator (CPC) is an SVM-based classifier that distinguishes between coding and ncRNAs using six sequence features (Kong et al. 2007). The first three features assess the quality, integrity and coverage of the ORF in the transcript. In contrast, the other three (number of hits, high-scoring segment pairs and score) assess the sequence homology against UniProt Reference Clusters (https://www.uniprot.org/) using BLASTX. The model was trained using 5610 coding cDNAs extracted from the Swiss-Prot database, while 2670 ncRNA sets were compiled from the RNAdb (Pang et al. 2005) and NONCODE (http://www.noncode.org/) databases. CPC showed significant improvement over previously available state-of-art tools with 95.77% accuracy.

CPC2 is the updated version of CPC with higher speed (~1000 times) and accuracy achieved by using an SVM model with the standard radial basis kernel function which can quickly predict the coding ability of the sequences (Kang et al. 2017). The model is trained based on four intrinsic sequence features, including ORF length, integrity, Fickett score and isoelectric point (pI). The Fickett score captures compositional bias and position asymmetry between protein coding and ncRNAs, while the pI captures distinct chemical properties of hypothetical peptides from non-coding sequences compared to natural peptides encoded by bona fide coding sequences. Though CPC and CPC2 have been trained using human data, the performance of CPC2 has been evaluated with human, mouse, zebrafish, fly, worm and *Arabidopsis* datasets, demonstrating its applicability to both animals and plants. Therefore, CPC2 is species-neutral and can be used for lncRNA predictions from non-model organism transcriptomes. A mobile-friendly web server (http://cpc2.gao-lab.org/) and a standalone package (http://cpc2.gao-lab.org/download.php) are available for download.

11.2.2 Coding-Non-Coding Index (CNCI)

This SVM-based tool was developed to deal with the challenge of incomplete transcripts from poorly annotated genomes. The Coding-Non-Coding Index (CNCI) profiles the adjoining nucleotide triplets (ANT) to distinguish between coding transcripts and lncRNA sequences independent of the annotations and is, therefore, effective with incomplete transcripts and sense–antisense pairs (Sun et al. 2013). The authors used five intrinsic sequence features, including length and score of most-like CDS (MLCDS), length percentage, score distance and codon bias of single nucleotide triplets in the MLCDS. CNCI was trained using human data extracted from the RefSeq (https://www.ncbi.nlm.nih.gov/refseq/) and GENCODE (https://www.gencodegenes.org/) databases. Its performance was assessed using vertebrate, invertebrate and plant datasets. When benchmarked against CPC and phylogenetic codon substitution frequencies (phyloCSF), CNCI performed better for vertebrate species, with relatively poor performance with invertebrates and plant datasets.

11.2.3 Predictor of lncRNAs and Messenger RNAs Based on an Improved *k*-mer Scheme (PLEK)

Predictor of lncRNAs and messenger RNAs based on an improved *k*-mer scheme (PLEK) is an alignment-free tool that uses an improved *k*-mer scheme to differentiate lncRNAs from mRNAs (Li, Zhang, and Zhou 2014). PLEK takes calibrated *k*-mer frequencies of transcript sequences into consideration to build a binary classification model using an SVM algorithm. Like CNCI, PLEK was also trained on human data and performed well on vertebrate data. Its performance was assessed with well-annotated mRNAs and lncRNA transcripts with an accuracy of up to 95.6%. Further, it demonstrated >94% accuracy for transcripts containing indel sequencing errors and was faster than CNCI and CPC. PLEK is a fast, robust and accurate tool that performs well with the PacBio or 454 sequencing data containing high levels of indels due to sequencing errors. It can be accessed through a web server (http://202.200.112.245/plek/).

11.2.4 Coding-Potential Assessment Tool (CPAT)

Like PLEK, the coding-potential assessment tool (CPAT) is also an alignment-free and species-neutral tool developed using a logistic regression model (Wang, Park, Dasari, Wang, Kocher, and Li 2013). CPAT outperformed both CPC and PhyloCSF in sensitivity and specificity. It uses four sequence features, including Fickett score, ORF size, coverage and hexamer usage bias for lncRNA prediction. While the other three features have been used in CPC as well, the hexamer score that considers the degree of hexamer usage bias between protein coding and ncRNAs, further improved the discriminatory power of CPAT. The model was trained using 10,000 ncRNA sequences downloaded from the GenCODE database. A web interface for CPAT was also developed that enables easy access to users (http://lilab.research.bcm.edu/). CPAT is clearly more robust, faster and easier to use compared to other popular tools, but the users have to define a threshold to discern ncRNAs from protein-coding transcripts.

11.3 TOOLS DEVELOPED FOR lncRNA IDENTIFICATION IN PLANTS

Most of the previously mentioned ML tools were trained using human and animal datasets and exhibit suboptimal performance when tested with the plant datasets. To facilitate better predictions from plant data, several new tools have been developed for lncRNA identification using hybrid (plant and animal) or exclusively plant data in the past few years. Here, we have reviewed a selection of tools available for lncRNA prediction in plants. These have been classified based on the ML algorithms used for generating the model (Figure 11.2; Table 11.2).

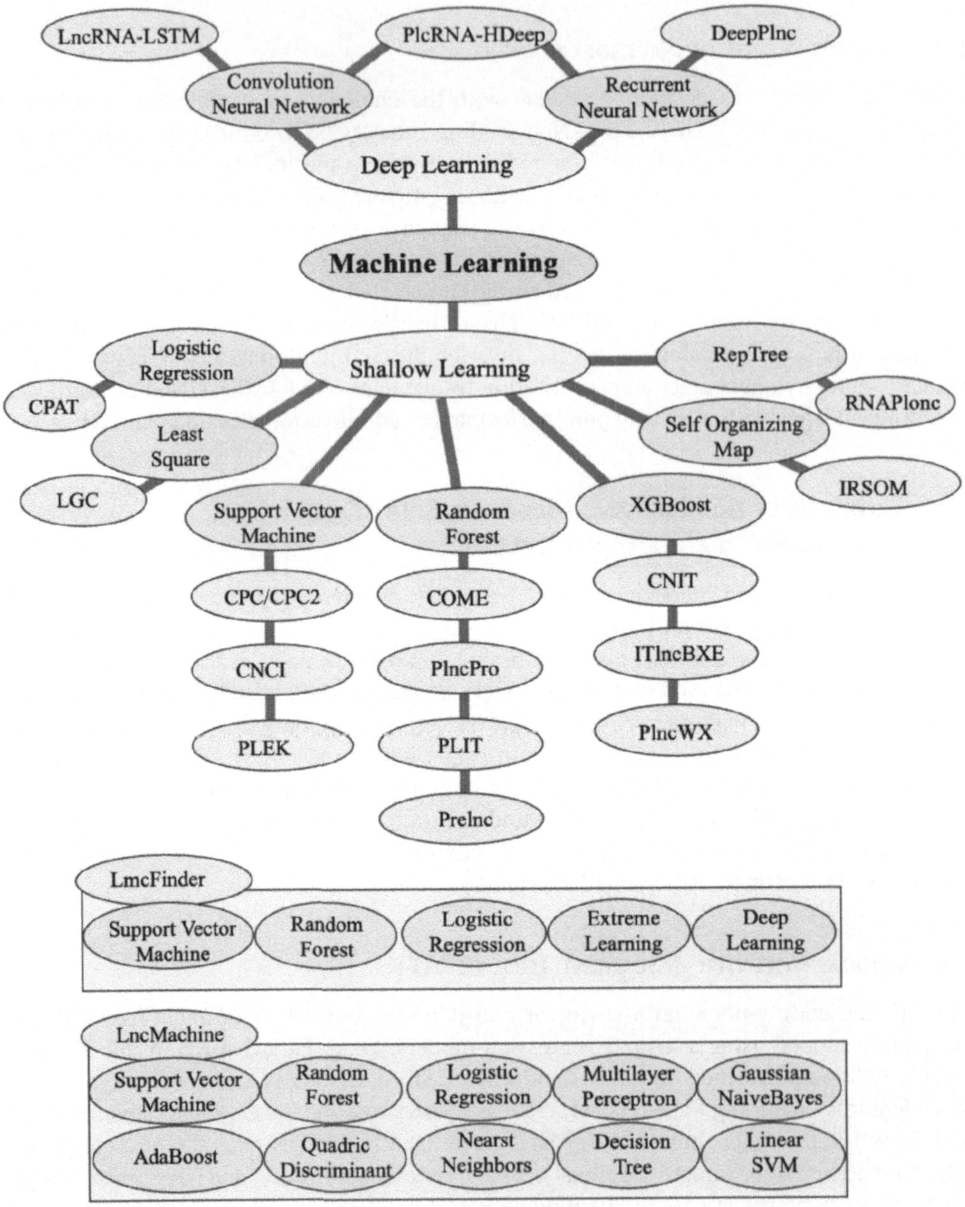

FIGURE 11.2 Flowchart of the machine learning algorithms implemented in the lncRNA identification tools reviewed here. The machine learning algorithms implemented in popular lncRNA identification tools are highlighted. LncFinder and LncMachine allow implementation of multiple algorithms.

TABLE 11.2
List of Tools Developed for lncRNA Identification in Plants

S. No.	Name of the Tool	Training Data	Feature Category	ML Algorithm	Tools used for Benchmarking	Strength/s	Reference
A. List of Tools Developed using Shallow Learning-Based Classifiers							
1	COME	Animal and plant	Sequence and experiment-based	Random forest	CNCI, RNAcode, HMMER, CPAT and PhyloCSF, CPC and PLEK	Uses multiple sequence and experimental features with decompose–compose method that improves prediction scores with incomplete transcripts.	Hu et al. 2017
2	PLncPRO	Plants	Sequence	Random forest	PLEK, CPAT, CNCI and CPC	Provides dicot- and monocot-specific models.	Singh et al. 2017
3	PLIT	Plants	Sequence	Iterative random forests	CPAT, CPC2, PLEK and LncScore	Alignment-free; uses L1 regularization for feature selection.	Deshpande et al. 2019
4	RNAPlonc	Plants	Sequence	RepTree	CPC, CPC2, CPAT and PLncPRO	Tested 5468 features with three methods before selecting 16 optimal features; uses pattern recognition approach; low computing cost.	Negri et al. 2019
5	Prelnc	Plants	Sequence	Random forest	CPC2, CPAT, PLEK, LncFinder, PLncPRO and RNAplonc	Includes ranking feature with Pearson correlation coefficient; IFS method used; compared multiple classifiers.	Cao et al. 2020
6	ItLnc-BXE	Plants	Sequence	XGBoost	PLIT and PLncPRO	Adopts the Relief-GA feature selection method; several base learners trained and combined into a single meta-learner by ensemble learning to improve performance and robustness; constructs species-specific models.	Zhang et al. 2020
7	CNIT	Plants	Sequence	XGBoost	CNCI, CPC2, CPAT and PLEK	An upgraded version of CNCI, with ~200 times higher speed and increased accuracy.	Guo et al. 2019
8	PlncWX	Plants	Sequence and structural	WOA-XGBoost	KNN, Naïve Bayes, SVM, decision tree, random forest and AdaBoost	Uses both sequence and structural features, compared 20 feature selection methods and 7 classifiers for accuracy.	Guo et al. 2021
9	IRSOM	Animal and plant	Sequence	Supervised SOMs	CPC2, CPAT, CNCI and PLEK	Provides a rejection option with user-defined threshold, alignment-free, highlights ambiguous transcripts; rejection threshold needs to be optimized.	Platon et al. 2018

(Continued)

TABLE 11.2 (*Continued*)
List of Tools Developed for lncRNA Identification in Plants

S. No.	Name of the Tool	Training Data	Feature Category	ML Algorithm	Tools used for Benchmarking	Strength/s	Reference
10	LGC	Animal and plant	Relationship between sequence features	Least square method and root mean square error	CPC, CPAT, CNCI, PLEK	Considers feature relationship between ORF and GC content.	Wang et al. 2019
				B. List of Tools Developed using Deep Learning-Based Classifiers			
11	LncRNA-LSTM	*Zea mays*		RNN-LSTM	CPC2, CNCI, PLEK and LncADeep	First plant-based deep learning classifier; training dataset is very small.	Meng et al. 2019
12	PlncRNA-HDeep	*Z. mays*		RNN-LSTM-CNN	CNCI, PLEK, CPC2, LncADeep and lncRNAnet	Implements hybrid deep learning based on two encoding styles; training dataset is very small.	Meng et al. 2021
13	DeepPlnc	Plants		Bimodal CNN	PLncPRO, CNIT, PreLnc, PlcRNA-HDeep and RNAplonc	Performs well with partial sequences from *de novo* assembled transcriptome; uses both sequence and structural features; training dataset is very small.	Ritu et al. 2022
				C. Customizable Tools with Option to Test Multiple Classifiers			
14	LncFinder	Human data	Sequence, structure and physiochemical property	Multiple algorithms	CPC, CPAT, CNCI, PLEK and CPC2	Provides a user-friendly platform to extract features, build classifiers with different ML algorithms and evaluate their performance.	Han et al. 2019
15	LncMachine	Wheat	Sequence	Multiple algorithms	CPC2, CPAT and CNIT	Highly customizable. Plant training data limited to wheat, but user can train the model using their own training dataset; crop-specific model.	Cagirici et al. 2021

11.3.1 Tools Trained Using Shallow ML Algorithms

Shallow ML algorithms use the pre-defined features to create a prediction model. Most of the tools developed in this category for lncRNA identification in plants use random forest or XGBoost classifiers.

11.3.1.1 Coding Potential Calculation Tool Based on Multiple Features (COME)

The coding potential calculation tool based on multiple features (COME) integrates various sequence and experiment-based features such as DNA conservation, protein conservation, GC content, RNA structure conservation, expression and histone methylation to predict lncRNAs by using a balanced random forest algorithm (Hu, Xu, Hu, and Lu 2017). It annotates each predicted lncRNA with supporting evidence that could help researchers conduct functional and mechanistic studies. The model has been trained on animal and plant datasets, and its performance was evaluated on the mouse, worm, fruit fly, and *Arabidopsis* datasets. On benchmarking, COME performed better than CNCI, RNAcode, HMMER, CPAT, PhyloCSF, CPC and PLEK.

11.3.1.2 Plant lncRNA Prediction by Random Forest (PLncPRO)

This random forest algorithm was developed specifically for plants (Singh, Khemka, Rajkumar, Garg, and Jain 2017). The use of consensus models for dicots and monocots facilitated the prediction of lncRNAs in non-model/orphan plants. The model was trained using datasets from 10 plant species including *Arabidopsis, Amborella*, *Chlamydomonas*, soybean, rice, *Physcomitrella*, *Selaginella*, potato, grapes and maize by using 71 sequence features and tested on datasets from rice and chickpea. PLncPRO is a multi-threaded memory-efficient standalone program that can run with Python on any environment (http://ccbb.jnu.ac.in/plncpro/).

11.3.1.3 RNAPlonc (Pattern Recognition Analysis on lncRNAs)

RNAPlonc is a pattern recognition tool which uses the Reduced Error Pruning Tree (RepTree) algorithm for identifying ncRNAs in plants (Negri, Alves, Bugatti, Saito, Domingues, and Paschoal 2019). The feature prioritization considered 5468 features but a set of only 16 features was used for prediction. RNAPlonc has a relatively low computational cost without compromising the performance. The algorithm was trained using data from five plant species including *Arabidopsis*, cucumber, poplar, rice and soybean. On benchmarking with CPC, CPC2 and CPAT, PLncPRO outperformed them for lncRNA identification in diverse plant genomes like *Populus*, cotton and *Brachypodium*.

11.3.1.4 Plant lncRNA Identification Tool (PLIT)

The Plant lncRNA Identification Tool (PLIT) is an alignment-free tool for identification of lncRNAs from plant transcriptomic datasets (Deshpande, Shuttleworth, Yang, Taramonli, and England 2019). PLIT uses L1 regularization and iterative random forests (IRF) for the selection of optimal features. A total of 31 optimal features were selected based on lncRNA and protein-coding data from eight plant species. The least absolute shrinkage and selection operator (LASSO) with the IRF algorithm were used for model building, and performance of the model was assessed on seven plant datasets. PLIT exhibited higher prediction accuracy, sensitivity and specificity compared to CPAT, CPC2, PLEK and LncScore (Zhao, Song, and Wang 2016).

11.3.1.5 Predicting lncRNAs (PreLnc)

PreLnc is another model developed using the random forest-based classifier (Cao, Wang, Bi, Ye, Yin, and Ye 2020). It uses high-confidence lncRNAs and mRNA transcripts to build prediction models. Authors combined ranking features with the Pearson correlation coefficient and used incremental feature selection (IFS) method to build the models with various classifiers including SVM, logistic regression, decision tree, random forests, and K-nearest neighbor (KNN). Finally, the model built using balanced random forest was used for classification. PreLnc was trained using data from three plant (rice, maize and *Arabidopsis*) and three animal species (human, mouse and cow). A set of eleven important features were used for lncRNA prediction. PreLnc was benchmarked against CPC2, CPAT, PLEK, PLncPRO, RNAplonc and LncFinder. Although PreLnc is a user-friendly tool, it still needs improvement in computing speed.

11.3.1.6 ItLnc-BXE (Identification of Plant lncRNAs by Bagging-XGBoost-Ensemble)

ItLnc-BXE uses extreme gradient boosting (XGBoost) for training the model using plant transcripts extracted from CANTATAdb (http://cantata.amu.edu.pl/), EnsemblPlants (https://plants.ensembl.org/index.html) and RNAcentral (https://rnacentral.org/) (Zhang, Liu, Dai, Yu, Liu, and Zhang 2020). CANTATAdb is a database dedicated to plant lncRNAs. Sequences from *Arabidopsis*, potato, rice, barley, *Physcomitrella* and *Chlamydomonas* were used for training and evaluation of the model. Four groups of features, including sequence, ORF, codon and alignment-based features were used for model building. The dicot and monocot-based models were created and the models performed with high accuracy with lower plant species as well. When benchmarked against PLIT and PLncPRO, ItLnc-BXE showed better overall performance in terms of accuracy and sensitivity.

11.3.1.7 Coding-Non-Coding Identifying Tool (CNIT)

A fast and accurate web-based tool, the Coding-Non-Coding Identifying Tool (CNIT), is an updated version of CNCI. It can be used with species lacking the whole-genome sequence or with poorly annotated transcripts. Like CNCI, CNIT also uses adjoining nucleotide triplets (ANT) features but with a more powerful ensemble ML model, XGBoost (extreme gradient boosting), used for creating the model (Guo et al. 2019). The model was trained using human and *Arabidopsis* lncRNAs with 67 intrinsic sequence features. CNIT has been benchmarked against, CNCI, CPC2, CPAT and PLEK and tested on 11 animal and 26 plant species. The area under the curve (AUC) values demonstrate the relatively accurate identification from diverse eukaryotic species. The fast computation time is another advantage of CNIT.

11.3.1.8 PLncWX (Plant lncRNA Identification Based on WOA-XGBoost)

PLncWX uses a whale optimization algorithm (WOA)-XGBoost-based model to identify lncRNAs in plants (Guo, Yin, Zhou, and Li 2021). XGBoost builds a decision tree using a gradient boosting framework, while WOA is used for the selection of features. The model was trained using data from five plant species, namely, *Arabidopsis*, *Brachypodium*, *Chlamydomonas*, *Physcomitrella* and *Selaginella,* representing distinct taxonomic categories. A total of 23 features were selected for model building. The performance of XGBoost was compared with seven other classifiers, including SVM, Naïve Bayes, KNN, decision tree, random forest, and AdaBoost, with XGBoost showing the highest overall performance.

11.3.1.9 IRSOM

IRSOM is a supervised classifier that implements a self-organizing map (SOM) with a rejection option that increases the reliability of prediction by discarding ambiguous classifications below a certain threshold (Platon, Zehraoui, Bendahmane, and Tahi 2018). The rejection technique lets users visualize and further analyze the rejected sequences. The overall architecture of the method has three layers: an input layer, a hidden layer containing the SOM with the rejection option, and an output layer. IRSOM uses several features associated with sequence statistics (such as, codon position bias, GC content, frequency of *k*-mer motifs and nucleotides), and putative ORFs (such as, ORF length, coverage, frequency, distribution, frame bias and distribution of start and end codons). The datasets used for non-coding sequences were extracted from RNAcentral, which combines data from several ncRNA databases. The accuracy of this alignment-free tool was greater than 0.95 for all the species tested with an average sensitivity of 0.99 and an average specificity of 0.98. IRSOM is also fast, with the ability to handle large datasets. IRSOM has been tested on several species, including human, mouse, *Escherichia coli*, *Saccharomyces*, *Drosophila*, zebrafish and *Arabidopsis*. The model has been benchmarked against several existing tools including CPAT, CPC2, CNCI and PLEK, with better or equivalent performance and accuracy. It can be accessed through a user-friendly web version (https://evryrna.ibisc.univ-evry.fr/evryrna/IRSOM/irsom_predict_form) or installed locally.

11.3.1.10 ORF Length and GC Content (LGC)

All the tools discussed previously use multiple sequence features as independent variables overlooking the biological relationship between them. A novel algorithm ORF Length and GC content (LGC) fills this caveat by taking the relationship between ORF length and GC content into account for differentiating between protein-coding and ncRNAs from diverse range of species without prior knowledge (Wang et al. 2019). The data from six representative species, including one vertebrate (zebrafish), one invertebrate (worm), two mammals (human and mouse) and two plants (rice and tomato), were downloaded from the GENCODE and Ensembl databases and used to examine the relationship between ORF length and GC content. The heterogeneity observed was used to distinguish lncRNAs from mRNAs. The LGC model, built on human data, was tested with plants and animals for lncRNA identification. Although LGC can be used in a cross-specific manner; species-specific models achieve higher accuracy. On benchmarking against CPC, PLEK, CNCI and CPAT, LGC showed better accuracy, specificity and robustness across different species ranging from mammals to plants.

11.3.2 Tools Trained Using Deep Learning Algorithms

With the increase in training data and lack of information about the most relevant features for prediction, deep learning methods have yielded promising results by leveraging complex neural network (CNNs) such as recurrent neural networks (RNNs), deep neural networks (DNNs) and deep belief networks (DBNs). These combine artificial neural networks with several hidden computational layers to make decisions.

So far, the deep learning methods developed for lncRNA identification in plants mainly rely on CNNs and RNNs (Figure 11.2). CNN is a popular neural network-based technique aggregating a complex feature using convolution and pooling layers. It has been used in various applications, such as predicting DNA binding sites, predicting the specificity of DNA and RNA binding proteins, CRISPR guide RNA design, etc. RNN is another popular neural network in which neurons draw insights from previous data during training. CNN considers all data points independent; on the contrary, data points in RNN are dependent on each other.

11.3.2.1 Long Short-Term Memory Networks (lncRNA-LSTM)

A pipeline developed for predicting lncRNAs in plants, long short-term memory networks (LncRNA-LSTM) uses K-means clustering to solve the challenge of unbalanced sample sizes. It uses an RNN-LSTM framework for building the model (Meng, Chang, Zhang, Shi, and Luan 2019). LSTM is a class of RNNs that retains the context drawn from previous data points for longer durations and is, therefore, useful for complex problems. Architecture-wise, the input first moves through the embedding layer, followed by the LSTM layer. With the *Zea mays* dataset, LncRNA-LSTM showed better performance when benchmarked against CPC2, CNCI, PLEK and LncADeep (Yang et al. 2018).

11.3.2.2 PlncRNA-HDeep

PlncRNA-HDeep is a hybrid deep learning model that comprises a layer each of RNN-LSTM and CNN stacked parallelly with two different encodings (Meng et al. 2021). LSTM and CNN automatically extract and learn the relevant information from the training data provided. Since this hybrid ensemble model learns from multiple perspectives, it performs better than a single model. Also, the model does not require prior knowledge for predicting plant lncRNAs. Like LncRNA-LSTM, it was also trained using *Z. mays* sequences. The dataset was extracted from GreeNC (Paytuvi-Gallart et al. 2019) and RefSeq, and the model was trained using *k*-mer frequency, ORF information and structural information of lncRNAs for predictions. Benchmarking with CPC2, PLEK, CNCI, LncADeep and lncRNAnet (Baek, Lee, Kwon, and Yoon 2018) showed that PlncRNA-HDeep outperformed all of them with 95.1% precision, 96.5% accuracy, 97.9% sensitivity and a 96.5% F1 score on the maize dataset. Since PlncRNA-HDeep has only been trained and tested on one plant species, its performance with other plant species remains to be investigated. Also, PlncRNA-HDeep does not provide any prebuilt models; therefore, one has to build the model.

11.3.2.3 DeepPlnc

Recently, a bi-modal CNN-based deep learning system, DeepPlnc, has been reported to identify plant lncRNAs by employing sequence and structural features (Ritu, Gupta, Sharma, and Shankar 2022). It is freely available as a web server and stand-alone program (https://scbb.ihbt.res.in/DeepPlnc/). Two different instances of positive and negative datasets were used. The lncRNA dataset is comprised of experimentally validated lncRNAs from 51 plant species extracted from EnsemblPlants (v51) with known status, PncSTress (http://bis.zju.edu.cn/pncstress/) and PLncDB v2.0 (https://www.tobaccodb.org/plncdb/) repositories. The non-lncRNA dataset comprised fully annotated mRNAs downloaded from EnsemblPlants (v51). The performance of DeepPlnc was further evaluated using another dataset derived from 16 plant species, based on known and predicted lncRNAs. The authors used two hidden layers and one-hot encoding to convert sequential and structural features into vectors. The DeepPlnc was compared with five tools (CNIT, PLncPRO, PreLnc, PlcRNA-HDeep and RNAplonc) to evaluate its performance and was found to work accurately with full-length as well as incomplete sequences with ambiguous boundaries. The authors demonstrated the application of this method to annotate lncRNAs from *de novo* assembled transcriptome of an endangered Himalayan medicinal plant species, *Rheum australe*. However, since the training data used are very small, the broad applicability of the model to other plant species remains to be tested.

11.3.3 Customizable Tools with the Option to Test Multiple Classifiers

11.3.3.1 LncFinder

LncFinder leverages sequence-intrinsic composition, structural attributes, and physicochemical properties to predict lncRNAs. It was released as an R package and a web server (http://bmbl.sdstate.edu/lncfinder/). While most available tools display unstable performance on datasets from different species and cannot be re-trained or customized by users, LncFinder is a highly flexible platform that uses a heterologous lncRNA predictor (Han et al. 2019). It can be used for feature extraction, selection, model validation, classifier construction and performance evaluation using different ML algorithms. This would allow the users to assess the impact of different ML algorithms on the performance of the classifier. So far, it allows customization of ML algorithms with five classifiers, including SVM, logistic regression, random forest, extreme learning machine and deep learning (Figure 11.2). Trained using ncRNAs extracted from GENCODE and Ensembl, the model has been evaluated with data from multiple species, including chicken, human, mouse, zebrafish and wheat. Compared to CPC, CPAT, CNCI, PLEK and CPC2, LncFinder performed better. LncFinder can also reveal some insights into lncRNA–protein interactions and evolution.

11.3.3.2 LncMachine

Recently, an alignment-free coding potential prediction tool, LncMachine has been reported for lncRNA annotation in plants by employing a comprehensive feature selection (Cagirici, Galvez, Sen, and Budak 2021). LncMachine claims to deploy several user-specified algorithms in real-time and, therefore, provides high flexibility to users. The authors extracted 93 sequence features subjected to several feature selection methods to finally shortlist a set of four features, including ORF length, sequence length, %GC and pI. Users can build prediction models in LncMachine by using nine different ML algorithms, including logistic regression, random forest, multi-layer perceptron, nearest neighbors, decision tree, AdaBoost, Gaussian Naïve Bayes, quadric discriminant analysis (QDA) and linear SVMs (Figure 11.2). Among these, the authors found random forest as the best predictor for the wheat dataset, whereas QDA and Naïve Bayes provided higher sensitivity for the rice dataset. Two independent transcriptomic datasets from the hexaploidy and tetraploid species of wheat were tested to evaluate the performance of the model. LncMachine was benchmarked against CPC2, CPAT and CNIT. The sensitivity of the model varied with different plant datasets, indicating that no single tool or algorithm is likely suitable for all plant datasets. Even when the same ML algorithm is used, the prediction accuracy can vary due to differences in parameters, platforms, feature sets and training data.

11.4 PROS AND CONS OF VARIOUS TOOLS AVAILABLE FOR lncRNA IDENTIFICATION IN PLANTS

With the exponential increase in high-throughput genomic and transcriptomic data and the mounting evidence of involvement of lncRNAs in regulating key aspects of plant development and stress response, it is essential to devise a robust strategy for accurate identification and annotation of lncRNAs (Waseem, Liu, and Xia 2020). At present, there are several prediction tools available for lncRNA identification. The prediction accuracy of these tools mainly depends on the training data, selection of features, classifiers and the quality of input data used for prediction. While the quality of training and input data and availability of features for selection depends on experimental techniques, the choice of an appropriate classifier can be ensured by training and evaluating different models.

Tools developed using alignment-based SVM classifiers have been widely used for lncRNA identification and still used for benchmarking the recent lncRNA predictors. CPC, CPC2, CNCI and PLEK are some popular tools that used SVM-based classifiers, while CPAT uses logistic regression for lncRNA prediction. However, species/lineage-specific training datasets and a limited number of features used for developing these models limit their applicability in plants. CPC and CPC2 rely on the quality and coverage of ORFs or homology against protein datasets. Therefore, these cannot identify lncRNAs containing long ORFs or those encoding for short peptide-like sequences. Also, these tools cannot differentiate lncRNAs from coding transcripts falling entirely within the UTR regions or lacking a conserved ORF. In addition to biological limitations, suboptimal alignments obtained using heuristic methods and long computation time further limit their utilization to predict lncRNAs from massive-scale RNA sequencing data. Moreover, these tools cannot be used to identify lncRNAs from incomplete transcripts obtained from poorly annotated *de novo*-assembled transcriptomes or identify plant species/lineage-specific lncRNAs. CNCI partially manages this challenge by leveraging intrinsic sequence features.

PLEK and CPAT are alignment-free methods that improve overall performance and computational speed. SVM-based PLEK considers *k*-mer frequencies, thereby improving the prediction performance with the NGS data containing a high frequency of insertions and deletions (indels). At the same time, the logistic regression-based CPAT shows an improved discriminatory power by taking hexamer usage into account. Overall, the alignment-free tools, CPAT, CNCI and PLEK, are more efficient and faster than CPCs but are highly influenced by the training data and, therefore, do not perform well with the plant data.

To manage this problem, several species-neutral or plant-specific tools have been developed in the past few years by using hybrid (plant and animal) or exclusively plant data, respectively. These tools primarily leverage Ensembl-based classifiers. Ensembl methods are ideal for heterogeneously distributed data. They avoid overfitting the data by combining multiple learners to produce one optimal model. Furthermore, their ability to use multiple classifiers combines the strengths of individual classifiers, thereby enhancing the accuracy of the prediction. For example, while using decision trees, bootstrap aggregating (BAGGing) or random forest models can be applied to produce the final tree. COME, PLncPRO, PLIT and Prelnc use random forest algorithms, while RNAPlonc takes advantage of Reptree, which has a relatively lower overall computational cost. COME is unique in leveraging experimental features such as expression and ribosome profiling data for lncRNA prediction.

Further, ItLnc-BXE, CNIT and PlncWX have been trained exclusively on plant data using XGBoost. Boosting is an iterative method with error reduction practice after each training. For example, the second model will try to fix the errors made by the first model, while the third model will try to fix the errors by the second model, and so on. The output of boosting is a weighted average of the predictions. Furthermore, ItLnc-BXE and CNIT rely on sequence features, while PlncWX also considers structural features.

Another recently developed tool, IRSOM, uses SOMs with a rejection option, enabling users to reject or consider a prediction based on probability scores. At the same time, LGC presents another advancement by considering the relationship between ORF length and GC content to differentiate between protein-coding RNAs and lncRNAs.

The deep learning methods exhibit better performance and adaptability with the ability to learn the rules itself to train the model, thereby ruling out the need for complex feature engineering based on prior knowledge and expertise. However, deep learning models developed for plant lncRNA identification are still scarce. LncRNA-LSTM uses RNN-LSTM, which was superseded by an improved model PlncRNA-HDeep by taking advantage of both the LSTM and CNN framework. Both these tools have been trained using maize data. More recently, DeepPlonc trained using bimodal CNN has exhibited enhanced performance with the ability to accurately identify lncRNAs from partial sequences derived from *de novo*-assembled transcriptomes. However, the training data used for DeepPlonc is also limited to maize.

11.5 CONCLUSIONS

Despite continuous efforts to accurately identify lncRNAs from plant species, methods developed to date are far from satisfactory, primarily due to the complex and ambiguous features of the lncRNA sequences. The scarcity of annotated and experimental data available for the plant species further adds to this challenge. A comparative study of 15 tools representing various methods available for lncRNA prediction using human data reported deep learning tools as top performers (Ammunet, Wang, Khan, and Elo 2022). However, the studies on plants are not yet enough to derive similar conclusions. Also, with the species and tissue-specific idiosyncrasies of lncRNAs, no single tool can likely outperform all others for diverse species. Therefore, customizable tools such as LncFinder and LncMachine would probably be better for training species-specific models by comparing different algorithms.

There are still immense opportunities to improve the performance and applicability of the existing tools. Recent studies have started unveiling unique aspects of lncRNA biogenesis, processing, subcellular localization, expression, and functions which must be incorporated to enhance prediction accuracy (Quinn and Chang 2016). For example, LGC takes the relationship between ORF length and GC content into consideration for lncRNA prediction; future updates may also take the correlation between other similar features such as codon usage bias, gene length and GC content, etc., into consideration for generating more robust models (Plotkin and Kudla 2011). Though most of the studies mentioned in this chapter reported development of a web-based interface for researchers from non-bioinformatics backgrounds, these interfaces can only handle a limited amount of data at a time, and some are not accessible anymore. Moreover, most prediction tools are developed based on the Linux operating system. Developing user-friendly Windows versions with the ability to append experimental features will help experimental biologists design specific experiments for empirical validations and input the results for better accuracy. Several databases containing lncRNAs from diverse or specific plant species have emerged recently which can serve as a good reference for training new models; however, the quality of the data in these databases needs to be ensured, and regular updates would be required to keep pace with the mounting transcriptomic data (Bhatia, Goyal, Sharma, Upadhyay, and Singh 2017; Paytuvi-Gallart et al. 2019; Shukla, Gupta, Srivastava, Sharma, Shukla, and Shasany 2021; Singh, Vivek, and Kumar 2021).

Since, the discovery of *Early Nodulin 40* (*Enod40*) that encodes short peptides in soybean, several transcripts, previously identified as lncRNAs, have been reported to encode small peptides, also known as micropeptides, with important biological implications (Rohrig, Schmidt, Miklashevichs, Schell, and John 2002; Yeasmin, Yada, and Akimitsu 2018). For example, micropeptides encoded by *PLS* (*POLARIS*), *ROT4* (*ROTUNDIFOLIA*), *DLV1* (*DEVIL1*) and *KOD* (*Kiss of Death*) genes of *Arabidopsis* regulate root growth, leaf morphogenesis, plant organogenesis and programmed cell death, respectively (Casson, Chilley, Topping, Evans, Souter, and Lindsey 2002; Narita et al. 2004; Wen, Lease, and Walker 2004; Chilley et al. 2006; Blanvillain et al. 2011). Similarly in maize, micropeptides encoded by *Brk* (*Brick1*), *Zm401* and *Zm908* regulate polarized leaf growth, tapetum development and pollen tube growth, respectively (Frank and Smith 2002; Ma et al. 2008; Wang et al. 2009). Some of these likely correspond to young proteins with evolutionary significance and were missed during annotation due to very short ORFs (Ruiz-Orera, Messeguer, Subirana, and Alba 2014). Therefore, incorporating appropriate strategies in the lncRNA identification pipelines would be required to delineate such unannotated micropeptide-encoding genes from true lncRNAs.

Another important caveat in lncRNA identification is the lack of standard guidelines for lncRNA annotation. Researchers focused on developing new tools select varied parameters and thresholds, which contribute to the variability in the performance, hence, benchmarking against previously available tools leads to conflicting results (Budak, Kaya, and Cagirici 2020). With more revelations about the structural and functional attributes using experimental studies and the development of better algorithms, we will hopefully be able to overcome these challenges soon.

REFERENCES

Ammunet, T., N. Wang, S. Khan, and L. L. Elo 2022. "Deep learning tools are top performers in long non-coding RNA prediction." *Brief Funct Genomics* 21 (3):230–41. doi: 10.1093/bfgp/elab045

Baek, J., B. Lee, S. Kwon, and S. Yoon 2018. "LncRNAnet: long non-coding RNA identification using deep learning." *Bioinformatics* 34 (22):3889–97. doi: 10.1093/bioinformatics/bty418

Bhatia, G., N. Goyal, S. Sharma, S. K. Upadhyay, and K. Singh. 2017. "Present scenario of long non-coding RNAs in plants." *Noncoding RNA* 3 (2). doi: 10.3390/ncrna3020016

Blanvillain, R., B. Young, Y. M. Cai, V. Hecht, F. Varoquaux, V. Delorme, J. M. Lancelin, M. Delseny, and P. Gallois 2011. "The *Arabidopsis* peptide kiss of death is an inducer of programmed cell death." *EMBO J* 30 (6):1173–83. doi: 10.1038/emboj.2011.14

Budak, H., S. B. Kaya, and H. B. Cagirici 2020. "Long non-coding RNA in plants in the era of reference sequences." *Front Plant Sci* 11:276. doi: 10.3389/fpls.2020.00276

Cabili, M. N., C. Trapnell, L. Goff, M. Koziol, B. Tazon-Vega, A. Regev, and J. L. Rinn 2011. "Integrative annotation of human large intergenic noncoding RNAs reveals global properties and specific subclasses." *Genes Dev* 25 (18):1915–27. doi: 10.1101/gad.17446611

Cagirici, H. B., S. Galvez, T. Z. Sen, and H. Budak 2021. "LncMachine: a machine learning algorithm for long noncoding RNA annotation in plants." *Funct Integr Genomics* 21 (2):195–204. doi: 10.1007/s10142-021-00769-w

Cao, L., Y. Wang, C. Bi, Q. Ye, T. Yin, and N. Ye. 2020. "PreLnc: an accurate tool for predicting lncRNAs based on multiple features." *Genes (Basel)* 11 (9). doi: 10.3390/genes11090981

Casson, S. A., P. M. Chilley, J. F. Topping, I. M. Evans, M. A. Souter, and K. Lindsey 2002. "The POLARIS gene of *Arabidopsis* encodes a predicted peptide required for correct root growth and leaf vascular patterning." *Plant Cell* 14 (8):1705–21. doi: 10.1105/tpc.002618

Chen, L. L., and G. G. Carmichael 2010. "Decoding the function of nuclear long non-coding RNAs." *Curr Opin Cell Biol* 22 (3):357–64. doi: 10.1016/j.ceb.2010.03.003

Chilley, P. M., S. A. Casson, P. Tarkowski, N. Hawkins, K. L. Wang, P. J. Hussey, M. Beale, J. R. Ecker, G. K. Sandberg, and K. Lindsey 2006. "The POLARIS peptide of *Arabidopsis* regulates auxin transport and root growth via effects on ethylene signaling." *Plant Cell* 18 (11):3058–72. doi: 10.1105/tpc.106.040790

Costa Negri, T. D., A. Rossi Paschoal, and W. A. Luz Alves 2020. "Comparison tools for lncRNA identification: analysis among plants and humans." *2020 IEEE Conference on Computational Intelligence in Bioinformatics and Computational Biology (CIBCB)*:1–8. doi: 10.1109/CIBCB48159.2020.9277716

Deshpande, S., J. Shuttleworth, J. Yang, S. Taramonli, and M. England 2019. "PLIT: an alignment-free computational tool for identification of long non-coding RNAs in plant transcriptomic datasets." *Comput Biol Med* 105:169–81. doi: 10.1016/j.compbiomed.2018.12.014

Dinger, M. E., K. C. Pang, T. R. Mercer, and J. S. Mattick 2008. "Differentiating protein-coding and noncoding RNA: challenges and ambiguities." *PLoS Comput Biol* 4 (11):e1000176. doi: 10.1371/journal.pcbi.1000176

Fickett, J. W. 1982. "Recognition of protein coding regions in DNA sequences." *Nucleic Acids Res* 10 (17):5303–18. doi: 10.1093/nar/10.17.5303

Frank, M. J., and L. G. Smith 2002. "A small, novel protein highly conserved in plants and animals promotes the polarized growth and division of maize leaf epidermal cells." *Curr Biol* 12 (10):849–53. doi: 10.1016/s0960-9822(02)00819-9

Guo, J. C., S. S. Fang, Y. Wu, J. H. Zhang, Y. Chen, J. Liu, B. Wu, J. R. Wu, E. M. Li, L. Y. Xu, L. Sun, and Y. Zhao 2019. "CNIT: a fast and accurate web tool for identifying protein-coding and long noncoding transcripts based on intrinsic sequence composition." *Nucleic Acids Res* 47 (W1):W516–22. doi: 10.1093/nar/gkz400

Guo, F., Z. Yin, K. Zhou, and J. Li. 2021. "PLncWX: a machine-learning algorithm for plant lncRNA identification based on WOA-XGBoost." *J Chem* 2021. doi: 10.1155/2021/6256021

Han, S., Y. Liang, Q. Ma, Y. Xu, Y. Zhang, W. Du, C. Wang, and Y. Li 2019. "LncFinder: an integrated platform for long non-coding RNA identification utilizing sequence intrinsic composition, structural information and physicochemical property." *Brief Bioinform* 20 (6):2009–27. doi: 10.1093/bib/bby065

Hu, L., Z. Xu, B. Hu, and Z. J. Lu 2017. "COME: a robust coding potential calculation tool for lncRNA identification and characterization based on multiple features." *Nucleic Acids Res* 45 (1):e2. doi: 10.1093/nar/gkw798

Kandpal, M., N. Dhaka, and R. Sharma 2021. "Genome-wide in silico analysis of long intergenic non-coding RNAs from rice peduncles at the heading stage." *Physiol Mol Biol Plants* 27 (10):2389–2406. doi: 10.1007/s12298-021-01059-2

Kang, Y. J., D. C. Yang, L. Kong, M. Hou, Y. Q. Meng, L. Wei, and G. Gao 2017. "CPC2: a fast and accurate coding potential calculator based on sequence intrinsic features." *Nucleic Acids Res* 45 (W1):W12–16. doi: 10.1093/nar/gkx428

Kong, L., Y. Zhang, Z. Q. Ye, X. Q. Liu, S. Q. Zhao, L. Wei, and G. Gao 2007. "CPC: assess the protein-coding potential of transcripts using sequence features and support vector machine." *Nucleic Acids Res* 35 (Web Server issue):W345–49. doi: 10.1093/nar/gkm391

Kung, J. T., D. Colognori, and J. T. Lee 2013. "Long noncoding RNAs: past, present, and future." *Genetics* 193 (3):651–69. doi: 10.1534/genetics.112.146704

Li, A., J. Zhang, and Z. Zhou 2014. "PLEK: a tool for predicting long non-coding RNAs and messenger RNAs based on an improved *k*-mer scheme." *BMC Bioinformatics* 15:311. doi: 10.1186/1471-2105-15-311

Ma, J., B. Yan, Y. Qu, F. Qin, Y. Yang, X. Hao, J. Yu, Q. Zhao, D. Zhu, and G. Ao 2008. "Zm401, a short-open reading-frame mRNA or noncoding RNA, is essential for tapetum and microspore development and can regulate the floret formation in maize." *J Cell Biochem* 105 (1):136–46. doi: 10.1002/jcb.21807

Meng, J., Z. Chang, P. Zhang, W. Shi, and Y. Luan. 2019. "lncRNA-LSTM: prediction of plant long noncoding RNAs using long short-term memory based on p-nts encoding." In Huang, D. S., Huang, Z. K., and Hussain, A. (eds.), *Intelligent Computing Methodologies. ICIC 2019. Lecture Notes in Computer Science( )*, vol 11645. Cham: Springer. doi: https://doi.org/10.1007/978-3-030-26766-7_32

Meng, J., Q. Kang, Z. Chang, and Y. Luan 2021. "PlncRNA-HDeep: plant long noncoding RNA prediction using hybrid deep learning based on two encoding styles." *BMC Bioinform* 22 (Suppl 3):242. doi: 10.1186/s12859-020-03870-2

Militti, C., S. Maenner, P. B. Becker, and F. Gebauer 2014. "UNR facilitates the interaction of MLE with the lncRNA roX2 during Drosophila dosage compensation." *Nat Commun* 5:4762. doi: 10.1038/ncomms5762

Morris, K. V., and J. S. Mattick 2014. "The rise of regulatory RNA." *Nat Rev Genet* 15 (6):423–37. doi: 10.1038/nrg3722

Narita, N. N., S. Moore, G. Horiguchi, M. Kubo, T. Demura, H. Fukuda, J. Goodrich, and H. Tsukaya 2004. "Overexpression of a novel small peptide ROTUNDIFOLIA4 decreases cell proliferation and alters leaf shape in *Arabidopsis* thaliana." *Plant J* 38 (4):699–713. doi: 10.1111/j.1365-313X.2004.02078.x

Negri, T. D. C., W. A. L. Alves, P. H. Bugatti, P. T. M. Saito, D. S. Domingues, and A. R. Paschoal 2019. "Pattern recognition analysis on long noncoding RNAs: a tool for prediction in plants." *Brief Bioinform* 20 (2):682–89. doi: 10.1093/bib/bby034

Pang, K. C., S. Stephen, P. G. Engstrom, K. Tajul-Arifin, W. Chen, C. Wahlestedt, B. Lenhard, Y. Hayashizaki, and J. S. Mattick 2005. "RNAdb–a comprehensive mammalian noncoding RNA database." *Nucleic Acids Res* 33 (Database issue):D125–30. doi: 10.1093/nar/gki089

Paytuvi-Gallart, A., W. Sanseverino, and R. Aiese Cigliano. 2019. "A walkthrough to the use of GreeNC: the plant lncRNA database." Chekanova, J. A., and Wang, H.-L. V. (eds.), *Plant Long Non-Coding RNAs: Methods and Protocols*. New York, NY: Springer. doi: 10.1007/978-1-4939-9045-0_25

Platon, L., F. Zehraoui, A. Bendahmane, and F. Tahi 2018. "IRSOM, a reliable identifier of ncRNAs based on supervised self-organizing maps with rejection." *Bioinformatics* 34 (17):i620–28. doi: 10.1093/bioinformatics/bty572

Plotkin, J. B., and G. Kudla 2011. "Synonymous but not the same: the causes and consequences of codon bias." *Nat Rev Genet* 12 (1):32–42. doi: 10.1038/nrg2899

Ponting, C. P., P. L. Oliver, and W. Reik 2009. "Evolution and functions of long noncoding RNAs." *Cell* 136 (4):629–41. doi: 10.1016/j.cell.2009.02.006

Quinn, J. J., and H. Y. Chang 2016. "Unique features of long non-coding RNA biogenesis and function." *Nat Rev Genet* 17 (1):47–62. doi: 10.1038/nrg.2015.10

Ritu, S. Gupta, N. K. Sharma, and R. Shankar 2022. "DeepPlnc: Bi-modal deep learning for highly accurate plant lncRNA discovery." *Genomics* 114 (5):110443. doi: 10.1016/j.ygeno.2022.110443

Rohrig, H., J. Schmidt, E. Miklashevichs, J. Schell, and M. John 2002. "Soybean ENOD40 encodes two peptides that bind to sucrose synthase." *Proc Natl Acad Sci USA* 99 (4):1915–20. doi: 10.1073/pnas.022664799

Ruiz-Orera, J., X. Messeguer, J. A. Subirana, and M. M. Alba 2014. "Long non-coding RNAs as a source of new peptides." *eLife* 3:e03523.

Shukla, B., S. Gupta, G. Srivastava, A. Sharma, A. K. Shukla, and A. K. Shasany 2021. "lncRNADetector: a bioinformatics pipeline for long non-coding RNA identification and MAPslnc: a repository of medicinal and aromatic plant lncRNAs." *RNA Biol* 18 (12):2290–95. doi: 10.1080/15476286.2021.1899673

Singh, U., N. Khemka, M. S. Rajkumar, R. Garg, and M. Jain 2017. "PLncPRO for prediction of long non-coding RNAs (lncRNAs) in plants and its application for discovery of abiotic stress-responsive lncRNAs in rice and chickpea." *Nucleic Acids Res* 45 (22):e183. doi: 10.1093/nar/gkx866

Singh, A., A. T. Vivek, and S. Kumar 2021. "AlnC: an extensive database of long non-coding RNAs in angiosperms." *PLoS One* 16 (4):e0247215. doi: 10.1371/journal.pone.0247215

Sun, L., H. Luo, D. Bu, G. Zhao, K. Yu, C. Zhang, Y. Liu, R. Chen, and Y. Zhao 2013. "Utilizing sequence intrinsic composition to classify protein-coding and long non-coding transcripts." *Nucleic Acids Res* 41 (17):e166. doi: 10.1093/nar/gkt646

Ulitsky, I., A. Shkumatava, C. H. Jan, H. Sive, and D. P. Bartel 2011. "Conserved function of lincRNAs in vertebrate embryonic development despite rapid sequence evolution." *Cell* 147 (7):1537–50. doi: 10.1016/j.cell.2011.11.055

Wang, D., C. Li, Q. Zhao, L. Zhao, M. Wang, D. Zhu, G. Ao, and J. Yu 2009. "Zm401p10, encoded by an anther-specific gene with short open reading frames, is essential for tapetum degeneration and anther development in maize." *Funct Plant Biol* 36 (1):73–85. doi: 10.1071/FP08154

Wang, L., H. J. Park, S. Dasari, S. Wang, J. P. Kocher, and W. Li 2013. "CPAT: Coding-Potential Assessment Tool using an alignment-free logistic regression model." *Nucleic Acids Res* 41 (6):e74. doi: 10.1093/nar/gkt006

Wang, G., H. Yin, B. Li, C. Yu, F. Wang, X. Xu, J. Cao, Y. Bao, L. Wang, A. A. Abbasi, V. B. Bajic, L. Ma, and Z. Zhang 2019. "Characterization and identification of long non-coding RNAs based on feature relationship." *Bioinformatics* 35 (17):2949–56. doi: 10.1093/bioinformatics/btz008

Waseem, M., Y. Liu, and R. Xi 2020. "Long non-coding RNAs, the dark matter: an emerging regulatory component in plants." *Int J Mol Sci* 22 (1):86. doi: 10.3390/ijms22010086

Wen, J., K. A. Lease, and J. C. Walker 2004. "DVL, a novel class of small polypeptides: overexpression alters *Arabidopsis* development." *Plant J* 37 (5):668–77. doi: 10.1111/j.1365-313x.2003.01994.x

Wu, L., S. Liu, H. Qi, H. Cai, and M. Xu 2020. "Research progress on plant long non-coding RNA." *Plants (Basel)* 9 (4):408. doi: 10.3390/plants9040408

Yang, C., L. Yang, M. Zhou, H. Xie, C. Zhang, M. D. Wang, and H. Zhu 2018. "LncADeep: an ab initio lncRNA identification and functional annotation tool based on deep learning." *Bioinformatics* 34 (22):3825–34. doi: 10.1093/bioinformatics/bty428

Yeasmin, F., T. Yada, and N. Akimitsu 2018. "Micropeptides encoded in transcripts previously identified as long noncoding RNAs: a new chapter in transcriptomics and proteomics." *Front Genet* 9:144. doi: 10.3389/fgene.2018.00144

Zhang, G., Z. Liu, J. Dai, Z. Yu, S. Liu, and W. Zhang 2020. "ItLnc-BXE: a bagging-XGBoost-ensemble method with comprehensive sequence features for identification of plant lncRNAs." *IEEE Access* 8:68811–19. doi: 10.1109/ACCESS.2020.2985114

Zhao, J., X. Song, and K. Wang 2016. "lncScore: alignment-free identification of long noncoding RNA from assembled novel transcripts." *Sci Rep* 6:34838. doi: 10.1038/srep34838

Index

Note: Page numbers in **bold** refer to tables and those in *italic* refer to figures.

J

K

L

M

Q

R

S

T

U

V

W

X

Y

Z